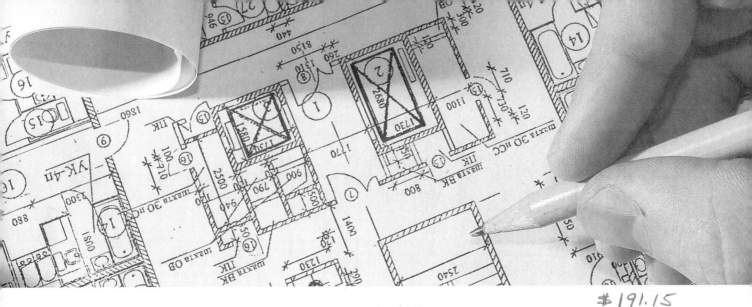

Estimating in Building Construction

SEVENTH EDITION

Frank R. Dagostino

Steven J. Peterson
Weber State University

Prentice Hall

Boston   Columbus   Indianapolis   New York   San Francisco   Upper Saddle River
Amsterdam   Cape Town   Dubai   London   Madrid   Milan   Munich   Paris   Montreal   Toronto
Delhi   Mexico City   Sao Paulo   Sydney   Hong Kong   Seoul   Singapore   Taipei   Tokyo

Vice President and Editor in Chief: Vernon R. Anthony
Acquisitions Editor: David Ploskonka
Editorial Assistant: Nancy Kesterson
Director of Marketing: David Gesell
Executive Marketing Manager: Derril Trakalo
Senior Marketing Coordinator: Alicia Wozniak
Project Manager: Susan Hannahs
Production Editor: Maren L. Miller
Associate Managing Editor: Alexandrina Benedicto Wolf

Senior Managing Editor: JoEllen Gohr
Art Director: Jayne Conte
Cover Image: Shutterstock Images
AV Project Manager: Janet Portisch
Full-Service Project Management: Niraj Bhatt/ Aptara®, Inc.
Composition: Aptara®, Inc.
Printer/Binder: Courier Kendallville
Cover Printer: Lehigh/Phoenix Color, Inc.
Text Font: Minion

Credits and acknowledgments borrowed from other sources and reproduced, with permission, in this textbook appear on the appropriate page within text.

Many of the designations by manufacturers and seller to distinguish their products are claimed as trademarks. Where those designations appear in this book, and the publisher was aware of a trademark claim, the designations have been printed in initial caps or all caps.

Library of Congress Cataloging-in-Publication Data

Dagostino, Frank R.
  Estimating in building construction / Frank R. Dagostino, Steven J. Peterson.—7th ed.
    p. cm.
  Includes bibliographical references and index.
  ISBN-13: 978-0-13-119952-1
  ISBN-10: 0-13-119952-8
  1. Building—Estimates. 2. Building—Estimates—Data processing. I. Peterson, Steven J.
II. Title.
  TH435.D18 2011
  692'.5—dc22

                                    2009038196

10 9 8 7 6 5 4 3 2 1

Prentice Hall
is an imprint of

www.pearsonhighered.com

ISBN 10:   0-13-119952-8
ISBN 13: 978-0-13-119952-1

*To my father for encouraging me to get an education and
my mother for her loving support. SP*

PREFACE

The seventh edition continues to build on the strong foundation of the previous editions. The need for estimators to understand the theory behind quantification is critical and must be fully understood prior to performing any computerized estimating. This underlying premise has been one of the guiding principles that began with Mr. Dagostino and continues with the current author. This edition uses extensive examples and exercises to demonstrate methodology and to the organization of the estimate. Estimating is an art that relies heavily on the judgment of the person performing the takeoff. A person's estimating skills can only be developed with practice; therefore, the reader is encouraged to work the example problems and apply the skills taught in this book. Since the estimate is used throughout the project, the assumptions and methodologies assumed must be documented and organized so that subsequent users will have access to this knowledge.

NEW TO THIS EDITION

The intent of this revision is to expand the estimating material covered by this book and to bring other material in line with current industry practices. The following is a list of key changes and additions that have been made to this edition:

- The discussion of the different types of estimates (e.g., square foot and parametric estimates) has been expanded in Chapter 1.

- A chapter discussing the project comparison method, square-foot estimating, and assembly estimating has been added at the end of the book as Chapter 21.

- The term "specifications" has been replaced with "project manual" when referring to the book that accompanies the plans and includes the contract documents and other information as well as the technical specifications. The term "specifications" is used to refer to the technical specification. This was done to be consistent with practices of the Construction Specifications Institute.

- A chapter providing an overview of the use of computers in construction estimating has been added as Chapter 5.

- A discussion of how to determine labor burden has been added to Chapter 7 (formerly Chapter 6).

- A discussion of how to determine labor productivity has been added to Chapter 7 (formerly Chapter 6).

- The term "work hour" has been replaced with the more commonly used term "labor hour" throughout the book.

- The use of published estimating data, such as RS Means, has been added to Chapters 7 and 21.

I thank the following for their insightful reviews: Frederick E. Gould, Roger Williams University; Donald E. Mulligan, Professor Emeritus at Arizona State University; and Wayne Reynolds, Eastern Kentucky State University.

CONTENTS

CHAPTER ONE

INTRODUCTION
TO ESTIMATING

1-1 GENERAL INTRODUCTION

Building construction estimating is the determination of probable construction costs of any given project. Many items influence and contribute to the cost of a project; each item must be analyzed, quantified, and priced. Because the estimate is prepared before the actual construction, much study and thought must be put into the construction documents. The estimator who can visualize the project and accurately determine its cost will become one of the most important persons in any construction company.

For projects constructed with the design-bid-build (DBB) delivery system, it is necessary for contractors to submit a competitive cost estimate for the project. The competition in construction bidding is intense, with multiple firms vying for a single project. To stay in business, a contractor must be the lowest-qualified bidder on a certain number of projects, while maintaining an acceptable profit margin. This profit margin must provide the general contractor an acceptable rate of return and compensation for the risk associated with the project. Because the estimate is prepared from the working drawings and the project manual for a building, the ability of the estimator to visualize all of the different phases of the construction project becomes a prime ingredient in successful bidding.

The working drawings usually contain information relative to the design, location, dimensions, and construction of the project, while the project manual is a written supplement to the drawings and includes information pertaining to materials and workmanship, as well as information about the bidding process. The working drawings and the project manual constitute the majority of the contract documents, define the scope of work, and *must* be considered together when preparing an estimate. The two complement each other, and they often overlap in the information they convey. The bid submitted must be based on the scope work provided by the owner or the architect. The estimator is responsible for including everything con-tained in the drawings and the project manual in the submitted bid. Because of the complexity of the drawings and the project manual, coupled with the potential cost of an error, the estimator must read everything thoroughly and recheck all items. Initially, the plans and the project manual must be checked to ensure that they are complete. Then the estimator can begin the process of quantifying all of the materials presented. Every item included in the estimate must contain as much information as possible. The quantities determined for the estimate will ultimately be used to order and purchase the needed materials. The estimated quantities and their associated projected costs will become the basis of project controls in the field.

Estimating the ultimate cost of a project requires the integration of many variables. These variables fall into either direct field costs or indirect field costs. The indirect field costs are also referred to as general conditions or project overhead costs in building construction. The direct field costs are the material, labor, equipment, or subcontracted items that are permanently and physically integrated into the building. For example, the labor and materials for the foundation of the building would be a direct field cost. The indirect field costs are the cost for the items that are required to support the field construction efforts. For example, the project site office would be a general conditions cost. In addition, factors such as weather, transportation, soil conditions, labor strikes, material availability, and subcontractor availability need to be integrated into the estimate. Regardless of the variables involved, the estimator must strive to prepare as accurate an estimate as possible. Since subcontractors or specialty contractors may perform much of the work in the field, the estimator must be able to articulate the scope of work in order for these companies to furnish a price quote. The complexity of an estimate requires organization, estimator's best judgment, complete specialty contractors' (subcontractors') bids, accurate quantity takeoffs, and accurate records of completed projects.

The design-build (DB) and construction-manager (CM) project delivery systems are gaining in popularity. In the design-build delivery system, the contractor acts as both the designer and the general contractor. In the construction-manager delivery system, the contractor is involved in the design process, providing expertise in construction methods and costs, as well as managing the construction process. Both of these delivery systems require the contractor to provide cost estimates for the proposed project throughout the design process.

At the conceptual stage of the project, the contractor prepares a cost estimate based on the project's concept. This is known as a conceptual estimate. When performing a conceptual estimate, typically, drawings are not available or they are very limited. What exists is often a vague verbal or written description of the project scope, which may be accompanied by a few sketches. When preparing this type of estimate, the contractor makes assumptions about virtually every aspect of the project. The conceptual estimate is used early in the design process to check to see if the owner's wants are in line with their budget and is often used as a starting point to begin contract negotiations.

During the design process, the contractor prepares and maintains a cost estimate based on the current, but incomplete, design. This is often referred to as a preliminary estimate. In addition, the contractor may prepare estimates that are used to select between building materials and to determine whether the cost to upgrade the materials is justified. What all these estimates have in common is that the design is incomplete. Once the design is complete, the contractor can prepare a detailed estimated for the project.

1–2 TYPES OF ESTIMATES

The required level of accuracy coupled with the amount of information about the project that is available will dictate the type of estimate that can be prepared. The different estimating methods are discussed below.

Detailed Estimate

The detailed estimate includes determination of the quantities and costs of everything that is required to complete the project. This includes materials, labor, equipment, insurance, bonds, and overhead, as well as an estimate of the profit. To perform this type of estimate, the contractor must have a complete set of contract documents. Each item of the project should be broken down into its parts and estimated. Each piece of work that is to be performed by the contractor has a distinct labor requirement that must be estimated. The items that are to be installed by others need to be defined and priced. Caution needs to be exercised to ensure that there is agreement between the contractor and the specialty contractor as to what they are to do and whether they are to install or supply and install the items. In addition, there needs to be an agreement about who is providing support items such as cranes and scaffolding. The contractor is responsible for making sure that the scope of work is divided among the contractor and subcontractors so that there are no overlaps in the individual scope of works and that everything has been included in someone's scope of work.

The detailed estimate must establish the estimated quantities and costs of the materials, the time required for and costs of labor, the equipment required and its cost, the items required for overhead and the cost of each item, and the percentage of profit desired, considering the investment, the time to complete, and the complexity of the project. The principles used to prepare the detailed estimates are covered in Chapters 4 and 6 through 20.

Assembly Estimating

In assembly estimating, rather than bidding each of the individual components of the project, the estimator bids the components in groups known as assemblies. The installation of the components of an assembly may be limited to a single trade or may be installed by many different trades. An example of a simple assembly would be a residential light switch, which includes a single-gang box, a single-pole switch, cover plate, two wire nuts, and an allowance of 20 feet of NM-B 12 gage wire. The entire assembly would be installed by an electrician. A residential electrical estimate could be prepared using assemblies for the switches, outlets, lights, power panels, and so forth rather than determining the individual components. An example of a complex assembly would be a metal-stud, gypsum-board partition wall, which would include bottom track, metal studs, top track, drywall, screws, tape, joint compound, insulation, primer, paint, and other miscellaneous items needed to construct the wall. This assembly would be installed by multiple trades.

Many high-end estimating computer programs, such as WinEst and Timberline, allow the user to prepare detailed estimates by taking off assemblies. For the switch assembly, the estimator would take off the number of switch assemblies needed for the project, and the software would add one single-gang box, one single-pole, one cover plate, two wire nuts, and 20 feet of NM-B 12-gage wire to the detailed estimate for each switch assembly. This simplifies the estimating process and increases the productivity of the estimator.

Assembly estimating is also useful for conceptual and preliminary estimates. By using broad assemblies, an estimate can be prepared quickly for an entire building. For example, an estimate for a warehouse can be prepared by using assembles for the spot footings, the continuous footings, the foundation wall, the floor slab (slab, reinforcement, granular base, vapor barrier, and fine grading), the exterior wall, personnel doors, overhead doors, joist and deck roof structure (including supports), roof insulation, roofing, wall cap, skylights, bathrooms, fire sprinklers, heating, lighting, and power distribution. This type of estimate can be prepared in hours instead of spending days preparing a detail estimate. The trade-off is that this type of estimate has many broad assumptions and is less accurate. This type of assembly estimating is good for estimates prepared with limited drawings, to compare design approaches, and as a check of a detailed estimate that seems way off. If the assembly price comes from previously completed projects, it is assumed that this project is identical to the completed projects. That assumption is clearly not valid in the construction of buildings. Weather conditions, building materials, and systems as well as design and construction team members change from project to project, all adding to the uniqueness of every project. Skill and judgment must be used while preparing this type of assembly estimate to ensure that proper adjustments are made by taking into account the varying conditions of each project. Companies such as R.S. Means publish annual guides (such as *Square Foot Costs*) that contain pricing for assemblies. Assembly estimating is discussed in Chapter 21.

Square-Foot Estimates

Square-foot estimates are prepared by multiplying the square footage of a building by a cost per square foot and then adjusting the price to compensate for differences in the building heights, length of the building perimeter, and other building components. In some cases, a unit other than square footage is used to measure the size of the building. For example, the size of a parking garage may be measured by the number of parking stalls in the garage. The information required to produce a square-foot estimate is much less than is needed to prepare a detailed estimate. For example, a preliminary set of design drawings (a single-line floor plan and key elevations) would have the dimensions that are necessary to prepare a square-foot estimate. Square-foot estimates are helpful to check whether the project, as designed, is within the owner's budget. Like an assembly estimate that uses broad assemblies, care must be exercised while preparing a square-foot estimate to ensure that the projects used to determine the cost per square foot are similar to the proposed project. Companies such as R.S. Means publish annual guides (such as *Square Foot Costs*) that contain a range of unit costs for a wide variety of building types. These guides provide a number of adjustments to compensate for varying building components, including the city where the project is located. Square-foot estimating is discussed in Chapter 21.

Parametric Estimates

Parametric estimates use equations that express the statistical relationship between building parameters and the cost of the building. The building parameters used in the equation may include the gross square footage, number of floors, length of perimeter, percentage of the building that is common space, and so forth. For an equation to be usable, the parameters used in the equation must be parameters that can be determined early in the design process; otherwise the equation is useless. Parametric estimates are similar to square-foot estimates; however, the equations used in parametric estimates are more complex and may use log functions, ratios of parameters, and multiplication of parameters. Parametric estimating is useful for preparing conceptual estimates based on assumptions of key building

parameters or estimates based upon early designs. As with square-foot estimates and assembly estimates that use broad assemblies, care must be taken to ensure that the proposed project is similar to the projects from which the equation has been derived.

Model Estimating

Model estimating uses computer models to prepare an estimate based on a number of questions answered by the estimator. Model estimating is similar to assembly estimating, but it requires less input from the estimator. For example, an estimate may be prepared for a warehouse by answering the following questions:

- What is the length of the building?
- How many bays are along the length of the building?
- What is the width of the building?
- How many bays are along the width of the building?
- What is the wall height above the grade?
- What is the depth (from the grade) to the top of the footing?
- What is the floor thickness?
- Do you want wire mesh in the slab?
- How many roof hatches do you want?
- How many personnel doors do you want?
- How many and what size of overhead doors do you want?
- How many and what size of skylights do you want?
- Do you want fire sprinklers?
- What bathroom facilities do you want (separate male and female, unisex, or none)?

On the basis of the answers to these questions, the model prepares a preliminary estimate for the project. Logic is built into the model, such that the model selects the necessary components for the estimate based upon the answers to the questions. For example, the size of the spot footings in the center of the building that support the roof and their costs are selected based on the area of the roof the footings support, which is equal to the width of a bay multiplied by the length of a bay. The length and width of the bays are calculated from the first four questions. A simple model estimate (Warehouse.xls) for a warehouse is provided on the companion disk. This model makes many assumptions as to the design of the warehouse, such as assuming the exterior wall is constructed of concrete masonry units (CMU). The model ignores the site and excavation cost, which needs to be added to the estimate from the model to get a complete estimate.

Estimating models may be complex and may prepare a detailed estimate for the entire project, or the models may be simple and prepare a preliminary estimate for part of a project. As with square-foot estimates, assembly estimates that use broad assemblies, and parametric estimates, care must be taken to make sure that the proposed project is similar to the projects from which the model was developed.

Project Comparison Estimates

Project comparison estimates are prepared by comparing the cost of a proposed project to a completed project. When preparing an estimate using this method, the estimator starts with the costs of a comparable project and then makes adjustments for differences in the project. For example, an estimate for the buildings in an apartment project may be prepared from a project built using the same plans during the previous year in a nearby city. In this example, the prices from the completed project need to be adjusted for inflation, changes in the availability and cost of labor, changes in the plans made to meet city codes, and so forth. In most cases, the site should be estimated using another method because of the many differences in site conditions. As with other estimating methods that do not prepare a detailed list of materials, care must be taken to ensure that the proposed project is similar to the completed project. The project comparison method is discussed in Chapter 21.

1–3 ESTIMATING OPPORTUNITIES

For anyone who is not aware of the many opportunities in the estimating field, this section will review some of the areas in which knowledge of estimating is necessary. Generally, knowledge of the procedures for estimating is required by almost everyone involved in or associated with the field of construction. From the estimator, who may be involved solely with the estimating of quantities of materials and the pricing of the project, to the carpenter, who must order the material required to build the framing for a home, this knowledge is needed to do the best job possible at the most competitive cost. Others involved include the project designer, drafters, engineers, contractors, subcontractors, material suppliers, and material representatives. In the following sections, a few of the estimating opportunities are described.

Architectural Offices.   The architectural office will require estimates at three design stages: preliminary (rough square foot or project comparison costs), cost evaluation during drawing preparation (usually more accurate square foot or assembly costs), and a final estimate (usually based on material and installation costs, to be as accurate as possible). For projects built using the design-build or construction-manager deliver systems, the preliminary estimate is often used during negotiation with the general contactor. Once the general contactor is hired, the general contractor's estimator will prepare the remaining estimates.

In large offices, the estimating may be done by an estimator hired primarily to do all the required estimating. In many offices, the estimating may be done by the chief drafter, head or lead architect, or perhaps someone else in the office who has developed the required estimating skills. There are also estimating services or consultants who perform estimates on a for-fee basis.

Engineering Offices.   The engineering offices involved in the design of building construction projects include civil, structural, mechanical (plumbing, heating, air-conditioning), electrical, and soil analysis. All of these engineering design phases require preliminary estimates, estimates while the drawings are being prepared, and final estimates as the drawings are completed. They are prepared in the same way estimates are prepared by the architects.

General Contractors.   For design-bid-build projects, the general contractor makes *detailed* estimates that are used to determine what the company will charge to do the required work.

The estimator will have to take off the quantities (amounts) of each material; determine the cost to furnish (buy and get to the site) and install each material in the project; assemble the bids (prices) of subcontractors; as well as determine all of the costs of insurance, permits, office staff, and the like. In smaller companies, one person may do the estimating, whereas in larger companies several people may work to negotiate a final price with an owner or to provide a competitive bid. On projects built using the design-build or construction manager deliver systems, the contractor's scope of work involves providing assistance to the owners, beginning with the planning stage, and continuing through the actual construction of the project. In this type of business, the estimators will also provide preliminary estimates and then update them periodically until a final price is set.

Estimating with Quantities Provided.   Estimating for projects with quantity surveys involves reviewing the specifications for the contract and material requirements, reviewing the drawings for the type of construction used, and assembling the materials used. The estimator will spend part of the time getting prices from subcontractors and material suppliers and the rest of the time deciding on how the work may be most economically accomplished.

Subcontractors.   Subcontractors may be individuals, companies, or corporations hired by the general contractor to do a particular portion of the work on the project. Subcontractors are available for all the different types of work required to build any project and include excavation, concrete, masonry (block, brick, stone), interior partitions, drywall, acoustical ceilings, painting, steel and precast concrete, erection, windows, metal and glass curtain walls, roofing, flooring (resilient, ceramic and quarry tile, carpeting, wood, terrazzo), and interior wall finishes (wallpaper, wood paneling, and sprayed-on finishes). The list continues to include all materials, equipment, and finishes required.

The use of subcontractors to perform all of the work on the project is becoming an acceptable model in building construction. The advantage of this model is that the general contractor can distribute the risk associated with the project to a number of different entities. In addition, the subcontractors and craft personnel perform the same type of work on a repetitive basis and are therefore quasi experts in their niche.

However, the general contractor relinquishes a substantial amount of control over the project when this method is employed. The more that the contractor subcontracts out, the more the field operation becomes involved in coordination rather than direct supervision of craft personnel.

The subcontractor carefully checks the drawings and project manual and submits a price to the construction companies that will be bidding on the project. The price given may be a unit or lump sum price. If a subcontractor's bid is presented as what he or she would charge per unit, then it is a *unit price* (such as per square foot, per block, per thousand brick, per cubic yard of concrete) bid. For example, the bid might be $5.25 per linear foot (lf) of concrete curbing. Even with unit price bids, the subcontractors need to perform a quantity takeoff in order to have an idea of what is involved in the project, at what stages they will be needed, how long it will take to complete their work, and how many workers and how much equipment will be required. The subcontractor needs the completed estimate to determine what is a reasonable amount for overhead and profit. Typically, as the quantity of work increases, the associated unit cost of jobsite overhead decreases. For example, the cost of mobilization for a 100 lf of curb is $1,000 or $10 per lf; if the quantity had been 1,000 lf, it would have been $1 per lf. The subcontractor would not know how much to add to the direct field cost unit price for overhead unless a quantity takeoff had been performed. If the subcontractor submits a lump-sum bid, then he or she is proposing to install, or furnish and install, a portion of work: For example, the bid might state, "agrees to furnish and install all Type I concrete curbing for a sum of $12,785.00."

Each subcontractor will need someone (or several people) to check specifications, review the drawings, determine the quantities required, and put the proposal together. It may be a full-time estimating position or part of the duties assumed, perhaps in addition to purchasing materials, helping to schedule projects, working on required shop drawings, or marketing.

Material Suppliers. Suppliers submit price quotes to the contractors (and subcontractors) to supply the materials required for the construction of the project. Virtually every material used in the project will be estimated, and multiple price quotes will be sought. Estimators will have to check the specifications and drawings to be certain that the materials offered will meet all of the requirements of the contract and required delivery dates.

Manufacturers' Representatives. Manufacturers' representatives represent certain materials, product suppliers, or manufacturers. They spend part of their time visiting contractors, architects, engineers, subcontractors, owners, and developers to be certain that they are aware of the availability of the material, its uses, and approximate costs. In a sense they are salespeople, but their services and the expertise they develop in their product lines make good manufacturers' representatives welcome not as salespersons, but as needed sources of information concerning the materials and products they represent. Representatives may work for one company, or they may represent two or more.

Manufacturers' representatives will carefully check the specifications and drawings to be certain that their materials meet all requirements. If some aspect of the specifications or drawings tends to exclude their product, or if they feel there may be a mistake or misunderstanding in these documents, they may call the architects/engineers and discuss it with them. In addition, many times they will be involved in working up various cost analyses of what the materials' or products' installed cost will be and in devising new uses for the materials, alternate construction techniques, and even the development of new products.

Project Management. Project management companies specialize in providing professional assistance in planning the construction of a project and keeping accurate and updated information about the financial status of the project. Owners who are coordinating large projects often hire such companies. Among the various types of owners are private individuals, corporations, municipal government agencies (such as public works and engineering departments), and various public utility companies.

Both the firms involved in project management, as well as someone on the staff of the owner being represented, must be knowledgeable in estimating and scheduling projects.

Government. When a government agency is involved in any phase of construction, personnel with experience in construction and estimating are required. Included are local, state or province, and nationwide agencies, including those involved in highways, roads, sewage treatment, schools, courthouses, nursing homes, hospitals, and single and multifamily dwellings financed or qualifying for financing by the government.

Employees may be involved in preparing or assisting to prepare preliminary and final estimates; reviewing estimates from architects, engineers, and contractors; the design and drawing of the project; and preparation of the specifications.

Professional Quantity Surveyors. Professional quantity surveyors are for-hire firms or individuals who make unit quantity takeoffs of materials required to build a project. They are available to provide this service to all who need it, including governmental agencies.

Freelance Estimators. Freelance estimators will do a material takeoff of a portion or entire project for whoever may want a job done. This estimator may work for the owner, architect, engineer, contractor, subcontractor, material supplier, or manufacturer. In some areas, the estimator will do a material takeoff of a project being competitively bid and then sell the quantity list to one or more contractors who intend to submit a bid on the project.

Many times a talented individual has a combined drafting and estimating business. Part of the drafting business may include preparing shop drawings (drawings that show sizes of materials and installation details) for subcontractors, material suppliers, and manufacturers' representatives.

Residential Construction.   Estimators are also required for the contractors, material suppliers, manufacturers' representatives, and most of the subcontractors involved in residential construction. From the designer who plans the house and the drafter who draws the plans and elevations to the carpenters who put up the rough framing and the roofers who install the roofing material, knowledge of estimating is necessary.

The designer and drafter should plan and draw the house plans using standard material sizes when possible (or being aware of it when they are not using standard sizes). In addition, they will need to give preliminary and final estimates to the owner. Workers need to have a basic knowledge of estimating so they can be certain that adequate material has been ordered and will be delivered by the time it is needed.

Computer Software.   The use of computers throughout the world of construction offers many different types of opportunities to the estimator. Job opportunities in all the areas mentioned earlier will be centered on the ability to understand, use, and manipulate computer software. The software available today integrates the construction drawings, estimating, bidding, purchasing, and management controls of the project. Some construction consultants specialize in building databases for computerized estimating systems and training estimators in the use these systems.

1–4 THE ESTIMATOR

Most estimators begin their career doing quantity takeoff; as they develop experience and judgment, they develop into estimators. A list of the abilities most important to the success of an estimator follows, but it should be more than simply read through. Any weaknesses affect the estimator's ability to produce complete and accurate estimates. If individuals lack any of these abilities, they must (1) be able to admit it and (2) begin to acquire the abilities they lack. Those with construction experience, who are subsequently trained as estimators, are often most successful in this field.

To be able to do quantity takeoffs, the estimator must

1. Be able to read and quantify plans.
2. Have knowledge of mathematics and a keen understanding of geometry. Most measurements and computations are made in linear feet, square feet, square yards, cubic feet, and cubic yards. The quantities are usually multiplied by a unit price to calculate material costs.
3. Have the patience and ability to do careful, thorough work.
4. Be computer literate and use computer takeoff programs such as On-Screen Takeoff or Paydirt.

To be an estimator, an individual needs to go a step further. He or she must

1. Be able, from looking at the drawings, to visualize the project through its various phases of construction. In addition, an estimator must be able to foresee problems, such as the placement of equipment or material storage, then develop a solution and determine its estimated cost.
2. Have enough construction experience to possess a good knowledge of job conditions, including methods of handling materials on the job, the most economical methods of construction, and labor productivity. With this experience, the estimator will be able to visualize the construction of the project and thus get the most accurate estimate on paper.
3. Have sufficient knowledge of labor operations and productivity to thus convert them into costs on a project. The estimator must understand how much work can be accomplished under given conditions by given crafts. Experience in construction and a study of projects that have been completed are required to develop this ability.
4. Be able to keep a database of information on costs of all kinds, including those of labor, material, project overhead, and equipment, as well as knowledge of the availability of all the required items.
5. Be computer literate and know how to manipulate and build various databases and use spreadsheet programs and other estimating software.
6. Be able to meet bid deadlines and still remain calm. Even in the rush of last-minute phone calls and the competitive feeling that seems to electrify the atmosphere just before the bids are due, estimators must "keep their cool."
7. Have good writing and presentation skills. With more bids being awarded to the best bid, rather than the lowest bid, being able to communicate what your company has to offer, what is included in the bid, and selling your services is very important. It is also important to communicate to the project superintendent what is included in the bid, how the estimator planned to construct the project, and any potential pitfalls.

People cannot be taught experience and judgment, but they can be taught an acceptable method of preparing an estimate, items to include in the estimate, calculations required, and how to make them. They can also be warned against possible errors and alerted to certain problems and dangers, but the practical experience and use of good judgment required cannot be taught and must be obtained over time.

How closely the estimated cost will agree with the actual cost depends, to a large extent, on the estimators' skill and judgment. Their skill enables them to use accurate estimating methods, while their judgment enables them to visualize the construction of the project throughout the stages of construction.

No.	Quantity	Unit	Item
02525	715	l.f.	Curb, type A
02525	75	l.f.	Curb, type B
04220	12,500	each	8" concrete block
04220	5,280	each	12" concrete block
04220	3,700	each	4" concrete block

FIGURE 1.1. Quantity Survey.

1–5   QUANTITY SURVEYING

In Canada, parts of Europe, and on most road construction projects in the United States, the estimated quantities of materials required on the project are determined by a professional quantity surveyor or engineer and provided to the interested bidders on the project. Figure 1.1 is an example of the quantities that would be provided by a quantity surveyor or engineer. This is often referred to as a unit price bid.

In this method of bidding, the contractors are all bidding based on the same quantities, and the estimator spends time developing the unit prices. For example, the bid may be $47.32 per cubic yard (cy) of concrete. Because all of the contractors are bidding on the same quantities, they will work on keeping the cost of purchasing and installing the materials as low as possible.

As the project is built, the actual number of units required is checked against the original number of units on which the estimates were made. For example, in Figure 1.1, the original quantity survey called for 715 linear feet (lf) of concrete curbing. If 722 lf were actually installed, then the contractor would be paid for the additional 7 lf. If 706 lf were used, then the owner would pay only for the 706 lf installed and not the 715 lf in the original quantity survey. This type of adjustment is quite common. When errors do occur and there is a large difference between the original quantity survey and the actual number of units, an adjustment to the unit price is made. Small adjustments are usually made at the same unit rate as the contractor bid. Large errors may require that the unit price be renegotiated.

If the contractor is aware of potential discrepancies between the estimated quantities and those that will be required, the contractor may price his or her bid to take advantage of this situation. With a belief that the estimated quantities are low, the contractor may reduce his or her unit price to be the low bidder. If the assumption is true, the contractor has the potential to make the same profit by distributing the project overhead over a greater number of units.

1–6   TYPES OF BIDS

Basically, the two bidding procedures by which the contractor gets to build a project for owners are as follows:

1. Competitive bidding
2. Negotiated bidding

Competitive bidding involves each contractor submitting a lump-sum bid or a proposal in competition with other contractors to build the project. The project may be awarded based on the price or best value. When the project is awarded based on the price, the lowest lump-sum bidder is awarded the contract to build the project as long as the bid form and proper procedures have been followed and this bidder is able to attain the required bonds and insurance. When the project is awarded based upon the best value, the proposals from the contractors are rated based on specified criteria with each criterion given a certain percentage of the possible points. The criteria may include review of the capabilities of the assigned project team, the company's capabilities and its approach to the project (including schedule), proposed innovation, method of mitigating risk, and price. The price is often withheld from the reviewers until the other criteria have been evaluated to prevent the price from affecting the ratings of the other criteria. Most commonly, the bids must be delivered to the person or place specified by a time stated in the instruction to bidders.

The basic underlying difference between negotiated work and competitive bidding is that the parties arrive at a mutually agreed upon price, terms and conditions, and contractual relationship. This arrangement often entails negotiations back and forth on virtually all aspects of the project, such as materials used, sizes, finishes, and other items that affect the price of the project. Owners may negotiate with as many contractors as they wish. This type of bidding is often used when owners know which contractor they would like to build the project, in which case competitive bidding would waste time. The biggest disadvantage of this arrangement is that the contractor may not feel the need to work quite as hard to get the lowest possible prices as when a competitive bidding process is used.

1–7   CONTRACT DOCUMENTS

The bid submitted for any construction project is based on the contract documents. If an estimator is to prepare a complete and accurate estimate, he or she must become familiar with all of the documents. The documents are listed and briefly described in this section. Further explanations of the portions and how to bid them are contained in later chapters.

For design-bid-build projects, the contract documents consist of the *owner-contractor agreement*, the *general conditions of the contract,* the *supplementary general conditions,* the *working drawings,* and *specifications,* including all *addenda* incorporated in the documents before their execution. All of these documents become part of the *contract.*

Agreement.   The agreement is the document that formalizes the construction contract, and it is the basic contract. It incorporates by reference all of the other documents and makes them part of the contract. It also states the contract sum and time allowed to construct the project.

General Conditions.   The general conditions define the rights, responsibilities, and relations of all parties to the construction contract.

Supplementary General Conditions (Special Conditions).   Because conditions vary by locality and project, the supplementary general conditions are used to amend or supplement portions of the general conditions.

Working Drawings.   The actual plans (drawings, illustrations) from which the project is to be built are the working drawings. They contain the dimensions and locations of building elements and materials required, and delineate how they fit together.

Specifications.   Specifications are written instructions concerning project requirements that describe the quality of materials to be used and their performance.

Addenda.   The addenda statement is a drawing or information that modifies the basic contract documents after they have been issued to the bidder, but prior to the taking of bids. They may provide clarification, correction, or changes in the other documents.

For projects built with the design-build and construction-manager deliver systems, the contract documents are more limited than for project built with the design-bid-build delivery system because the contractor is involved in the design and selection of the specifications for the project. These documents can be as simple as an agreement with a conceptual description of the project.

1-8   BIDDING INFORMATION

There are several sources of information pertaining to the projects available for bidding. Public advertising (advertisement for bids) is required for many public contracts. The advertisement is generally placed in newspapers, trade magazines, and journals, and notices are posted in public places and on the Internet. Private owners often advertise in the same manner to attract a large cross section of bidders (Figure 1.2). Included in the advertisement is a description of the nature, extent, and location of the project; the owner; the availability of bidding documents; bond requirements; and the time, manner, and place that the bids will be received.

Reporting services, such as *Dodge Reports* and *Engineering News Record* (ENR), provide information about projects that are accepting bids or proposals. The *Dodge Reports* are issued for particular, defined localities throughout the country, and separate bulletins are included that announce new projects within the defined area and provide a constant updating on jobs previously reported. The updating may include a listing of bidders, low bidders, awards of contracts, or abandonment of projects. In short, the updates provide information that is of concern to the contractors.

Other reporting services are the *Associated General Contractors* (AGC) and local building contractor groups. They generally perform the same type of service as the *Dodge Reports* but are not quite as thorough or as widely distributed. In most locales, the reporting services provide plan rooms where interested parties may review the drawings and project manual of current projects. While most general contractors will obtain several sets of contract documents for bidding, the various subcontractors and material suppliers make extensive use of such plan rooms.

1-9   AVAILABILITY OF CONTRACT DOCUMENTS

When paper copies of the plans and the project manual are used, there is usually a limit on the number of sets of contract documents a general contractor may obtain from the architect/engineer, and this limitation is generally found in the invitation to bid or instructions to bidders. Subcontractors, material suppliers, and manufacturers' representatives can usually obtain prints of individual drawings and specification sheets for a fee from the architect/engineer, but it should be noted that this fee is rarely refundable. The architect/engineer will require a deposit for each set of contract documents obtained by the prime contractors. The deposit, which acts as a guarantee for the safe return of the contract documents, usually ranges from $10 to over $200 per set and is usually refundable. It should be realized that the shorter the bidding period, the greater the number of sets that would be required. Also, a large complex job requires extra sets of contract documents to make an accurate bid.

To obtain the most competitive prices on a project, a substantial number of subcontractors and material suppliers must bid the job. To obtain the most thorough coverage, there should be no undue restrictions on the number of sets of contract documents available. If this situation occurs, it is best to call the architect/engineer and discuss the problem. For many projects, the owner makes drawings available in computer files, which can be printed or used in estimating software (such as On-screen Takeoff). This reduces the cost of reproducing the drawings and project manual, making it economical to distribute them to numerous contractors and subcontractors. Often electronic copies of the plans and the project manual can be downloaded via the Internet.

INVITATION TO BID FROM NEWSPAPER

REQUEST FOR BID PROPOSAL: Administration Building Annex, Project No. 1-2796, All American Independent School District, Littleville, Texas 77777.

RECEIPT OF BIDS: Sealed Proposals will be received by the All American Independent School District at the central administration building, 2005 Sarah Lane Littleville, Texas, until 2:00 p.m., April 24, 20__ and then publicly opened and read aloud. Bids mailed should be addressed to Mr. Ryan Smith, Director of Facilities, All American Independent School District, 2005 Sarah Lane, Littleville, Texas 77777, and should be clearly marked "HOLD FOR BID OPENING – PROJECT NO. 1-2796." Bid proposals cannot be withdrawn for sixty days from the date of bid opening.

A certified check or cashier's check on a state or national bank or a bidders bond from an acceptable surety company authorized to transact business in the state of Texas, in the amount of not less than five percent (5%) of the greatest total amount of the bidder's proposal must accompany each proposal as a guarantee that, if awarded the contract, the bidder will within ten (10)

calendar days after award of contract enter into contract and execute performance and payment bonds on the forms provided in the contract documents.

Proposals must be completed and submitted on the forms provided by the architect. Incomplete bid proposals will invalidate the bid proposal and the bid will be rejected and returned to the bidder. The right to accept any bid, or to reject any or all bids and to waive all formalities is hereby reserved by the All American Independent School District.

SCOPE OF WORK: This project consists of constructing a new Administration building. The site for this building is located at 123 Ryan Lane, Littleville, Texas.

PRIME CONTRACT: All work will be awarded under a single prime contract.
Information and bidding documents can be obtained from: A. B. Architects & Associates, Architects / Engineers, 7920 Anita Circle, Littleville, Texas 77777.
Plan Deposit: $50.00 per set. The deposit will be returned if the documents are returned in good condition within three weeks after bid opening; otherwise, no refund will

be made. Checks to be made payable to: "A. B. Architects & Associates, Architects / Engineers." Bid documents are available to established plan rooms without charge. It is the intent of the All American Independent School District that historically underutilized businesses be afforded every opportunity to participate in its construction projects as prime contractors, subcontractors and/or suppliers.

Plans and specifications are on file at the following locations and may be examined without charge:
AGC of America
10806 Gulfdale
San Antonio, Texas 78216

F. W. Dodge
333 Eastside Street
Houston, Texas 77098

BVCA Plan Room
2828 Finfeather
Bryan, Texas 77801

Houston Minority Business Development Center
1200 Smith Street, Ste. 2870
Houston, Texas 77002
PRE-BID CONFERENCE: A pre-Bid Conference will be held on Friday, April 11, 20__ at 1:00 p.m. on the site of the proposed project. All bidders are required to attend this conference.

FIGURE 1.2. Advertisement for Bids.

During the bidding period, the lead estimator needs to be certain that the contract documents are kept together. Never lend out portions of the documents. This practice will eliminate subcontractors' and material suppliers' complaints that they did not submit a complete proposal because they lacked part of the information required for a complete bid.

Some subcontractors and suppliers still prefer to work with paper copies of the plans. The general contractors often set aside space in their offices where the subcontractors' and material suppliers' estimators may work. In this manner, the contract documents never leave the contractor's office and are available to serve a large number of bidders who want to use the paper copies.

1–10 SOURCES OF ESTIMATING INFORMATION

For matters relevant to estimating and costs, the best source of information is your historical data. These figures allow for the pricing of the project to match how the company actually performs its construction. This information takes into account the talent and training of the craft personnel and the management abilities of the field staff personnel. In addition, it integrates the construction companies' practices and methodologies. This is why a careful, accurate accounting system combined with accuracy in field reports is so important. If all of the information relating to the job is tracked and analyzed, it will be available for future reference. Computerized cost accounting systems are very helpful in gathering this information and making it readily available for future reference. See *Construction Accounting and Financial Management* by Steven J. Peterson for more information on managing construction accounting systems.

There are several "guides to construction cost" manuals available; however, a word of extreme caution is offered regarding the use of these manuals. They are only *guides;* the figures should *rarely* be used to prepare an actual estimate. The manuals may be used as a guide in checking current prices and should enable the estimator to follow a more uniform system and save valuable time. The actual pricing in the manuals is most appropriately used in helping architects check approximate current prices and facilitate their preliminary estimate. In addition to these printed guides, many of these companies provide electronic databases that can be utilized by estimating software packages. However, the same caution needs to be observed as with the printed version. These databases represent an average of the methodologies of a few contractors. There is no simple way to convert this generalized information to match the specifics of the construction companies' methodologies.

WEB RESOURCES

www.fwdodge.com

enr.construction.com

REVIEW QUESTIONS

1. What information is contained in the working drawings?
2. What information is contained in the specifications?
3. What is the relationship between the working drawings and the specifications?
4. How does the work involved in being an estimator for a general contractor differ from that of an estimator who works for a subcontractor?
5. What is the difference between doing a quantity takeoff and doing a full detailed estimate?
6. What additional skills must the estimator have to be able to take a quantity survey and turn it into a detailed estimate?
7. What is the difference between competitive and negotiated bidding?
8. What is the difference between a detailed estimate and a square-foot estimate?
9. What are the contract documents, and why are they so important?
10. Why is it important to bid only from a full set of contract documents?

CHAPTER TWO

CONTRACTS, BONDS, AND INSURANCE

2-1 THE CONTRACT SYSTEM

Contracts may be awarded either by a single contract for the entire project or by separate contracts for the various phases required for the completion of the project. The single contract comprises all work required for the completion of a project and is the responsibility of a single, prime contractor. This centralization of responsibility provides that one of the distinctive functions of the prime contractor is to plan, direct, and coordinate all parties involved in completing the project. The subcontractors (including mechanical and electrical) and material suppliers involved in the project are responsible directly to the prime contractor, who in turn is responsible directly to the owner. The prime contractor must ensure that all work is completed in accordance with the contract documents, that the work is completed on time, and that all subcontractors and vendors have been paid. Under the system of separate contracts, the owner signs separate agreements for the construction of various portions of a project. The separate awards are often broken into the following phases:

1. General construction
2. Plumbing
3. Heating (ventilating, air-conditioning)
4. Electrical
5. Sewage disposal (if applicable)
6. Elevators (if applicable)
7. Specialties
8. Other

In this manner, the owner retains the opportunity to select the contractors for the various important phases of the project. Also, the responsibility for the installation and operation of these phases is directly between the owner and contractors rather than through the general contractor. In this contracting scheme, the owner or the owner's agents provide the coordination between the contractors. There is disagreement as to which system provides the owner with the best and the most cost-effective project. In certain states, laws require the award of separate contracts when public money is involved. Most general contractor trade organizations favor single contracts, but in contrast, most large specialty contract groups favor separate contracts. Owners, however, must critically evaluate their needs and talents and decide which method will provide them with the best product.

Under the single contract, the prime contractor will include a markup on the subcontracted items as compensation for the coordination effort and associated risk. If one of the subcontractors is unable to perform, the prime contractor absorbs the added cost of finding a replacement and any associated delays. It is this markup that encourages the owner to use separate contracts. If no general contractor assumes the responsibility for the management and coordination of the project, then the owner must shoulder this responsibility and its associated risk. If the owner does not have the talents or personnel to accomplish these tasks, he or she must hire them. The architect, for an added fee, may provide this service, or a construction management firm that specializes in project coordination may be hired.

2-2 TYPES OF AGREEMENTS

The owner-contractor agreement formalizes the construction contract. It incorporates, by reference, all other contract documents. The owner selects the type of agreement that will be signed: It may be a standard form of agreement such as those promulgated by the American Institute of Architects (AIA) or by other professional or trade organizations.

The agreement generally includes a description of the project and contract sum. Other clauses pertaining to alternates accepted, completion date, bonus and penalty clauses, and any other items that should be amplified, are included. No contract should ever be signed until the attorneys for all parties have had a chance to review the document. Each party's attorney will normally give attention only to matters

> *... agrees to build the project in accordance with the contract documents herein described for the lump sum of $275,375.00 ....*

FIGURE 2.1. Lump-Sum Agreement.

that pertain to his or her client's welfare. All contractors should employ the services of an attorney who understands the nuances of the construction industry and property law.

Types of agreements generally used are as follows:

1. Lump-sum agreement (stipulated sum, fixed price)
2. Unit-price agreement
3. Cost-plus-fee agreements

Lump-Sum Agreement (Stipulated Sum, Fixed Price)

In a lump-sum agreement, the contractor agrees to construct the project in accordance with the contract documents, for a set price arrived at through competitive bidding or negotiation. The contractor agrees that the work will be satisfactorily completed regardless of the difficulties encountered. This type of agreement (Figure 2.1) provides the owner advance knowledge of construction costs and requires the contractor to accept the bulk of the risk associated with the project. The accounting process is simple, and it creates centralization of responsibility in single contract projects. It is also flexible with regard to alternates and changes required on the project. However, the cost of these changes may be high. When the owner issues a change order, the contractor is entitled to additional monies for the actual work and for additional overhead, as well as additional time. If the original work is already in place, then the cost of the change order includes not only the cost of the new work but also the cost of removing the work that has already been completed. The later in the project that change orders are issued the greater their cost. Therefore, changes need to be identified as early as possible to minimize their impact on the construction cost and completion date. Seeking input from the contractor on the constructability of the project, adequacy of the drawings, and any recommended changes during the design process helps reduce the number of change orders. Contractors should help their clients understand that changes in the design during the construction phase are more expensive and increase the construction time more than if they were made during the design process. In addition, the contractor should not begin work on any change orders prior to receiving written authorization from the owner.

The major disadvantages to the contractor of lump-sum agreements are that the majority of the risk is placed upon the general contractor and they have to guarantee a price even though all of the costs are estimated.

Because of the very nature and risks associated with the lump-sum price, it is important that the contractor be able to accurately understand the scope of the project work required at the time of bidding.

Unit-Price Agreement

In a *unit-price agreement,* the contractor bases the bid on estimated quantities of work and on the completion of the work in accordance with the contract documents. The owner of the contracting agency typically provides the quantity takeoff. This type of contracting is most prevalent in road construction. Because of the many variables associated with earthwork, the main component of road projects, it is virtually impossible to develop exact quantities. The owner, therefore, provides the estimated quantities, and the contractors are in competition over their ability to complete the work rather than their quantity estimating ability. Figure 2.2 is an example of unit price quantities.

Bidders will base their bids on the quantities provided or will use their estimate of the quantities to determine their unit price bids. If contractors have insight into the quantities, they can use that information to their competitive advantage. The contractor's overhead is either directly or indirectly applied to each of the unit price items. If the contractor believes that the stated quantities are low, the overhead can be spread over a greater quantity rather than the quantity provided by the owner. This allows the contractor to submit a lower bid while making the same or more profit. In government agency projects, the low bidder will be determined based on the owner-provided quantities. In Figure 2.3, the illustration shows the unit price bid tabulation for a portion of the project.

Payments are made based on the price that the contractor bids for each unit of work and field measurements of the work actually completed. A field crew that represents the owner must make the verification of the in-place units, meaning that neither the owner nor the contractor will

No.	Quantity	Unit	Item
025-254-0300	1000	L.F.	Curb, Straight
025-254-0400	75	L.F.	Curb, Radius
022-304-0100	600	C.Y.	Compacted crushed stone base
025-104-0851	290	Tons	1 ½" thick asphalt

FIGURE 2.2. Typical Quantity Survey.

				Contractor 1		Contractor 2		Contractor 3	
				Bid Unit Price	Estimated Item Cost	Bid Unit Price	Estimated Item Cost	Bid Unit Price	Estimated Item Cost
No.	Quantity	Unit	Item						
025-254-0300	1000	L.F.	Curbs, Straight	$5.35	$5,350.00	$5.50	$5,500.00	$6.25	$6,250.00
025-254-0400	75	L.F.	Curbs, Radius	$9.25	$693.75	$9.36	$702.00	$8.00	$600.00
022-304-0100	600	C.Y.	Compacted Base	$31.50	$18,900.00	$32.50	$19,500.00	$38.50	$23,100.00
025-104-0851	290	Tons	1 1/2" Thick Asphalt	$42.50	$12,325.00	$43.75	$12,687.50	$45.00	$13,050.00
Total					$37,268.75		$38,389.50		$43,000.00

Unit Price Bid Tabulation

FIGURE 2.3. Unit Price Bid Tabulation.

know the exact cost of the project until its completion. The biggest advantages of the unit-price agreement are that

1. It allows the contractors to spend most of their time working on pricing the labor and materials required for the project while checking for the most economical approach to handle the construction process.
2. Under lump-sum contracts, each contractor does a quantity takeoff, which considerably increases the chances for quantity errors and adds overhead to all the contractors.

Cost-Plus-Fee Agreements

In *cost-plus-fee agreements,* the contractor is reimbursed for the construction costs as defined in the agreement. However, the contractor is not reimbursed for all items, and a complete understanding of reimbursable and nonreimbursable items is required. This arrangement is often used when speed, uniqueness of the project, and quality take precedence. This contract arrangement allows for construction to begin before all the drawings and specifications are completed, thus reducing the time required to complete the project. The contract should detail accounting requirements, record keeping, and purchasing procedures. There are many types of fee arrangements, any of which may be best in a given situation. The important point is that whatever the arrangement, not only the amount of the fee, but also how and when it will be paid to the contractor must be clearly understood by all parties.

Cost-plus-fee contracts include a project budget developed by the members of the project team. Although the owner typically is responsible for any expenditure over the project budget, all team members have an intrinsic motivation to maintain the project budget. The members of the project team put their professional reputation at risk. It is unlikely that an owner would repeatedly hire a contractor who does not complete projects within budget. The same holds true for architects if they design projects that are typically over budget; they most likely will not get repeat business. When dealing with owners/developers, there is little elasticity in the project budget. Their financing, equity partners, and rental rates are based on a construction budget, and few sources for additional funds are available.

Percentage Fee.   The advantage of the percentage fee is that it allows the owner to save fees paid to the contractor when construction costs go down. The major disadvantage is

that the fee increases with construction costs, so there is little incentive on the contractor's part to keep costs low. The primary incentive for a contractor to keep costs under control is the maintenance of her reputation.

Fixed Fee.   The advantage of the fixed fee is that it removes the temptation for the contractor to increase construction costs to increase his fee. The disadvantage is that the contractor has little incentive to keep the costs low, because the fee is the same if the project is over budget as it is if the project is under budget.

Fixed Fee with Guaranteed Maximum Cost. Advantages of this fixed fee with a guaranteed maximum cost (g-max) are that a guaranteed maximum cost is assured to the owner, and it generally provides an incentive to contractors to keep the costs down since they share in any savings. Again, the contractor assumes a professional status. Disadvantages are that drawings and specifications must be complete enough to allow the contractor to set a realistic maximum cost.

Sliding Scale Fee.   The sliding scale fee provides an answer to the disadvantages of the percentage fee, because as the cost of the project increases, the percent fee of construction decreases. The contractor is motivated to provide strong leadership so that the project will be completed swiftly at a low cost. The disadvantage is that extensive changes may require modifications of the scale.

Fixed Fee with a Bonus and Penalty.   With this type of fixed fee, the contractor is reimbursed the actual cost of construction plus a fee. A target cost estimate is set up; and if the cost is less than the target amount, the contractor receives a bonus in the form of a percentage of the savings. If the cost goes over the target figure, there is a penalty (reduction of percentage).

2-3 AGREEMENT PROVISIONS

Although the exact type and form of agreement may vary, certain provisions are included in all of them. Contractors must check each of those items carefully before signing the agreement.

Scope of the Work. The project, drawings, and specifications are identified; the architect is listed. The contractor agrees to furnish all material and perform all of the work for the project in accordance with the contract documents.

Time of Completion. The agreement should specify the starting and completion time. Starting time should never precede the execution date of the contract. The completion date is expressed either as a number of days or as a specific date. If the number of days is used, it should be expressed in calendar days and not working days to avoid subsequent disagreements about the completion date. Any liquidated damages or penalty and bonus clauses are usually included here; they should be clearly written and understood by all parties concerned.

Contract Sum. Under a lump-sum agreement, the *contract sum* is the amount of the accepted bid or negotiated amount. The accepted bid amount may be adjusted by the acceptance of alternates or by minor revisions that were negotiated with the contractor after receiving the bid. In agreements that involve cost-plus conditions, there are generally articles concerning the costs for which the owner reimburses the contractor. Customarily, not all costs paid by the contractor are reimbursed by the owner; reimbursable and nonreimbursable items should be listed. The contractor should be certain that all costs incurred in the construction are included somewhere. Also, in cost-plus-fee agreements, the exact type of compensation should be stipulated.

Progress Payments. Because of the cost and duration of construction projects, contractors must receive payments as work is completed. These payments are based on the completed work and stored materials. However, the owner typically retains a portion of all progress payments as security to ensure project completion and payment of all contractor's financial obligations.

The *due date* for payments is any date mutually acceptable to all concerned. In addition, the agreement needs to spell out the maximum time the architect/engineer can hold the contractor's application for payment and how soon the owner must pay the contractor after the architect makes out the certificate of payment. There should also be some mention of possible contractor action if these dates are not met. Generally, the contractor has the option of stopping the work. Some contracts state that if the contractor is not paid when due, the owner must also pay interest at the legal rate in force in the locale of the building.

Retained Percentage. It is customary for the owner to withhold a certain percentage of the payments, which is referred to as *retainage,* and is protection for the owner to ensure the completion of the contract and payment of the contractor's financial obligations. The most typical retainage is 10 percent, but other percentages are also used. On some projects, this retainage is continued through the first half of the project, but not through the last half.

In some states, the retainage is set by statute and limits the owner's liability for the nonpayment to subcontractors and suppliers to the amount retained. In these states, if the owner retains less than the percentage specified, liability is still the amount set by statute.

Schedule of Values. The contractor furnishes the architect/engineer with a statement, called a *schedule of values,* that shows sales prices for specific items within the project. This statement breaks the project into quantifiable components. Contractors typically overvalue the initial items on the project. This practice is referred to as *front-end loading.*

Work in Place and Stored Materials. The *work in place* is usually calculated as a percentage of the work that has been completed. The amounts allowed for each item in the schedule of values are used as the base amounts due on each item. The value of the work completed is equal to the work in place for each line item in the schedule of values multiplied by the sales price or the value for that line item. The contractor may also receive payment for materials stored on the site or some other mutually agreed upon location. The contractor may have to present proof of purchase, bills of sale, or other assurances to receive payment for materials stored off the job site.

Acceptance and Final Payment. The acceptance and final payment sets a time for the final payment to the contractor. When the final inspection, certification of completion, acceptance of the work, and required lien releases are completed, the contractor will receive the *final payment,* which is the amount of retainage withheld throughout the construction period. Many agreements are set up so that if full completion is held up through no fault of the contractor, the architect can issue a certificate for part of the final retainage.

2-4 BONDS

Often referred to as *surety bonds,* bonds are written documents that describe the conditions and obligations relating to the agreement. (In law, a surety is one who guarantees payment of another party's obligations.) The bond is not a financial loan or insurance policy, but serves as an endorsement of the contractor. The bond guarantees that the contract documents will be complied with, and all costs relative to the project will be paid. If the contractor is in breach of contract, the surety must complete the terms of the contract. Contractors most commonly use a corporate surety that specializes in construction bonds. The owner will reserve the right to approve the surety company and form of bond, as the bond is worth no more than the company's ability to pay.

To eliminate the risk of nonpayment, the contract documents will, on occasion, require that the bonds be obtained from one specified company. To contractors, this may mean doing business with an unfamiliar company, and they may be required to submit financial reports, experience records, projects (in progress and completed), as well as other material that could create a long delay before the bonds are approved. It is up to the owner to decide whether the surety obtained by the contractor is acceptable or to specify a company. In the latter case, the contractor has the option of complying with the contract documents or not submitting a bid on the project. No standard form of surety bond is applicable to every project. Statutory bonds are bond forms that conform to a particular governing statute; they vary from one jurisdiction to another. Nonstatutory bonds are used when a statutory form is not required. There is no standard form of bond that is nationally accepted. The customary bond forms used by the surety companies are generally employed.

Bid Bond

The *bid bond* ensures that if a contractor is awarded the bid within the time specified, the contractor will enter into the contract and provide all other specified bonds. If the contractor fails to do so without justification, the bond will be forfeited to the owner. The amount forfeited will in no case exceed the amount of the bond or the difference between the original bid and the next highest bid that the owner may, in good faith, accept. The contractor's surety usually provides these bonds free or for a small annual service charge of from $25 to $100. The usual contract requirements for bid bonds specify that they must be 5 to 10 percent of the bid price, but higher percentages are sometimes used. Contractors should inform the surety company once the decision to bid a project is made, especially if it is a larger amount than they usually bid or if they already have a great deal of work. Once a surety writes a bid bond for a contractor, that company is typically obligated to provide the other bonds required for the project. Surety companies therefore may do considerable investigation of contractors before they will write a bid bond for them, particularly if it is a contractor with whom they have not done business before or with whom they have never had a bid bond.

Performance Bond

The *performance bond* guarantees the owner that the contractor will perform all work in accordance with the contract documents and that the owner will receive the project built in substantial agreement with the documents. It protects the owner against default on the part of the contractor up to the amount of the bond penalty. The warranty period of one year is usually covered under the bond also. The contractor should check the documents to see if this bond is required and in what amount, and must also make the surety company aware of all requirements. Most commonly these bonds must be made out in the amount of 100 percent of the contract price. The rates vary according to the classification of work being bid. If the work required on the project comes under more than one classification, whichever premium rate is the highest is the one used. Almost all general construction work on buildings rates a "B" classification. The premium

rates are subject to change without notice, and it is possible to get lower rates from "preferred companies" if the contractor is acceptable to the company.

Labor and Material Bond

The *labor and material bond,* also referred to as a payment bond, guarantees the payment of the contractor's bill for labor and materials used or supplied on the project. It acts as protection for the third parties and the owner, who are exempted from any liabilities in connection with claims against the project. In public works, the statutes of that state or entity will determine whether a specific item of labor or material is covered. Claims must be filed in accordance with the requirements of the bond used. Most often a limitation is included in the bond stating that the claimant must give written notice to the general contractor, owner, or surety within 90 days after the last day the claimant performed any work on the project or supplied materials to it.

Subcontractor Bonds. *Performance,* and *labor and materials* (payment) bonds are those that the subcontractors must supply to the prime contractor. They protect the prime contractor against financial loss and litigation due to default by a subcontractor. Because these bonds vary considerably, prime contractors may require the use of their own bond forms or reserve the right of approval of both the surety and form of the bond. These types of bonds are often used when the general contractor is required to post a bond for the project. This arrangement protects the general contractor, reduces risk, and allows the general contractor greater bonding capacity.

License or Permit Bond. The license or permit bond is required of the prime contractor when a state law or municipal ordinance requires a contractor's license or permit. The bond guarantees compliance with statutes and ordinances.

Lien Bond. The lien bond is provided by the prime contractor and indemnifies the owner against any losses resulting from liens filed against the property.

2-5 OBTAINING BONDS

The surety company will thoroughly check out a contractor before it furnishes a bid bond. The surety checks such items as financial stability, integrity, experience, equipment, and professional ability of the firm. The contractor's relations with sources of credit will be reviewed, as will current and past financial statements. At the end of the surety company's investigations, it will establish a maximum bonding capacity for that particular contractor. The investigation often takes time to complete, so contractors should apply well in advance of the time at which they desire bonding; waits of two months are not uncommon. Each time the contractor requests a bid bond for a particular job, the application must be approved. If the contractor is below the workload limit and there is nothing unusual about the project, the application will be approved quickly. If a contractor's maximum bonding capacity is approached, or if the type of construction is new to the particular contractor or is not conventional, a considerably longer time may be required. The surety puts the contractor through investigations before giving a bond for a project to be sure that the contractor is not overextended.

To be successful, the contractor requires equipment, working capital, and an organization. None of these should be spread thin. The surety checks the contractor's availability of credit so that, if already overextended, the contractor will not take on a project that is too big. The surety will want to know if the contractor has done other work similar to that about to be bid upon. If so, the surety will want to know the size of the project. The surety will encourage contractors to stay with the type of work in which they have the most experience. The surety may also check progress payments and the amount of work to be subcontracted. If the surety refuses the contractor a bond, the contractor must first find out why and then attempt to demonstrate to the surety that the conditions questioned can be resolved.

The contractor must remember that the surety is in business to make money and can only do so if the contractor is successful. The surety is not going to take any unnecessary chances in the decision to bond a project. At the same time, some surety companies are more conservative than others. If contractors believe that their surety company is too conservative or not responsive enough to their needs, they should shop around, talk with other sureties, and try to find one that will work with their organization. If contractors are approaching a surety for the first time, they should pay particular attention to what services the company provides. Some companies provide a reporting service that includes projects being bid and low bidders. Also, when contractors are doing public work, the surety company can find out when the particular contractor can expect to get payment and what stage the job is in at a given time. Contractors need to select the company that seems to be the most flexible in its approach and offers the greatest service.

2-6 INSURANCE

Contractors must carry insurance for the protection of the assets of their business, and because it is often required by the contract documents. The contractor's selection of an insurance broker is of utmost importance, because the broker must be familiar with the risks and problems associated with construction projects. The broker also must protect the contractor against the wasteful overlapping of protection, yet there can be no gaps in the insurance coverage that might cause the contractor serious financial loss. Copies of the insurance requirements in the contract documents should be forwarded immediately to the insurance broker. The broker should be under strict instructions from the contractor that all insurance must be supplied in accordance with the contract documents. The broker will then supply the cost of the required insurance to the contractor for inclusion in the bidding proposal.

Insurance is not the same as a bond. With an insurance policy, the responsibility for specified losses is shouldered by the insurance company. In contrast, with a bond, the bonding companies will fulfill the obligations of the bond and turn to the contractor to reimburse them for all the money that they expended on their behalf. In addition to the insurance required by the contract documents, the contractor also has insurance requirements. Certain types of insurance are required by state statute. For example, some states require all employers to obtain workers' compensation and motor vehicle insurance. In addition, there is other insurance that is required but provided by governmental agencies. Examples of this type of insurance are unemployment and social security. Other insurance that is usually carried includes fire, liability, accident, life, hospitalization, and business interruption. No attempt will be made in this book to describe all of the various types of insurance that are available. A few of the most common types are described here.

Workers' Compensation Insurance. A workers' compensation insurance policy provides benefits to employees or their families if they are killed or injured during the course of work. The rates charged for this insurance vary by state, type of work, and the contractor. The contractor's experience rating depends on the company's work record with regard to accidents and claims. Contractors with the fewest claims enjoy lower premiums. Workers should be classified correctly to keep rates as low as possible. The rate charged is expressed as a percentage of payroll and will vary considerably. The rates may range from less than 1 percent to over 30 percent, depending on the location of the project and the type of work being performed. The contractor pays the cost of the policy in full.

Builder's Risk Fire Insurance. Builder's risk fire insurance protects projects under construction against direct loss due to fire and lightning. This insurance also covers temporary structures, sheds, materials, and equipment stored at the site. The cost usually ranges from $0.40 to $1.05 per $100 of valuation, depending on the project location, type of construction assembly, and the company's past experience with a contractor. If desirable, the policy may be extended to all direct loss causes, including windstorms, hail, explosions, riots, civil commotion, vandalism, and malicious mischief. Also available are endorsements that cover earthquakes and sprinkler leakage. Other policies that fall under the category of project and property insurance are as follows:

1. Fire insurance on the contractor's buildings
2. Equipment insurance
3. Burglary, theft, and robbery insurance
4. Fidelity insurance, which protects the contractor against loss caused by any dishonesty on the part of employees

WEB RESOURCES

www.sio.org

REVIEW QUESTIONS

1. What is a single contract, and what are its principal advantages and disadvantages for the owner?

2. What are separate contracts, and what are the principal advantages and disadvantages for the owner?

3. With separate contracts, describe three options available to the owner for managing the contractor's work on the project.

4. List and briefly define the types of agreements that may be used for the owner's payment to the contractor.

5. What is the "time of completion," and why must it be clearly stated in the contract agreement provisions?

6. What are progress payments, and why are they important to the contractor?

7. What is retainage, where is the amount specified, and why is it used?

8. What is a bid bond, and how does it protect the owner?

9. Where would information be found on whether a bid bond was required and, if so, its amount?

10. What are performance bonds? Are they required on all proposals?

11. How are the various surety bonds that may be required on a specific project obtained?

12. How does insurance differ from a surety bond?

CHAPTER THREE

PROJECT MANUAL

3-1 INTRODUCTION

The project manual, often referred to as the specifications, is a document that accompanies the drawings and includes information on how to bid the project, the contractual obligations of the successful contractor, and the specifications for the materials used in the construction. In this book, the term "project manual" is used when referring to the complete written document or a set of documents that accompanies the plans. The term "specifications" is used when referring to the material or technical specifications.

The contractor submits a bid or proposal based on the drawings and the project manual. The contractor is responsible for everything contained in the project manual and what is covered on the drawings. The project manual should be read thoroughly and reviewed when necessary. Contractors have a tendency to read only the portions of the project manual that refer to materials and workmanship; however, they are also responsible for anything stated in the proposal (or bid form), the information to bidders, the general conditions, and the supplementary general conditions.

There is a tendency among estimators to simply skim over the project manual. Reading the average project manual is time-consuming, but many important items are mentioned only in the project manual and not on the drawings. Because the project manual is part of the contract documents, the general contractor is responsible for the work and materials mentioned in it.

The project manual contains items ranging from the types of bonds and insurance required to the type, quality, and color of materials used on the job. A thorough understanding of the materials contained in the specification portion of the project manual may make the difference between being the low bidder and not.

There is no question that skimming the project manual is risky. Either the bids will be too high, because of contingency allowances added to cover uncertainty in the bid, or too low, from not including required items.

The project manual is generally presented in the following sequence:

1. Invitation to bid (advertisement for bidders)
2. Instructions to bidders
3. Bid (or proposal) forms
4. Form of owner/contractor agreement
5. Form of bid bond
6. Forms of performance bonds
7. General conditions of the contract
8. Supplementary general conditions
9. Specifications (technical specifications)

Separate contracts and many large projects often have a separate project manual for the mechanical and electrical trades.

3-2 CONSTRUCTION SPECIFICATIONS INSTITUTE

The Construction Specifications Institute (CSI) has developed a standard format for organizing the specification known as the MasterFormat. Prior to 2004, the MasterFormat consisted of 17 divisions (0 through 16). In 2004, the MasterFormat was revised to include 50 divisions (0 through 49), with many of the divisions being reserved for future use. Each division is subdivided into specific areas; for example, division 8 covers openings (doors, windows, and skylights), while the subdivision 08 50 00 deals specifically with windows. Subdivision 08 50 00 is further divided, and subdivision 08 51 00 deals with all types of metal windows. Subdivision 05 51 00 is subdivided by type of metal window, with 08 51 13 dealing with aluminum windows. The MasterFormat has found wide acceptance in the construction industry. The first two levels of the 2004 MasterFormat are shown in Figure 3.1. The CSI MasterFormat also ties in easily with computer programs and cost accounting systems. It is not necessary to memorize the major

DIVISION 00 – PROCUREMENT AND CONTRACTING REQUIREMENTS

00 00 00	PROCUREMENT AND CONTRACTING REQUIREMENTS
00 10 00	SOLICITATION
00 20 00	INSTRUCTIONS FOR PROCUREMENT
00 30 00	AVAILABLE INFORMATION
00 40 00	PROCUREMENT FORMS AND SUPPLEMENTS
00 50 00	CONTRACTING FORMS AND SUPPLEMENTS
00 60 00	PROJECT FORMS
00 70 00	CONDITIONS OF THE CONTRACT
00 90 00	REVISIONS, CLARIFICATIONS, AND MODIFICATIONS

DIVISION 01 – GENERAL REQUIREMENTS

01 00 00	GENERAL REQUIREMENTS
01 10 00	SUMMARY
01 20 00	PRICE AND PAYMENT PROCEDURES
01 30 00	ADMINISTRATIVE REQUIREMENTS
01 40 00	QUALITY REQUIREMENTS
01 50 00	TEMPORARY FACILITIES AND CONTROLS
01 60 00	PRODUCT REQUIREMENTS
01 70 00	EXECUTION AND CLOSEOUT REQUIREMENTS
01 80 00	PERFORMANCE REQUIREMENTS
01 90 00	LIFE CYCLE ACTIVITIES

DIVISION 02 – EXISTING CONDITIONS

02 00 00	EXISTING CONDITIONS
02 20 00	ASSESSMENT
02 30 00	SUBSURFACE INVESTIGATION
02 40 00	DEMOLITION AND STRUCTURE MOVING
02 50 00	SITE REMEDIATION
02 60 00	CONTAMINATED SITE MATERIAL REMOVAL
02 70 00	WATER REMEDIATION
02 80 00	FACILITY REMEDIATION

DIVISION 03 – CONCRETE

03 00 00	CONCRETE
03 10 00	CONCRETE FORMING AND ACCESSORIES
03 30 00	CAST-IN-PLACE CONCRETE
03 40 00	PRECAST CONCRETE
03 50 00	CAST DECKS AND UNDERLAYMENT
03 60 00	GROUTING
03 70 00	MASS CONCRETE
03 80 00	CONCRETE CUTTING AND BORING

DIVISION 04 – MASONRY

04 00 00	MASONRY
04 20 00	UNIT MASONRY
04 40 00	STONE ASSEMBLIES
04 50 00	REFRACTORY MASONRY
04 60 00	CORROSION-RESISTANT MASONRY
04 70 00	MANUFACTURED MASONRY

DIVISION 05 – METALS

05 00 00	METALS
05 10 00	STRUCTURAL METAL FRAMING
05 20 00	METAL JOISTS
05 30 00	METAL DECKING
05 40 00	COLD-FORMED METAL FRAMING
05 50 00	METAL FABRICATIONS
05 70 00	DECORATIVE METAL

DIVISION 06 – WOOD, PLASTICS, AND COMPOSITES

06 00 00	WOOD, PLASTICS, AND COMPOSITES
06 10 00	ROUGH CARPENTRY
06 20 00	FINISH CARPENTRY
06 40 00	ARCHITECTURAL WOODWORK
06 50 00	STRUCTURAL PLASTICS
06 60 00	PLASTIC FABRICATIONS
06 70 00	STRUCTURAL COMPOSITES
06 80 00	COMPOSITE FABRICATIONS

DIVISION 07 – THERMAL AND MOISTURE PROTECTION

07 00 00	THERMAL AND MOISTURE PROTECTION
07 10 00	DAMPPROOFING AND WATERPROOFING
07 20 00	THERMAL PROTECTION
07 25 00	WEATHER BARRIERS
07 30 00	STEEP SLOPE ROOFING
07 40 00	ROOFING AND SIDING PANELS
07 50 00	MEMBRANE ROOFING
07 60 00	FLASHING AND SHEET METAL
07 70 00	ROOF AND WALL SPECIALTIES AND ACCESSORIES
07 80 00	FIRE AND SMOKE PROTECTION
07 90 00	JOINT PROTECTION

DIVISION 08 – OPENINGS

08 00 00	OPENINGS
08 10 00	DOORS AND FRAMES
08 30 00	SPECIALTY DOORS AND FRAMES
08 40 00	ENTRANCES, STOREFRONTS, AND CURTAIN WALLS
08 50 00	WINDOWS
08 60 00	ROOF WINDOWS AND SKYLIGHTS
08 70 00	HARDWARE
08 80 00	GLAZING
08 90 00	LOUVERS AND VENTS

DIVISION 09 – FINISHES

09 00 00	FINISHES
09 20 00	PLASTER AND GYPSUM BOARD
09 30 00	TILING
09 50 00	CEILINGS
09 60 00	FLOORING
09 70 00	WALL FINISHES
09 80 00	ACOUSTIC TREATMENT
09 90 00	PAINTING AND COATING

DIVISION 10 – SPECIALTIES

10 00 00	SPECIALTIES
10 10 00	INFORMATION SPECIALTIES
10 20 00	INTERIOR SPECIALTIES
10 30 00	FIREPLACES AND STOVES
10 40 00	SAFETY SPECIALTIES
10 50 00	STORAGE SPECIALTIES
10 70 00	EXTERIOR SPECIALTIES
10 80 00	OTHER SPECIALTIES

DIVISION 11 – EQUIPMENT

11 00 00	EQUIPMENT
11 10 00	VEHICLE AND PEDESTRIAN EQUIPMENT
11 15 00	SECURITY, DETENTION AND BANKING EQUIPMENT
11 20 00	COMMERCIAL EQUIPMENT
11 30 00	RESIDENTIAL EQUIPMENT
11 40 00	FOODSERVICE EQUIPMENT
11 50 00	EDUCATIONAL AND SCIENTIFIC EQUIPMENT
11 60 00	ENTERTAINMENT EQUIPMENT
11 65 00	ATHLETIC AND RECREATIONAL EQUIPMENT
11 70 00	HEALTHCARE EQUIPMENT
11 80 00	COLLECTION AND DISPOSAL EQUIPMENT
11 90 00	OTHER EQUIPMENT

DIVISION 12 – FURNISHINGS

12 00 00	FURNISHINGS
12 10 00	ART
12 20 00	WINDOW TREATMENTS
12 30 00	CASEWORK
12 40 00	FURNISHINGS AND ACCESSORIES
12 50 00	FURNITURE
12 60 00	MULTIPLE SEATING
12 90 00	OTHER FURNISHINGS

DIVISION 13 – SPECIAL CONSTRUCTION

13 00 00	SPECIAL CONSTRUCTION
13 10 00	SPECIAL FACILITY COMPONENTS
13 20 00	SPECIAL PURPOSE ROOMS
13 30 00	SPECIAL STRUCTURES
13 40 00	INTEGRATED CONSTRUCTION
13 50 00	SPECIAL INSTRUMENTATION

DIVISION 14 – CONVEYING EQUIPMENT

14 00 00	CONVEYING EQUIPMENT
14 10 00	DUMBWAITERS
14 20 00	ELEVATORS
14 30 00	ESCALATORS AND MOVING WALKS
14 40 00	LIFTS
14 70 00	TURNTABLES
14 80 00	SCAFFOLDING
14 90 00	OTHER CONVEYING EQUIPMENT

FIGURE 3.1. CSI MasterFormat Divisions (2004 edition).

The Numbers and Titles used in this product are from *MasterFormat*™ and is published by The Construction Specification Institute (CSI) and Construction Specifications Canada (CSC), and is used with permission from CSI, 2009.

The Construction Specifications Institute (CSI)

99 Canal Center Plaza, Suite 300

Alexandria, VA 22314

800-689-2900; 703-684-0300

CSINet URL: http://www.csinet.org

DIVISION 21 – FIRE SUPPRESSION

21 00 00	FIRE SUPPRESSION
21 10 00	WATER-BASED FIRE-SUPPRESSION SYSTEMS
21 20 00	FIRE-EXTINGUISHING SYSTEMS
21 30 00	FIRE PUMPS
21 40 00	FIRE-SUPPRESSION WATER STORAGE

DIVISION 22 – PLUMBING

22 00 00	PLUMBING
22 10 00	PLUMBING PIPING AND PUMPS
22 30 00	PLUMBING EQUIPMENT
22 40 00	PLUMBING FIXTURES
22 50 00	POOL AND FOUNTAIN PLUMBING SYSTEMS
22 60 00	GAS AND VACUUM SYSTEMS FOR LABORATORY AND HEALTHCARE FACILITIES

DIVISION 23 – HEATING, VENTILATING, AND AIR-CONDITIONING (HVAC)

23 00 00	HEATING, VENTILATING, AND AIR-CONDITIONING (HVAC)
23 10 00	FACILITY FUEL SYSTEMS
23 20 00	HVAC PIPING AND PUMPS
23 30 00	HVAC AIR DISTRIBUTION
23 40 00	HVAC AIR CLEANING DEVICES
23 50 00	CENTRAL HEATING EQUIPMENT
23 60 00	CENTRAL COOLING EQUIPMENT
23 70 00	CENTRAL HVAC EQUIPMENT
23 80 00	DECENTRALIZED HVAC EQUIPMENT

DIVISION 25 – INTEGRATED AUTOMATION

25 00 00	INTEGRATED AUTOMATION
25 10 00	INTEGRATED AUTOMATION NETWORK EQUIPMENT
25 30 00	INTEGRATED AUTOMATION INSTRUMENTATION AND TERMINAL DEVICES
25 50 00	INTEGRATED AUTOMATION FACILITY CONTROLS
25 90 00	INTEGRATED AUTOMATION CONTROL SEQUENCES

DIVISION 26 – ELECTRICAL

26 00 00	ELECTRICAL
26 10 00	MEDIUM-VOLTAGE ELECTRICAL DISTRIBUTION
26 20 00	LOW-VOLTAGE ELECTRICAL DISTRIBUTION
26 30 00	FACILITY ELECTRICAL POWER GENERATING AND STORING EQUIPMENT
26 40 00	ELECTRICAL AND CATHODIC PROTECTION
26 50 00	LIGHTING

DIVISION 27 – COMMUNICATIONS

27 00 00	COMMUNICATIONS
27 10 00	STRUCTURED CABLING
27 20 00	DATA COMMUNICATIONS
27 30 00	VOICE COMMUNICATIONS
27 40 00	AUDIO-VIDEO COMMUNICATIONS
27 50 00	DISTRIBUTED COMMUNICATIONS AND MONITORING SYSTEMS

DIVISION 28 – ELECTRONIC SAFETY AND SECURITY

28 00 00	ELECTRONIC SAFETY AND SECURITY
28 10 00	ELECTRONIC ACCESS CONTROL AND INTRUSION DETECTION
28 20 00	ELECTRONIC SURVEILLANCE
28 30 00	ELECTRONIC DETECTION AND ALARM
28 40 00	ELECTRONIC MONITORING AND CONTROL

DIVISION 31 – EARTHWORK

31 00 00	EARTHWORK
31 10 00	SITE CLEARING
31 20 00	EARTH MOVING
31 30 00	EARTHWORK METHODS
31 40 00	SHORING AND UNDERPINNING
31 50 00	EXCAVATION SUPPORT AND PROTECTION
31 60 00	SPECIAL FOUNDATIONS AND LOAD-BEARING ELEMENTS
31 70 00	TUNNELING AND MINING

DIVISION 32 – EXTERIOR IMPROVEMENTS

32 00 00	EXTERIOR IMPROVEMENTS
32 10 00	BASES, BALLASTS, AND PAVING
32 30 00	SITE IMPROVEMENTS
32 70 00	WETLANDS
32 80 00	IRRIGATION
32 90 00	PLANTING

DIVISION 33 – UTILITIES

33 00 00	UTILITIES
33 10 00	WATER UTILITIES
33 20 00	WELLS
33 30 00	SANITARY SEWERAGE UTILITIES
33 40 00	STORM DRAINAGE UTILITIES
33 50 00	FUEL DISTRIBUTION UTILITIES
33 60 00	HYDRONIC AND STEAM ENERGY UTILITIES
33 70 00	ELECTRICAL UTILITIES
33 80 00	COMMUNICATIONS UTILITIES

DIVISION 34 – TRANSPORTATION

34 00 00	TRANSPORTATION
34 10 00	GUIDEWAYS/RAILWAYS
34 20 00	TRACTION POWER
34 40 00	TRANSPORTATION SIGNALING AND CONTROL EQUIPMENT
34 50 00	TRANSPORTATION FARE COLLECTION EQUIPMENT
34 70 00	TRANSPORTATION CONSTRUCTION AND EQUIPMENT
34 80 00	BRIDGES

DIVISION 35 – WATERWAY AND MARINE CONSTRUCTION

35 00 00	WATERWAY AND MARINE CONSTRUCTION
35 10 00	WATERWAY AND MARINE SIGNALING AND CONTROL EQUIPMENT
35 20 00	WATERWAY AND MARINE CONSTRUCTION AND EQUIPMENT
35 30 00	COASTAL CONSTRUCTION
35 40 00	WATERWAY CONSTRUCTION AND EQUIPMENT
35 50 00	MARINE CONSTRUCTION AND EQUIPMENT
35 70 00	DAM CONSTRUCTION AND EQUIPMENT

DIVISION 40 – PROCESS INTEGRATION

40 00 00	PROCESS INTEGRATION
40 10 00	GAS AND VAPOR PROCESS PIPING
40 20 00	LIQUIDS PROCESS PIPING
40 30 00	SOLID AND MIXED MATERIALS PIPING AND CHUTES
40 40 00	PROCESS PIPING AND EQUIPMENT PROTECTION
40 80 00	COMMISSIONING OF PROCESS SYSTEMS
40 90 00	INSTRUMENTATION AND CONTROL FOR PROCESS SYSTEMS

DIVISION 41 – MATERIAL PROCESSING AND HANDLING EQUIPMENT

41 00 00	MATERIAL PROCESSING AND HANDLING EQUIPMENT
41 10 00	BULK MATERIAL PROCESSING EQUIPMENT
41 20 00	PIECE MATERIAL HANDLING EQUIPMENT
41 30 00	MANUFACTURING EQUIPMENT
41 40 00	CONTAINER PROCESSING AND PACKAGING
41 50 00	MATERIAL STORAGE
41 60 00	MOBILE PLANT EQUIPMENT

DIVISION 42 – PROCESS HEATING, COOLING, AND DRYING EQUIPMENT

42 00 00	PROCESS HEATING, COOLING, AND DRYING EQUIPMENT
42 10 00	PROCESS HEATING EQUIPMENT
42 20 00	PROCESS COOLING EQUIPMENT
42 30 00	PROCESS DRYING EQUIPMENT

DIVISION 43 – PROCESS GAS AND LIQUID HANDLING, PURIFICATION, AND STORAGE EQUIPMENT

43 00 00	PROCESS GAS AND LIQUID HANDLING, PURIFICATION, AND STORAGE EQUIPMENT
43 10 00	GAS HANDLING EQUIPMENT
43 20 00	LIQUID HANDLING EQUIPMENT
43 30 00	GAS AND LIQUID PURIFICATION EQUIPMENT
43 40 00	GAS AND LIQUID STORAGE

DIVISION 44 – POLLUTION CONTROL EQUIPMENT

44 00 00	POLLUTION CONTROL EQUIPMENT
44 10 00	AIR POLLUTION CONTROL
44 20 00	NOISE POLLUTION CONTROL
44 40 00	WATER TREATMENT EQUIPMENT
44 50 00	SOLID WASTE CONTROL

DIVISION 45 – INDUSTRY-SPECIFIC MANUFACTURING EQUIPMENT

45 00 00	INDUSTRY-SPECIFIC MANUFACTURING EQUIPMENT

DIVISION 48 – ELECTRICAL POWER GENERATION

48 00 00	ELECTRICAL POWER GENERATION
48 10 00	ELECTRICAL POWER GENERATION EQUIPMENT
48 70 00	ELECTRICAL POWER GENERATION TESTING

FIGURE 3.1. (*Continued*)

divisions of the format; their constant use will facilitate their memorization.

3–3 INVITATION TO BID (ADVERTISEMENT FOR BIDS)

In public construction, public agencies must conform to regulations that relate to the method they use in advertising for bids. Customarily, the notice of proposed bidding is posted in public places and on the Internet and by advertising for bids in newspapers (Figure 3.2), trade journals, and magazines. Where the advertisement is published, how often, and over what period of time it will be published vary considerably according to the jurisdictional regulations. An estimator must not be bashful. If contractors are interested in a certain project, they should never hesitate to call and ask when it will be bid, when and where the agencies will advertise for bids, or for any other information that may be of importance.

Generally, the advertisement describes the location, extent, and nature of the work. It will designate the authority under which the project originated. Concerning the bid, it will give the place where bidding documents are available and list the time, manner, and place where bids will be received. It will also list bond requirements and start and completion dates for the work.

In private construction, owners often do not advertise for bidders. They may choose to negotiate with a contractor of choice, put the job out to bid on an invitation basis, or put the project out for competitive bidding open to anyone who wants to bid. If the owner puts a job out for competitive bidding, the architect/engineer will call the construction reporting services, which will pass the information on to their members or subscribers.

INVITATION TO BID FROM NEWSPAPER

REQUEST FOR BID PROPOSAL:
Administration Building Annex, Project No. 1-2796, All American Independent School District, Littleville, Texas 77777.
RECEIPT OF BIDS:
Sealed Proposals will be received by the All American Independent School District at the Central Administration Building, 2005 Sarah Lane Littleville, Texas, until 2:00 p.m., April 24, 20__ and then publicly opened and read aloud. Bids mailed should be addressed to Mr. Ryan Smith, Director of Facilities, All American Independent School District, 2005 Sarah Lane, Littleville, Texas 77777, and should be clearly marked "HOLD FOR BID OPENING – PROJECT NO. 1-2796." Bid proposals cannot be withdrawn for sixty days from the date of bid opening.
A certified check or cashier's check on a State of National Bank or a Bidders Bond from an acceptable Surety Company authorized to transact business in the State of Texas, in the amount of not less than five percent (5%) of the greatest total amount of the Bidder's proposal must accompany each proposal as a guarantee that, if

awarded the contract, the Bidder will within ten (10) calendar days after award of contract enter into contract and execute Performance and Payment Bonds on the forms provided in the Contract Documents.
Proposals must be completed and submitted on the forms provided by the ARCHITECT. Incomplete bid proposals will invalidate the bid proposal and the bid will be rejected and returned to the bidder. The right to accept any bid, or to reject any or all bids and to waive all formalities is hereby reserved by the All American Independent School District.
SCOPE OF WORK: This project consists of constructing a new Administration building. The site for this building is located at 123 Ryan Lane, Littleville, Texas.
PRIME CONTRACT: All work will be awarded under a single prime contract.
Information and bidding documents can be obtained from: A. B. Architects & Associates, Architects / Engineers, 7920 Anita Circle, Littleville, Texas 77777.
Plan Deposit: $50.00 per set. The deposit will be returned if the documents are returned in good condition within three

weeks after bid opening; otherwise, no refund will be made. Checks to be made Payable to: "A. B. Architects & Associates, Architects / Engineers."
Bid documents are available to established plan rooms without charge. It is the intent of the All American Independent School District that historically underutilized businesses be afforded every opportunity to participate in its construction projects as prime contractors, subcontractors and/or suppliers.
Plans and specifications are on file at the following locations and may be examined without charge:
AGC of America
10806 Gulfdale
San Antonio, Texas 78216
F. W. Dodge
333 Eastside Street
Houston, Texas 77098
BVCA Plan Room
2828 Finfeather
Bryan, Texas 77801
Houston Minority Business Development Center
1200 Smith Street, Ste. 2870
Houston, Texas 77002
PRE-BID CONFERENCE:
A pre-Bid Conference will be held on Friday, April 11, 20__ at 1:00 p.m. on the site of the proposed project. All bidders are required to attend this conference.

FIGURE 3.2. Invitation to Bid.

3-4 INSTRUCTIONS TO BIDDERS (INFORMATION FOR BIDDERS)

Instructions to bidders is the document that states the procedures to be followed by all bidders. It states in what manner the bids must be delivered; the time, date, and location of bid opening; and whether it is a public opening. (Bids may be either opened publicly and read aloud or opened privately.)

The instructions to bidders states where the drawings and the project manual are available and the amount of the deposit required. It also lists the form of owner/contractor agreement to be used, bonds required, times of starting and project completion, and any other bidder requirements.

Each set of instructions is different and should be read carefully. In reading the instructions to bidders, contractors should note special items. Figure 3.3 is an example of a set of instructions to bidders.

GENERAL REQUIREMENTS

INFORMATION FOR BIDDERS

DEFINITIONS

All definitions in the General Conditions, AIA Document A201, apply to the Information for Bidders.

Owner is All American Independent School District, Littleville, Texas 77777
Architect is A. B. Architects & Associates, Architects / Engineers
Contractor will be the successful bidder with whom a contract is signed by the Owner.

RESPONSIBILITY

By making a bid, bidders represent that they have read and understand the bidding documents. It is also each bidder's responsibility to visit the site and become familiar with any local conditions which may affect the work.

BIDDING REQUIREMENTS

Submit all bids on prepared forms and at the time and place specified under "PROPOSALS."
All blanks pertinent to the contract proposal must be filled in.

BID SECURITY

Bid security, in the amount of 10% of the bid, must accompany each proposal. Bid bonds and certified checks, made payable to the Owner, are acceptable bid security. Bid security will be returned to all unsuccessful bidders after award of contract has been made by the Owner. If the successful bidder fails to enter into contract within ten (10) days after receiving written notice of the award of contract, that contractor's bid security shall be forfeited to the Owner.

PERFORMANCE BOND

The Owner reserves the right to require the bidder to furnish bonds covering the faithful performance of the Contract and the payment of all obligations arising from the performance of the Contract in such form and amount as the Owner may require. In the event the Owner elects to require said executed bond, the Owner shall cause the Contract sum to be increased by the cost to the Contractor of said bond.

AWARD OR REJECTION OF BID

The Owner reserves the right to reject any or all bids and to waive any informality or irregularity in any bid received. The Owner reserves the right to reject a bid if the data is not submitted as required by the bidding documents, if no bid security has been furnished, or if the bid is in any way incomplete or irregular.

POST-BID INFORMATION

Within seven (7) days after the Contract signing, the Contractor shall submit an itemized statement of costs for each major item of work and a list of the proposed subcontractors for approval.

PROPOSALS

Sealed proposals will be received by All American Independent School District, 2005 Sarah Lane, Littleville Texas 77777, until 2:00 p.m. on April 24, 20__

All proposals together with the bid security must be submitted in sealed envelope marked "PROPOSAL FOR PROJECT NO: 1-2796 and bear the title of the work and the name of the bidder.

All bidders are invited to be present at the bid opening. Proposals will be publicly opened and read aloud at that time.

SUBSTITUTIONS

All bids must be based on the materials and equipment described in the bidding documents.

Requests for material substitution must be submitted to the Architect, in writing, not less than ten (10) days prior to the bid date. Any approval of proposed substitution will be set forth in an addendum.

INTERPRETATIONS

All requests for interpretation or correction of any ambiguity, inconsistency, or error in the bidding documents must be made in writing not less than seven (7) days prior to the bid date.

Interpretations and corrections will be issued as an addendum by the Architect. Only written interpretations and corrections are binding. Interpretations and corrections are not binding when made verbally and not spelled out in writing.

FIGURE 3.3. Instructions to Bidders.

Proposals. Be sure to check exactly where the bids are being received. Be sure to check each addendum to see if the time or location has been changed. It is rather embarrassing (as well as unprofitable) to wind up in the wrong place at the right time or in the right place at the wrong time. Typically, bids will be returned unopened if they are submitted late. Figure 3.4 is an example of a proposal (bid) form.

Commencement and Completion. Work on the project will commence within a specified period after the execution of the contract. The contractor will have to determine the number of calendar days to complete the project. The completion date must be realistic, as most contractors have a tendency to be overly optimistic with work schedules. At the same time, they should not be overly conservative since it may cause the owner concern over their ability to expedite the work.

Responsibility of Bidders. Contractors should read the responsibilities to bidders section thoroughly, as it indicates the importance of checking for all of the drawings and a complete project manual. The contractor clearly has the burden to visit the site.

Award or Rejection of Bid. The owner may reserve the right to

1. Reject any or all bids
2. Accept a bid other than the lowest

GENERAL CONSTRUCTION PROPOSAL FORM
A. B. Architects & Associates, Architects / Engineers
7920 Anita Circle
Littleville, Texas 77777.

Proposals are due on April 24, 20__, at the Central Administration Building for All American Independent School District, 2005 Sarah Lane, Littleville, Texas 77777
Having carefully read the Instruction to Bidders, General Conditions, Supplementary General Conditions, the complete General Construction Specifications Divisions 1 through 17, Drawings A/101 through A/108, HVAC/1 through HVAC/3, PL/1 through PL/4 and E/1 through E/4, all dated March 15, 20__, for the new administration annex for All American School District and any addenda issued on the work, the undersigned submits proposals to furnish all labor and materials for the complete General Construction Work as follows:

BASE BID _____ ($_____) *Dollars*

ALTERNATE Proposals
Alternate No. One _____ ($_____) *Dollars*

Alternate No. Two _____ ($_____) *Dollars*

In accordance with the provision of the Information for Bidders, the estimated date of substantial completion is _____.

ADDENDA: Receipt of Addendum No. _____is hereby acknowledged.

The Undersigned certifies that the Contract Documents have been carefully examined and that the site of the work has been personally visited and that the amount and nature of the work to be done is understood and that at no time will a misunderstanding of the Contract Documents be pleaded.

ACCEPTANCE OF PROPOSAL: The Undersigned understands that the Owner reserves the right to accept or reject any or all Proposals, but that if written notice of the acceptance of the above Proposal is mailed, faxed, or delivered to the Undersigned within thirty (30) days after the formal opening of bids, or any time thereafter before this proposal is withdrawn, the Undersigned will enter into, execute, and deliver a contract within five (5) days after the date of said mailing, faxing, or delivering of such notice.

The above proposals are submitted by:

Firm: _____

By: _____

Titles: _____

Address: _____

Dated: _____, 20_____
Contractor to submit this Proposal in duplicate and retain the triplicate.

FIGURE 3.4. Bid or Proposal Form.

3. Reject any proposal not prepared and submitted in accordance with the contract documents.

These are common stipulations; in effect, owners may contract to any bidder they select. Needless to say, it causes some hard feelings when an owner does not accept the lowest bid.

Many owners are now selecting the contractor based on the best bid. The best bid is based upon a set of criteria established in the bid documents. Each proposal is rated for each criterion, and a final score is calculated by summing the score for each criterion multiplied by the weighting for that criterion.

3-5 BID (PROPOSAL) FORMS

If a prepared *proposal form* (Figure 3.4) is included in the project manual, the contractor must use this form to present the bid. By using a prepared bid form, the owner can evaluate all bids on the same basis.

The proposal form stipulates the price for which the contractor agrees to perform all of the work described in the contract documents. It also ensures that if the owner accepts the proposal within a certain time, the contractor must enter into an agreement or the owner may keep the bid security as liquidated damages.

The proposal must be submitted according to the requirements in the instructions to bidders. Any deviation from these requirements for the submission of the proposal may result in the proposal being rejected.

Contractors must fill in all blanks of the proposal, acknowledge receipt of all addenda, submit the required number of copies of the proposal, supply proper bid security, and be at the right place at the right time. Countless bids have never been opened, because they were delivered a few minutes late; others have been rejected, because a blank space was not filled in.

3-6 FORM OF OWNER/CONTRACTOR AGREEMENT

The owner/contractor agreement form spells out exactly the type or form of agreement between the owner and the contractor. The agreement may be a standard form published by the American Institute of Architects (AIA), in which case the form may be referenced rather than including a copy of the form in the project manual. Government agencies controlling the work usually have their own forms of agreement; the same is true for many corporations. In these cases, a copy of the agreement is included in the project manual. If the contract agreement is unfamiliar, the contractor must have his lawyer review it before submitting the bid. If the form of agreement is unacceptable, the contractor may prefer not to bid that particular project. Types of agreements are discussed in Chapter 2.

3-7 GENERAL CONDITIONS

The *general conditions* assembled and published by the AIA is the most commonly used standard form. It has found wide acceptance throughout the industry. Many branches of government as well as large corporations have also assembled their own versions of general conditions. The contractor must carefully read each article and make appropriate notes. In some cases, it is best for the contractor to give a copy of the general conditions to a lawyer for review and comment. In the event if contractors decide it is not in their best interest to work under the proposed set of general conditions, the architect/engineer should be informed of the reasons and asked if they would consider altering the conditions. If not, the contractor may decide it is best not to bid the project. Typically, all general conditions will include, in one form or another, the 14 topics included in the AIA general conditions. The general conditions clearly spells out the rights and responsibilities of all the parties. Obviously, the most stringent demands are placed on the contractors, because they are entrusted with the responsibility of actually building the project.

3-8 SUPPLEMENTARY GENERAL CONDITIONS

The *Supplementary General Conditions* of the project manual amends or supplements portions of the general conditions. It is through these supplemental conditions that the general conditions are geared to all the special requirements of geography, local requirements, and individual project needs. Part of the supplemental conditions cancels or amends the articles in the general conditions, while the remaining portion adds articles.

Contractors must carefully check the supplementary conditions, as each set is different. Items that may be covered in this section include insurance, bonds, and safety requirements. Also included may be comments concerning the following:

1. Pumping and shoring
2. Dust control
3. Temporary offices
4. Temporary enclosures
5. Temporary utilities
6. Temporary water
7. Material substitution
8. Soil conditions
9. Signs
10. Cleaning
11. Shop drawings—drawings that illustrate how specific portions of the work will be fabricated and/or installed
12. Surveys

As the contractor reviews the supplementary general conditions, notes must be made of the many requirements included in them, as they may be costly, and an amount to cover these items must be included in the estimate. When actually figuring the estimate, contractors will go through the entire supplementary general conditions carefully, noting all items that must be covered in the bid and deciding how much to allow for each item.

3-9 SPECIFICATIONS

Specifications, as defined by the AIA, are the written descriptions of materials, construction systems, and workmanship. The AIA further states that defining the quality of materials and the results to be provided by the application of construction methods is the purpose of the specifications.

The *specifications* (sometimes referred to as the technical specifications) generally follow the CSI MasterFormat. These specifications include the type of materials required, their required performance, and the method that must be used to obtain the specified result. When a particular method is specified, the contractor should base the bid on that methodology. Although deviations may be allowed once the contract has been signed, those items can be handled through the use of change orders. If an alternative method is assumed in the estimate and later denied by the architect, the contractor would have to shoulder any losses.

The material portion of the specifications usually mentions the physical properties, performance requirements, handling, and storage requirements. Often, specific brands or types of material are listed as the standard of quality required. Sometimes, two or three acceptable brands are specified, and the contractor has a choice of which to supply. If the contractor wishes to substitute another manufacturer's materials, it must be done in accordance with the contract documents.

Results that may be specified include items such as the texture of the material, appearance, noise reduction factors, allowable tolerances, heat loss factors, and colors.

3-10 ALTERNATES

In many projects, the owner requests prices for alternate methods or materials of construction (Figure 3.5). These alternates are generally spelled out on a separate listing in the project manual, and they are listed on the proposal form. The alternates may be either an *add price* or a *deduct price,* which means that contractors either add the price to the base bid (the price without any alternatives) or deduct it from the base bid. The price for any alternates must be complete and include all taxes, overhead, and profit. When an owner has a limited budget, the system of alternates allows a choice on how to best spend the available money.

Since lump-sum contracts are awarded on the basis of the total base bid, plus or minus any alternates accepted, there is always concern that the owner will select alternates in a way that will help a particular contractor become the low bidder. This concern has become so great that some contractors will not bid projects with a large number of alternates. To relieve the contractor of this concern, many architects include in the contract documents the order of acceptance of the alternates. Alternates deserve the same estimating care and consideration as the rest of the project, so contractors should not rush through them or leave them until the last minute.

3-11 ADDENDA

The period after the basic contract documents have been issued to the bidders and before the bids are due is known as the *bidding period.* Any amendments, modifications, revisions, corrections, and explanations issued by the architect/engineer during the bidding period are effected by issuing the *addenda* (Figure 3.6). The statements, and any drawings included, serve to revise the basic contract documents. They notify the bidder of any corrections in the documents, interpretations required, and any additional requirements, as well as other similar matters. The addenda must be in writing.

Because the addenda become part of the contract documents, it is important that all prime contractors promptly receive copies of them. Many architect/engineer offices send copies to all parties who have the plans and the project manual (include the plan rooms). The addenda are also of concern to the subcontractors, material suppliers, and manufacturers' representatives who are preparing proposals for the project, as the revisions may affect their bids.

The receipt of the addenda for a particular project by the reporting service will be noted in the information the service sends to their subscribers. It is suggested that the contractor call the architect/engineer's office once several days before bids are to be received, and again the day before to check that all addenda have been received. Most proposals have a space provided in which the contractor must list the addenda received. Failure to complete this space may result

---

ALTERNATES

ALTERNATE ONE

The Contractor shall state in the Proposal Form the amount to be added to the bid to furnish and install 4" face brick with 8" block exterior walls (as detailed in section A-1) in lieu of the painted 12" concrete block exterior walls.

FIGURE 3.5. Alternates.

A. B. Architects & Associates, Architects / Engineers
7920 Anita Circle
Littleville, Texas 77777

ADDENDUM NO. 1

All American Independent School District

April 1, 20--

GENERAL

Bid date and location has been revised to 2:00 p.m. May 1, 20--. Bid will be received at the Architect's office, 7920 Anita Circle, Littleville, Texas until that time. Bids will be publicly opened and read.

SPECIFICATIONS

1C   Article "ALTERNATE NO. ONE" Delete in full as written and substitute the following therefore:

The Contractor shall state in this Proposal Form the amount to be added to or deducted from the Base Bid to furnish and install all materials as shown on drawings A L/1 and in accordance with the other bidding documents.

3B   Article "MATERIALS" Amend by adding the following paragraph thereto as follows:

All ends of the double tee planks which rest on steel beams shall have a bearing plate 1/8 x 2-3/4 x 5 inches attached to the reinforcing bar and welded, on the job, to the steel beam.

FIGURE 3.6. Addendum.

in the bid being disqualified. For subcontractor bids that are used to bid the project, the contractors should verify that the addenda have been received by subcontractor and incorporated into his or her bid.

3–12 ERRORS IN THE SPECIFICATIONS

Ideally, the final draft of the specifications should be written concurrently with the preparation of the working drawings. The specification writer and production staff should keep each other posted on all items so that the written and graphic portions of the document complement and supplement each other.

Many architects/engineers have been highly successful in achieving this difficult balance. Unfortunately, there are still offices that view specification writing as dull and dreary. For this reason, they sometimes assign a person

not sufficiently skilled to this extremely important task. Also, many times, the specifications are put off until the last minute and then rushed so that they are published before they have been proofread. Some architects/engineers brush off the errors that arise with "we'll pick it up in an addendum."

Another practice that results in errors is when the architect/engineer uses the specifications from one job on a second job, which involves cutting a portion of the old specification out and inserting portions to cover the new job. Inadvertently, items are usually left out in such cases.

The real question is what to do when such an error is found. If the error is discovered early in the bidding period and no immediate answer is needed, the error is kept on a sheet that specifically lists all errors and omissions. Contractors are strongly urged to keep one sheet solely for the purpose of errors and omissions so that they can find the error when they need to. Most specifications require that all requests for interpretations must be made in writing and state how many days before the date set for opening bids they must be received. Check for this (often in instructions to bidders) and note it on the errors and omissions sheet. It is further stipulated in the project manual that all interpretations will be made in writing in the form of addenda and sent to all bidders. In actual practice, it is often accepted that estimators will telephone the architect's/engineer's office and request clarifications (interpretations, really). If the interpretation will materially affect the bid, the contractor must be certain to receive it in writing to avoid later problems. If there are contradictions between the drawings and the specifications, they should also attempt to get them resolved as early in the bidding period as possible.

In keeping a list of all discrepancies (errors) and any items not thoroughly understood, contractors should make notations about where on the drawings and in the project manual the problems occur. In this manner, when they ask for clarification, it is relatively easy to explain exactly what they want. The architect/engineer should not be contacted about each problem separately, but about a few at a time. Often the contractors will answer the questions themselves as they become more familiar with the drawings and the project manual. When calling architects/engineers, contractors need to be courteous, as everyone makes mistakes and being courteous will help keep them on your side. Besides, the information may have been included, but simply overlooked. Contractors should not wait until bids are due to call with questions, since verbal interpretations often will not be given; and even if they are, the person who knows the answers may not be available at the time. Regardless of the project or the type of estimate, the keynotes to success are cooperation and organization.

WEB RESOURCES

www.aia.org

www.csinet.org

REVIEW QUESTIONS

1. What types of information are found in the project manual?

2. Why is it important for the estimator to review carefully the entire project manual?

3. Describe the CSI MasterFormat and how it is used.

4. What types of information are provided in the invitation to bid?

5. Why is it important that the bids be delivered at the proper time and place?

6. How do the supplementary general conditions differ from the general conditions?

7. What information is contained in the specifications (technical specifications) of the project manual?

8. Explain what an alternate is and how it is handled during the bidding process.

9. Why is it important to prepare the alternate amounts carefully and thoroughly?

10. Explain what an addendum is and when it is used.

11. Why is it important that the estimator be certain that all addenda have been received before submitting a bid?

12. How should the estimator handle any errors or omissions that may be found in the contract documents?

CHAPTER FOUR

THE ESTIMATE

4–1 ORGANIZATION

The estimator must maintain a high degree of organization throughout the estimate development stage. A well-organized estimate improves the probability of getting the work, facilitating the actual work in the field, and completing the work within budget. The organization required includes a plan for completing the estimate and maintaining complete and up-to-date files. It must include a complete breakdown of costs for the project, both of work done by company forces (in-house) and of work done by subcontractors. Using appropriate software can be an effective way to keep organized. The estimate information should include quantities, material prices, labor conditions, costs, weather conditions, job conditions, delays, plant costs, overhead costs, and salaries of forepersons and superintendents. All data generated during the development of the estimate must be filed in an orderly manner. The estimating costs are often stored in spreadsheets, databases, or estimating software packages. The original paper documents may be stored in file cabinets and archived after a specified period, or they may be scanned and stored electronically.

The estimate of the project being bid must be systematically done, neat, clear, and easy to follow. The estimator's work must be kept organized to the extent that in an unforeseen circumstance (such as illness or accident), someone else might step in, complete the estimate, and submit a proposal on the project. If the estimate is not organized and easy to read and understand, then there is no possible way that anyone can pick up where the original estimator left off. The easiest way for you to judge the organization of a particular estimate is to ask yourself if someone else could pick it up, review it and the contract documents, and be able to complete the estimate. Ask yourself: Are the numbers labeled? Are calculations labeled? Where did the numbers come from? What materials are being estimated? An organized estimate is also important in case the estimator leaves the company before the project is complete. If he leaves, someone else needs to be able to order the materials and answer questions as to what was included in the bid.

4–2 PLANNING THE ESTIMATE

The need for organization during the estimating process is critical. There are many decisions that need to be made concerning the logistics of who will do which portion and when. Figure 4.1 is a diagrammatic representation of the steps that are required to complete an estimate. Another helpful tool when preparing an estimate is a bar chart schedule that details when the activities comprising the estimate will be completed. In addition, the persons who are responsible for those activities should be listed on the schedule. Figure 4.2 is a sample bar chart schedule for completing an estimate. The bars and milestones will be darkened as the activities are completed.

Since the preparation of an estimate is a corroborative effort, it is essential that all persons have input into when certain items are required and that they understand the interrelationships between the responsible parties. Therefore, one of the first things that needs to be done when preparing the estimate is to bring together all the estimate team members to develop the overall estimate schedule.

4–3 NOTEBOOK

A *notebook* should be kept for each estimate prepared. The notebook should be broken down into several areas: the workup sheets, summary sheets, errors and omissions sheets, proposals received from subcontractors, proposals received from material suppliers and manufacturers' representatives, and notes pertaining to the project. Also, a listing of all calls made to the architect/engineer should be kept together, specifying who called, who was contacted at the architect/engineer's office, the date of the call, and what was discussed. The notebook should be neat and easy to read and understand.

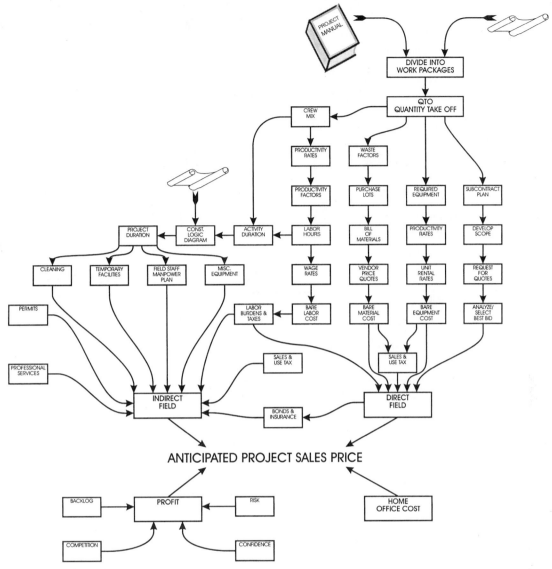

FIGURE 4.1. The Estimate Process.

Every page of the estimate should be numbered and initialed by the person who prepared that portion of the estimate. In addition, every page of the estimate should be checked and verified, and that person's initials should be placed on the page. This rather cumbersome procedure is required to help answer questions that may arise later. When construction begins and the estimate is used to purchase materials, if there are questions concerning a specific item, the estimator can be found and asked to clarify any questions.

4-4 TO BID OR NOT TO BID

It is impossible for a contractor to submit a proposal for every project that goes out to bid. Through personal contact and the reporting services, the contractor finds out what projects are out for bid and then must decide on which projects to submit a proposal. Many factors must be considered: the type of construction involved compared with the type of construction the contractor is usually involved in, the loca-

tion of the project, the size of the project in terms of total cost and in relation to bonding capacity, the architect/ engineer, the amount of work currently under construction, the equipment available, and whether qualified personnel are available to run the project.

There are also certain projects for which a contractor is not allowed to submit a proposal. The owners may accept proposals only from contractors who are invited to bid; other projects may have certain conditions pertaining to work experience or years in business that must be met.

4-5 THE ESTIMATE

Once the contractor has decided to bid on a particular project, arrangements need to be made to pick up the contract documents. The estimator should proceed with the estimate in a manner that will achieve the greatest accuracy and completeness possible. The accuracy required must be in the range of 98 to 99 percent for all major items on the estimate.

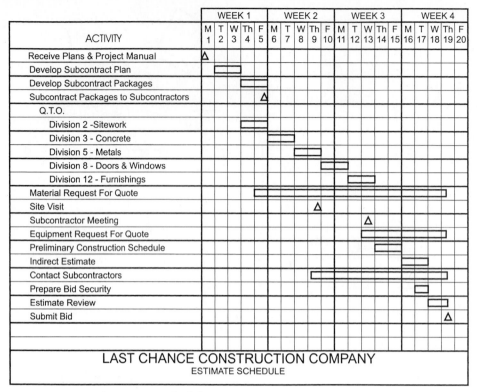

ACTIVITY	WEEK 1					WEEK 2					WEEK 3					WEEK 4				
	M 1	T 2	W 3	Th 4	F 5	M 6	T 7	W 8	Th 9	F 10	M 11	T 12	W 13	Th 14	F 15	M 16	T 17	W 18	Th 19	F 20
Receive Plans & Project Manual	△																			
Develop Subcontract Plan		▭																		
Develop Subcontract Packages				▭																
Subcontract Packages to Subcontractors					△															
Q.T.O.																				
Division 2 -Sitework				▭																
Division 3 - Concrete							▭													
Division 5 - Metals									▭											
Division 8 - Doors & Windows										▭										
Division 12 - Furnishings												▭								
Material Request For Quote						▭▭▭▭▭▭▭▭▭▭▭▭▭													
Site Visit								△												
Subcontractor Meeting												△								
Equipment Request For Quote												▭▭▭▭▭▭▭▭▭								
Preliminary Construction Schedule													▭							
Indirect Estimate															▭					
Contact Subcontractors										▭▭▭▭▭▭▭										
Prepare Bid Security																▭				
Estimate Review																		▭		
Submit Bid																			△	

LAST CHANCE CONSTRUCTION COMPANY
ESTIMATE SCHEDULE

FIGURE 4.2. Sample Estimate Schedule.

Listed below are the steps in working up a detailed estimate. These steps should form a basis for estimating, so it is important to read and understand them.

1. Carefully check the drawings and the project manual to be sure that you have everything, including all addenda. Not all architectural and engineering offices number their drawings in the same manner, so sometimes there can be confusion as to whether you have all the drawings. Architectural drawings are usually prefixed with the letter A. Structural drawings may be prefixed with the letter S, or they may be included with the architectural drawings. Mechanical drawings may be prefixed with M, P, or HVAC. Electrical drawings typically use the designation E. Some jobs have no prefixes before the numbers, but in these circumstances the pages are typically labeled Sheet 1 of 25. Typically, the front of the project manual or the drawings contains a list of all the drawings included in the set. Check all sources to ensure that you have received all of the drawings. If there are any discrepancies, check with the architect/engineer and complete your set. Follow the same procedure with the project manual. Check the list in the front of the project manual against what was received.

2. Scan the drawings to get a feel for the project. How large is it? What shape is it? What are the principal materials? Pay particular attention to the elevations. At this step, it is important that the estimator understands the project. Make a mental note of exterior finish materials, the amount of glass required, and any unusual features.

3. Review the floor plans, again getting the "feel" of the project. The estimator should begin to note all unusual plan features of the building. Look it over; follow through the rooms, starting at the front entrance. Again, make a mental note of what types of walls are used. Note whether enlarged floor plans show extra dimensions or whether special room layouts are required.

4. Begin to examine the wall sections for a general consideration of materials, assemblies, and makeup of the building. Take special note of any unfamiliar details and assemblies; circle them lightly with a red pencil or a highlighter so that you can refer to them readily.

5. Review the structural drawings. Note what types of structural systems are being used and what types of construction equipment will be required. Once again, if the structural system is unusual, the estimator should make a mental note to spend extra time on this area.

6. Review the mechanical drawings, paying particular attention to how they will affect the general construction, underground work requirements, outlet requirements, chases in walls, and other items of this sort. Even under separate contracts, the mechanical portions must be checked.

7. The submitted bid is based on the drawings and the project manual. You are responsible for everything contained in the project manual as well as what is covered on the drawings. Read and study the project manual thoroughly and review it when necessary. Take notes on all unusual items contained in the project manual.

8. Visit the site after making a preliminary examination of the drawings and the project manual. The visit should be made by the estimator or by other experienced persons, including members of the proposed project execution team. By including these persons on the site visit,

expertise and estimate ownership will be enhanced. The information that is obtained from the site visit will influence the bidding of the project. It is a good idea to take pictures of the site to reference when preparing the bid.

9. Even though estimators must rely on their own experience in construction, it is imperative that they create and maintain a close liaison with the other office personnel and field superintendents. After you have become familiar with the drawings and the project manual, call a meeting with the people who would most likely hold the key supervisory positions if you are the successful bidder. Be sure to allow these people time before the meeting to become familiar with the project. During this meeting, the project should be discussed in terms of the construction methods that could be followed, the most desirable equipment to use, the time schedules to be followed, and the personnel needed on the project.

10. Check carefully through the general conditions and supplementary general conditions, making a list of all items contained in the project manual that will affect the cost of the project.

11. Send a copy of all insurance requirements for the project to your insurance company and all bonding requirements for the project to your bonding company.

12. The estimator may now begin the takeoff of the quantities required. Each item must be accounted for, and the estimate itself must be as thorough and complete as possible. The items should be listed in the same manner and with the same units of measure in which the work will be constructed on the job. Whenever possible, the estimate should follow the general setup of the specifications. This work is done on a workup sheet. As each item is estimated, the type of equipment to be used for each phase should be listed. The list will vary depending on the equipment owned and what is available for rent. Prices on equipment to be purchased or rented must be included.

13. At the time the estimator is preparing the quantity takeoff on workup sheets, the following tasks can also be ongoing:
    (a) Notify subcontractors, material suppliers, and manufacturers' representatives that the company is preparing a proposal for the project and ask them if they intend to submit bids on the project.
    (b) Begin to make a list of all items of overhead that must be included in the project. This will speed up the future pricing of these items.

14. The information on the workup is carried over to the summary sheet. Work carefully; double-check all figures. If possible, have someone go over the figures with you. The most common error is the misplaced decimal point. Other common errors include the following:
    (a) Errors in addition, subtraction, multiplication, and division.
    (b) Omission of items such as materials, labor, equipment, or overhead.
    (c) Errors in estimating the length of time required to complete the project.
    (d) Errors in estimating construction waste.
    (e) Errors in estimating quantities of materials.
    (f) Errors in transferring numbers from one sheet to another.
    (g) Adding a line to a spreadsheet and not checking to make sure that the new line is included in the total.
    (h) Errors in setting up formulas, items, assemblies, markups, and so forth in estimating software.
    (i) Using typical productivity rates and costs from estimating software without adjusting them for individual project conditions.
    (j) Improper use of estimating software because the user does not understand the limits of the software or the inputs required by the software.

15. Having priced everything, make one last call to the architect/engineer's office to check the number of addenda issued to be sure that you have received them all. Double-check the time, date, and place that bids are being received. Double-check that all of the requirements for the submission of the proposal have been followed; be sure the proposal is complete.

4–6 SITE INVESTIGATION

It is often required by the contract documents that the contractor visit the site and attend a pre-bid conference. The importance of the visit and the items to be checked vary depending on the type of the project and its location. As a contractor expands to relatively new and unfamiliar areas, the importance of the preliminary site investigation increases, as does the list of items that must be checked. Examples of the type of information that should be collected are as follows:

1. Site access
2. Availability of utilities (electric, water, telephone)
3. Site drainage
4. Transportation facilities
5. Any required protection or underpinning of adjacent property
6. A rough layout of the site locating the proposed storage trailer and equipment locations
7. Subsurface soil conditions (bring a post hole digger to check this)
8. Local ordinances and regulations, and note any special requirements (permits, licenses, barricades, fences)
9. The local labor situation and local union rules
10. Availability of construction equipment rentals, the type and conditions of what is available as well as the cost
11. Prices and delivery information from local material suppliers (request proposals for the project)
12. The availability of local subcontractors (note their names, addresses, and what type of work they usually handle)
13. The conditions of the roads leading to the project, low bridges, and load limits on roads or bridges
14. Housing and feeding facilities if workers must be imported
15. Banking facilities

4-7 SPECIALTY CONTRACTORS

A *specialty contractor* or *subcontractor* is a separate contractor hired by the prime contractor to perform certain portions of the work. The amount of work that the prime contractor will subcontract varies from project to project. Some federal and state regulations limit the proportion of a project that may be subcontracted, but this is rarely the case in private work. There are advantages and disadvantages to using specialty contractors. Trades such as plumbing, electrical, and heating and air-conditioning have a tradition of being performed by specialty contractors, due to their specialized nature and licensing requirements. However, specialty contractors can now be found who are capable of performing every aspect of the construction project. Contractors today can construct entire projects without having any direct-hire craft personnel. The use of specialty contractors has gained popularity as a means to reduce risk and overhead; however, the contractor gives up a substantial amount of control when subcontracting the entire project.

If specialty contractors are to be used, the contractor must be certain to notify them early in the bidding period so that they have time to prepare a complete, accurate proposal. If rushed, the specialty contractor tends to bid high just for protection against what might have been missed.

The use of specialty contractors can be economical, but estimates still must be done for each portion of work. Even if the estimator intends to subcontract the work, an estimate of the work should be prepared. It is possible that the estimator will not receive proposals for a project before the bid date and will have to use an estimated cost of the work in totaling the proposal. All subcontractors' proposals (Figure 4.3) are compared with the estimator's price; it is important that a subcontractor's price is neither too high nor too low. If either situation exists, the estimator should call the subcontractor and discuss the proposal with him.

The specialty contractor's proposal is often phoned, faxed, or e-mailed into the general contractor's office at the last minute because of the subcontractor's fear that the contractor will tell other subcontractors the proposal price and encourage lower bids. This practice is commonly referred to as *bid peddling* or *bid shopping* and is highly unethical and should be discouraged. To prevent bid shopping, specialty contractors submit their final price only minutes before the bids close, which leads to confusion and makes it difficult for the estimator to analyze all bids carefully. This confusion is compounded by specialty contractors who submit unsolicited bids. These bids come from specialty contractors who were not contacted or invited to submit a bid, but who find out which contractors are bidding the project and submit a bid. Since these companies are not prequalified, there is an element of risk associated with accepting one of these bids. On the other hand, not using low bids from unsolicited subcontractors places the contractor at a price disadvantage.

In checking subcontractor proposals, note especially what is included and what is left out. Each subsequent proposal may add or delete items. Often the proposals set up certain conditions, such as use of water, heat, or hoisting facilities. The estimator must compare each proposal and select the one that is the most economical.

All costs must be included somewhere. If the subcontractor does not include an item in the proposal, it must be considered elsewhere. A tricky task for the prime contractor is the comparison of the individual subcontractor's price quotes. Throughout the estimating process, the prime contractor

---

McBill Precast
1215 Miriam Rd.
Littleville, Texas 77777

April 24, 20__

Ace Construction
501 Hightower St.
Littleville, Texas 77777
RE: All American Independent School District
               Administration Building Annex

Bids Due: April 24, 20__

Gentlemen:

We propose to furnish and install all precast double tees for the above mentioned project for the lump sum of: $13,250.00
Furnish double tee fillers (for ends of tees, literature enclosed). $3.75 each

All materials are bid in accordance with the contract documents, including Addendum #1.

Sincerely,

Charles McBill

FIGURE 4.3. Subcontractor's Proposal.

BID TABULATION

Project: _____
Location: _____
Architect: _____
Subcontract Package: _____

Estimate No. _____
Sheet No. _____
Date: _____
By: _____ Checked: _____

Scope of Work	Subcontractor 1	Subcontractor 2	Subcontractor 3	Subcontractor 4	Subcontractor 5
Base Bid					
Adjustments					
1					
2					
3					
4					
5					
6					
7					
8					
9					
10					
11					
Adjusted Bids					

Comments

FIGURE 4.4. Subcontract Bid Tabulation.

should be communicating with the specific subcontractors concerning the fact that they will submit a price quote and what scope of work is to be included within that quote. However, subcontractors will include items that they were not asked to bid and will exclude items that they were asked to bid. A "bid tabulation" or "bid tab" is used to equalize the scope between subcontractors so that the most advantageous subcontractor's bid can be included in the prime contractor's bid. Figure 4.4 is an example of a bid tabulation form. A spreadsheet version of Figure 4.4 (Bid-Tab.xls) is provided on the companion disk.

4–8 MATERIALS

For each project being bid, the contractor will request quotations from materials suppliers and manufacturers' representatives for all materials required. Although on occasion a manufacturer's price list may be used, it is more desirable to obtain written quotations that spell out the exact terms of the freight, taxes, time required for delivery, materials included in the price, and the terms of payment (Figure 4.5). The written proposals should be checked against the specifications to make certain that the specified material was bid.

All material costs entered on the workup and summary sheets (sections 4–9 and 4–10) must be based on delivery to the job site. The total will include all necessary freight, storage, transportation insurance, sales tax, and inspection costs. Remember that the sales tax that must be paid on a project is the tax in force in the area in which the project is being built; sometimes cities have different rates than the county in which they are located. Contractors should take time to check.

United Block Company
713 Charles Blvd.
Littleville, Texas 77777

April 20, 20____

Ace Construction
501 Hightower St.
Littleville, Texas 77777

Re: All American Independent School District
           Administration Building Annex

Gentlemen:

We are pleased to quote on the materials required for the above referenced project.

All of the materials listed below meet the requirements as specified in the drawings and specifications.

8 X 8 X 16	Concrete Block	$1.18 ea.
8 X 4 X 16	Concrete Block	$.82 ea.
8 X 6 X 16	Concrete Block	$.98 ea.

All prices quoted are delivered without sales tax.
Terms: 2% - 30 Days
FOB Jobsite

FIGURE 4.5. Materials Price Quote.

4–9 WORKUP SHEETS

The estimator uses two basic types of manual takeoff sheets: the workup sheet and the summary sheet. The *workup sheet* can be a variety of forms contingent upon what is being quantified. Figure 4.6 is an example of a workup sheet that could be

ESTIMATE WORK SHEET
REINFORCING STEEL

Project:	Little Office Building
Location	Littleville, Tx
Architect	U.R. Architects
Items	Foundation Concrete

Estimate No. 1234
Sheet No. 1 of 1
Date 11/11/20XX
By LHF   Checked JBC

Cost Code	Description	L (ft)	W (ft)	Space (/ft)	Count	Count	Bar Size	Linear Feet	Pounds / Foot	Quantity	Unit
	Continuous Footings										
	Perimeter - Long Bars	336	3.167	2		8	5	2688	1.043	2,804	Pounds
	Perimeter - Short Bars	2.83	336	2		673	5	1904.59	1.043	1,986	Pounds
	Interior - Long Bars	76	3	2		7	5	532	1.043	555	Pounds
	Interior - Short Bars	2.67	76	2		153	5	408.51	1.043	426	Pounds
	Dowels										
	Perimeter	4	336	1		337	5	1348	1.043	1,406	Pounds
	Interior	4	76	1		77	5	308	1.043	321	Pounds
	Foundation Walls										
	Perimeter - Long Bars	336	4	2		9	5	3024	1.043	3,154	Pounds
	Perimeter - Short Bars	3.67	336	2		673	5	2469.91	1.043	2,576	Pounds
	Interior - Long Bars	76	8	2		17	5	1292	1.043	1,348	Pounds
	Interior - Short Bars	8	76	2	Count	153	5	1224	1.043	1,277	Pounds
	Spread Footings	2.67	2.67	2	3	42	5	112.14	1.043	117	Pounds
	Dowels	4			3	12	5	48	1.043	50	Pounds
	Column Piers - Vertical Bars	3.67			3	12	5	44.04	1.043	46	Pounds
	Column Piers - Stirrups	3.33	4	1	3	15	3	49.95	0.376	19	Pounds
	Drilled Piers										
	Vertical Bars	20			3	18	5	360	1.043	375	Pounds
	Horizontal Bars	3.67	20	1.5	3	93	3	341.31	0.376	128	Pounds
	Grade Beams - Long Bars	69				6	5	414	1.043	432	Pounds
	Grade Beams - Stirrups	3.67	69	1		70	3	256.9	0.376	97	Pounds

FIGURE 4.6. Estimate Workup Sheet—Reinforcing Steel.

used to quantify reinforcing steel. A spreadsheet version of Figure 4.6 (Rebar.xls) is provided on the companion disk.

The workup sheet is used to make calculations and sketches and to generally "work up" the cost of each item. Material and labor costs should always be estimated separately. Labor costs vary more than material costs, and the labor costs will vary in different stages of the project. For example, a concrete block will cost less for its first 3 feet than for the balance of its height and the labor cost goes up as the scaffold goes up, yet material costs remain the same.

When beginning the estimate on workup sheets, the estimator must be certain to list the project name and location, the date that the sheet was worked on, and the estimator's name. All sheets must be numbered consecutively, and when completed, the total number of sheets is noted on each sheet (e.g., if the total number was 56, sheets would be marked "1 of 56" through "56 of 56"). The estimator must account for every sheet, because if one is lost, chances are that the costs on that sheet will never be included in the bid price. Few people can write so legibly that others may easily understand what they have written; it is, therefore, suggested that the work be printed. If a spreadsheet program is being used, contractors must be very careful to verify that all formulas are correct and that the page totals are correct. Errors in spreadsheet programs can be costly. Never alter or destroy calculations; if they need to be changed, simply draw a line through them and rewrite. Numbers that are written down must be clear beyond a shadow of a doubt. Too often a "4"

can be confused with a "9," or "2" with a "7," and so on. All work done in compiling the estimate must be totally clear and self-explanatory. It should be clear enough to allow another person to come in and follow all work completed and all computations made each step of the way.

When taking off the quantities, contractors must make a point to break down each item into different sizes, types, and materials, which involves checking the specifications for each item they are listing. For example, in listing concrete blocks, they must consider the different sizes required, the bond pattern, the color of the unit, and the color of the mortar joint. If any of these items varies, it should be listed separately. It is important that the takeoff be complete in all details; do not simply write "wire mesh," but "wire mesh 6×6 10/10"—the size and type are very important. If the mesh is galvanized, it will increase your material cost by about 20 percent, so this should be noted on the sheet. Following the CSI MasterFormat helps organize the estimate and acts as a checklist.

4-10 SUMMARY SHEET

All costs contained on the workup sheets are condensed, totaled, and included on the summary sheet. All items of labor, equipment, material, plant, overhead, and profit must likewise be included. The workup sheets are often summarized into summary sheets that cover a particular portion of the project. Figure 4.7 is an example of a summary sheet used to summarize the concrete contained in the project.

ESTIMATE SUMMARY SHEET

Project	Little Office Building
Location	Littleville, TX
Architect	U.R. Architect, Inc.
Items	Foundation Concrete

Estimate No.	1234
Sheet No.	1 of 1
Date	11/11/20XX
By	LHF Checked JBC

Cost Code	Description	Q.T.O.	Waste Factor	Purch. Quan.	Unit	Crew	Prod. Rate	Wage Rate	Work Hours	Unit Cost Labor	Unit Cost Material	Unit Cost Equipment	Labor	Material	Equipment	Total
	Continous Footing #5 Bar															
	(2804+1986+555+426)/2000	2.886	10.00%	3.17	Ton		15	$12.75	43.3	$191.25	$550.00		$607	$1,746	$0	$2,353
	Dowels From Ftg to Wall															
	(1406+321)/2000	0.864	10.00%	0.95	Ton		15	$12.75	13.0	$191.25	$550.00		$182	$523	$0	$704
	Foundation Walls															
	(3154+2576+1348+1277)/2000	4.178	10.00%	4.60	Ton		11	$12.75	46.0	$140.25	$550.00		$645	$2,528	$0	$3,172
	Spread Footings w/Dowels															
	(117+50)/2000	0.084	10.00%	0.09	Ton		15	$12.75	1.3	$191.25	$550.00		$18	$51	$0	$68
	Column Piers															
	#5 Bar	0.023	10.00%	0.03	Ton		24	$12.75	0.6	$306.00	$550.00		$8	$14	$0	$22
	#3 Bar	0.010	10.00%	0.01	Ton		24	$12.75	0.2	$306.00	$550.00		$3	$6	$0	$9
	Driller Piers															
	#5 Bar	0.188	10.00%	0.21	Ton		24	$12.75	4.5	$306.00	$550.00		$63	$114	$0	$177
	#3 Bar	0.064	10.00%	0.07	Ton		24	$12.75	1.5	$306.00	$550.00		$22	$39	$0	$60
	Grade Beam															
	#5 Bar	0.216	10.00%	0.24	Ton		22	$12.75	4.8	$280.50	$550.00		$67	$131	$0	$197
	#3 Bar	0.049	10.00%	0.05	Ton		22	$12.75	1.1	$280.50	$550.00		$15	$30	$0	$45
	Slab on Grade															
	(1348+1277)/2000	1.313	10.00%	1.44	Ton		13	$12.75	17.1	$165.75	$550.00		$239	$794	$0	$1,034
									133.21						9/15/20099/15/	
													$1,868	$5,974	$0	$7,843

FIGURE 4.7. Estimate Summary Sheet.

10:22 AM
PROJECT:_____
LOCATION:_____
BID DATE:_____

LAST CHANCE CONSTRUCTION COMPANY
ESTIMATE SUMMARY

2/23/2002
REVISION:_____
BY:_____
CHECKED:_____

DIV. DESCRIPTION	WORKHOURS	LABOR $	MATERIAL $	EQUIPMENT $	SUBCONTRACT $	TOTAL $
DIRECT FIELD COSTS						
2 SITEWORK						0
3 CONCRETE						0
4 MASONRY						0
5 METALS						0
6 WOODS & PLASTICS						0
7 MOISTURE - THERMAL CONTROL						0
8 DOORS, WINDOWS & GLASS						0
9 FINISHES						0
10 SPECIALTIES						0
11 EQUIPMENT						0
12 FURNISHINGS						0
13 SPECIAL CONSTRUCTION						0
14 CONVEYING SYSTEMS						0
15 MECHANICAL						0
16 ELECTRICAL						0
						0
						0
						0
TOTAL DIRECT FIELD COSTS	0	0	0	0	0	0
INDIRECT FIELD COSTS						
FIELD STAFF						0
TEMPORARY OFFICES						0
TEMPORARY FACILITIES						0
TEMPORARY UTILITIES						0
REPAIRS & PROTECTION						0
CLEANING						0
PERMITS						0
PROFESSIONAL SERVICES						0
BONDS						0
INSURANCE						0
MISC. EQUIPMENT						0
						0
LABOR BURDENS (STAFF)						0
LABOR BURDENS (CRAFT)						0
SALES TAX						0
TOTAL INDIRECT FIELD COSTS	0	0	0	0	0	0
HOME OFFICE COSTS						0
LAST MINUTE CHANGES						0
						0
						0
						0
						0
TOTAL LAST MINUTE CHANGES	0	0	0	0	0	0
PROFIT						0
TOTAL PROJECT COST	0	0	0	0	0	0
COMMENTS						

FIGURE 4.8. Overall Estimate Summary Sheet.

Every item on this page is supported by workup sheets. A spreadsheet version of Figure 4.7 (Recap.xls) is provided on the companion disk.

In addition to summarizing portions of the project, it is helpful to summarize the entire estimate onto a single page. Figure 4.8 is an example of a summary sheet that could be used to summarize all costs for the entire project. A spreadsheet version of Figure 4.8 (Est-Summary.xls) is provided on the companion disk.

The summary sheet should list all the information required, but none of the calculations and sketches that were used on the workup sheets. It should list only the essentials, yet still provide enough information for the person pricing the job not to have to continually look up required sizes, thicknesses, strengths, and similar types of information.

4–11 ERRORS AND OMISSIONS

No matter how much care is taken in the preparation of the contract documents, it is inevitable that certain errors will occur. Errors in the specifications were discussed in Section 3–12, and the same note-taking procedure is used for all other discrepancies, errors, and omissions. The instructions to bidders or supplementary general conditions ordinarily states that if there are discrepancies, the specifications take precedence over drawings and dimensioned figures, and detailed drawings take precedence over scaled measurements from drawings. All important discrepancies (those that affect the estimate) should be checked with the architect/engineer's office.

WEB RESOURCES

www.construction.com

www.e-builder.net

www.fwdodge.com

www.rsmeans.com

www.dcd.com

www.costbook.com

www.frankrwalker.com

www.bidshop.org

REVIEW QUESTIONS

1. Why should a notebook of the estimate be kept? What items should be kept in it?

2. Why might a contractor decide not to bid a particular project?

3. Why should estimators check carefully to be certain that they have all the contract documents before bidding a project?

4. One of the requirements of most contract documents is that the contractor visit the site. Why is the site investigation important?

5. Explain what a subcontractor is and the subcontractor's relationship with the owner and the general contractor.

6. Why should material quotes be in writing? What items must be checked in these proposals?

7. What is the difference between workup sheets and summary sheets?

8. Go to a local plan room, stop at the desk, and introduce yourself. Explain that you are taking a class in estimating and ask if you could be given a tour of their facilities.

CHAPTER FIVE

COMPUTERS IN ESTIMATING

5-1 OVERVIEW

During the last 20 years, computers have revolutionized estimating. With the advent of cheap personal computers, easy-to-use spreadsheet software, and specialized estimating software, the computer now takes a central role in the estimating process. With a portable computer, a portable printer, and a mobile Internet connection, an estimator can visit a client's office or home, review the project's requirements, access the company's database and estimating software, and, for simple jobs, give the client a typewritten estimate before leaving. The same estimator can return to the office, download a set of plans, perform the estimate, and e-mail a proposal to a client without using a single piece of paper.

If an estimator is to survive in today's competitive work environment, they must be computer savvy. Even old-time estimators have had to learn to use computers to remain competitive. In this chapter, we will look at how computers are being used in the estimating process. The purpose is to give the reader an overview of the types of software available, not to teach them how to use the software.

5-2 BENEFITS AND DANGERS OF COMPUTERIZED ESTIMATING

Computerized estimating offers many benefits to the estimator. When set up and used properly, computerized estimating can increase the efficiency of the estimating process. The following are some of the benefits of computerized estimating:

1. Computerized estimating can reduce calculation errors, which gives the estimator a more accurate cost for the project. This reduces the number of unprofitable jobs that are won because the job was inadvertently priced too cheap or jobs that are lost because the job was priced too high due to a computation error.

2. Computerized estimating increases the speed at which the estimate is prepared by performing the math that the estimator would have to do. This gives the estimator more time to focus on the critical aspects of the estimate (such as getting better pricing) and improving the quality of the estimate. It may also allow the estimator time to prepare more estimates, thus reducing the estimating cost to the company.

3. Many computerized estimating packages allow the estimator to track where the quantities came from. This is important when the estimator needs to answer questions about the estimate to the client or company's field personnel many months after the estimate was prepared.

4. Computerized estimating allows the estimator to quickly change a price or a productivity rate and get an instantaneous change in the project's cost. This allows the estimator to easily make last-minute price adjustments or to see how missing the target productivity would change the profitability of a job.

Alternately, if used incorrectly or carelessly, computerized estimating can cause many problems and generate many incorrect estimates. An inaccurate equation entered into a spreadsheet or specialized estimating software package can create errors in multiple estimates before it is found. There are two key dangers to avoid when using computerized estimating.

The first danger is to turn the thinking over to the computer, making the estimator nothing more than a data-entry person. Computers are very good at performing repetitive, mundane tasks, such as mathematical calculations. They can even be taught to make simple decisions, such as selecting the correct equation to use based on a response to a simple question. But the computer cannot think for the estimator. Computerized estimating is a tool, just like a calculator, to be used by the estimator to help her prepare the estimate. When the estimating program retrieves pricing or productivity rates from a database, the estimator must determine whether the retrieved data are appropriate for the current estimate. If job conditions are different for the job being estimated than typical conditions, the estimator must make the appropriate adjustment to the price or productivity. To help avoid these errors, the estimator must have a strong understanding of estimating concepts and processes. She must look at each number the software has prepared and ask herself, "Does this number make sense?"

The second danger is to use the software for a project that it was not designed to estimate. For example, one may prepare a spreadsheet to estimate the installation of a water-line up to a depth of four feet. If this spreadsheet were to be used on a project where the waterline was at eight feet deep, the software would underestimate the cost of the waterline because additional safety measures would be required for the additional depth. When using estimating software, it is important for estimators to understand the intended use and limitation of the software. It is also good to understand the methodology used by the software to calculate quantities and costs.

Estimating software can be divided into three broad categories: spreadsheets, specialized estimating software, and takeoff software. These are discussed in the following sections.

5-3 SPREADSHEETS

Spreadsheet software, such as Microsoft Excel, is a powerful tool for the construction estimator. Its usage ranges from simple spreadsheets that are used to add up quantities to complex spreadsheets that take hundreds of hours to develop and continue evolving over time. Spreadsheets are often used to augment specialized estimating software packages. In 2003, the American Society of Professional Estimators reported that 29 percent of construction companies used Excel as an estimating tool.[1]

Excel and other spreadsheets have many advantages. If they are not already available on the estimator's computer, they can be purchased cheaply. They are easy to use and can be readily adapted to the company's existing style. For a company that uses paper forms, the forms can simply be converted to a spreadsheet by creating the form's layout in the spreadsheet and then by adding the formulas needed to perform the mathematical calculations. Spreadsheets can make simple decisions by using IF functions. Simple databases can also be integrated into the spreadsheet.

There are many examples of estimating spreadsheets throughout this book, and sample spreadsheets have been included on the companion disk. The Warehouse.xls spreadsheet on the companion disk incorporates both IF functions to make simple decisions and a simple database.

5-4 SPECIALIZED ESTIMATING SOFTWARE

In 2003, the American Society of Professional Estimators reported that 47 percent of construction companies used specialized estimating software, such as WinEst or Timber-line.[2] These estimating software packages use a spreadsheet layout for the estimate, similar to what you would see in Excel (Figure 5.1), and combine it with a database (Figure 5.2). The spreadsheet layout displays the costs associated with each bid item in table format. These items are selected from items in the database. The database contains standard costs, productivity rates, labor rates, crews, and so on for each item. When selecting items from the database, the estimator must verify that the data selected are appropriate for the project he is bidding. The costs shown on the spreadsheet layout are summarized on a totals page (Figure 5.3). The features available vary from software to software, and some software allows you to select additional options or add on modules.

Specialized estimating software packages can increase the productivity of the estimator. They have the following advantages:

1. They can take off a group of items (an assembly) at the same time. For example, all of the components of an interior wall (studs, track, sound insulation, drywall, paint, and base) can be taken off by entering the length and the height of the wall. The software then uses formulas to determine the materials needed (the number of studs, length of track, rolls of insulation, sheets of drywall, gallons of drywall mud, pounds of screws, gallons of paint, etc.) and the labor and equipment needed to install them.

2. The estimate can be easily and quickly viewed in different formats. For example, the sales tax can be listed on the totals page or be spread out over the individual items. Different formats can be used to print the estimate. You can present one format to your boss and another to the client with little preparation time.

3. The software can prepare standard and custom reports, such as materials lists.

4. Changes can be made quickly to the estimates.

The disadvantage of specialized estimating software is that they are expensive and require a substantial commitment to set up and maintain them. They are useful for companies that do a lot of estimating. For companies that perform a low volume of estimating, the time it takes to maintain a specialized estimating software often exceeds the time it takes to perform the estimating, using Excel or paper forms. These companies often abandon specialized estimating software after a short while.

An educational version of WinEst is included on the companion disk.

---

[1]American Society of Professional Estimators, 2003 ASPE Cost Estimating Software Survey, posted on http://www.aspenational.com/cgi-bin/coranto/viewnews.cgi?id=epzkzkupuyhepgiait&tmpl=fullstory&style=fullstory, downloaded February 4, 2004.

[2]Ibid.

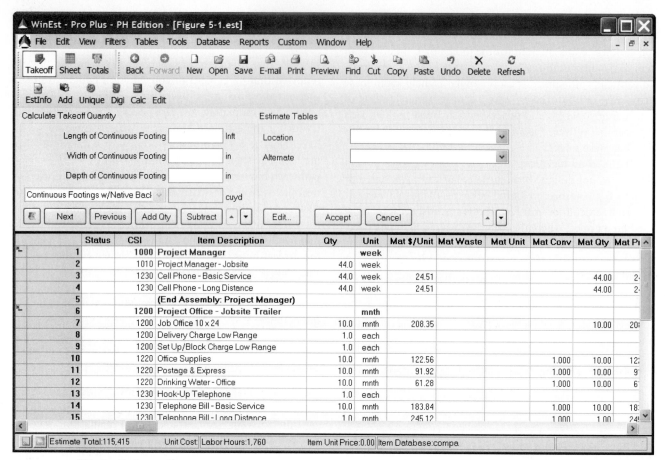

FIGURE 5.1. WinEst Estimate Sheet.

Courtesy of WinEstimator Inc.

5-5 TAKEOFF SOFTWARE

Takeoff software allows the user to determine the estimate quantities from an electronic set of drawings. Software such as On-Screen Takeoff allows the user to determine the lengths, areas, and volumes (Figure 5.4) of the different components in a building project. Other packages are designed for specific trades. For example, Trimble's Paydirt is specifically designed to estimate the quantities of earthwork,

concrete pavement, and asphalt for excavating contractors. These packages may be integrated into a specialized estimating program, or they may export the data to a spreadsheet or other specialized estimating package.

Takeoff software has the following advantages:

1. By using layers, it can keep track of where the quantities came from. This allows the estimator to review her takeoff

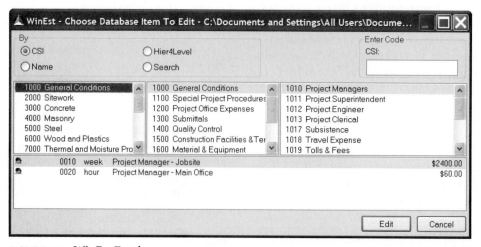

FIGURE 5.2. WinEst Database.

Courtesy of WinEstimator Inc.

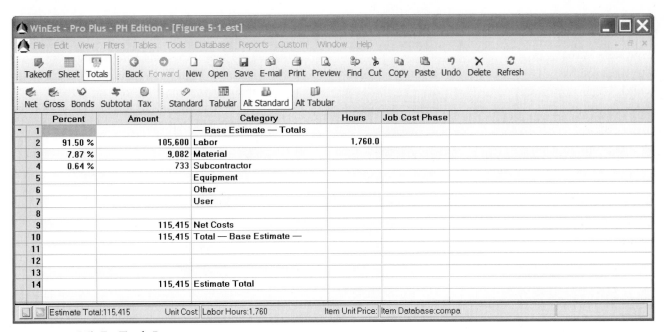

FIGURE 5.3. WinEst Totals Page.

Courtesy of WinEstimator Inc.

for missed items and recall where the individual quantities came from.

2. These software packages perform the math needed to convert linear dimensions or survey data into areas and volumes. For example, by tracing the perimeter of a concrete driveway, the software will determine the length of the perimeter to be formed, the area of the slab to be finished, and the volume of the concrete needed.

3. The drawings can be delivered electronically, reducing copying and delivery costs.

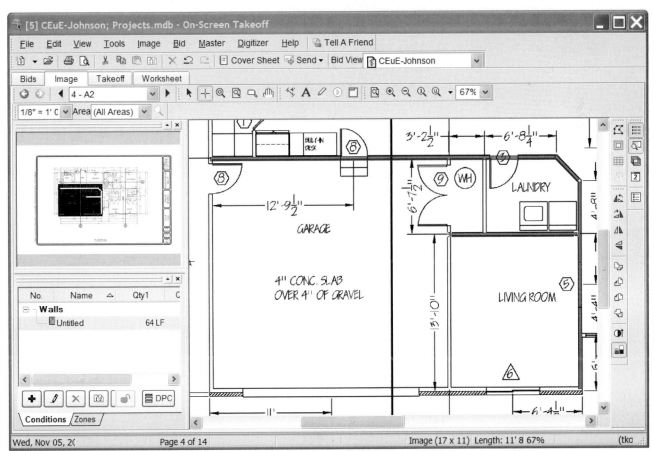

FIGURE 5.4. On-Screen Takeoff.

Courtesy of OnCenter Software

WEB RESOURCES

www.bid2win.com

www.buildsoft.com

www.hcss.com

www.mc2-ice.com

www.oncenter.com

www.sagesoftware.com

www.trimble.com/paydirt.shtml

www.winest.com

REVIEW QUESTIONS

1. What are the benefits of computerized estimating?

2. What are the two dangers that must be avoided when using computerized estimating?

3. What are the three categories of estimating software? How are they used in estimating?

CHAPTER SIX

OVERHEAD AND CONTINGENCIES

6-1 OVERHEAD

Overhead costs are generally divided into *home office overhead costs* and *job overhead costs.* The home office overhead costs include items that cannot be readily charged to any one project, but represent the cost of operating the construction company. The job overhead costs include all overhead expenses that will be incurred as a result of executing a specific project. The major difference between the two is that the home office overhead costs are incurred regardless of any specific project.

Job overhead costs constitute a large percentage of the total cost of a construction project. The job overhead costs can range from 15 to 40 percent of the total project cost. Because these costs are such a large portion of the total project costs, they must be estimated with the same diligence and precision as the direct cost. Simply applying a percentage for the project overhead degrades the overall accuracy of the estimate. If the direct portion of the estimate is quantified and priced out with diligence and precision and the project overhead is guessed, then the overall accuracy of the estimate is only as good as the guess of the project overhead. Estimators must consider overhead carefully, make a complete list of all required items, and estimate the cost for each of these items.

6-2 HOME OFFICE OVERHEAD

Home office overhead costs, also known as general overhead or indirect overhead costs, are costs that are not readily chargeable to one particular project. These costs are fixed expenses that must be paid by the contractor and are the costs of staying in business. These expenses must be shared proportionally among the projects undertaken; usually the home office cost items are estimated based on a fiscal year budget and reduced to a percentage of the anticipated annual revenue. The following items should be included in a home office overhead budget.

Office. Rent (if owned, the cost plus return on investment), electricity, heat, water, office supplies, postage, insurance (fire, theft, liability), taxes (property), telephone, office machines, and furnishings

Salaries. Office employees such as executives, accountants, estimators, purchasing agents, bookkeepers, and secretaries

Miscellaneous. Advertising, literature (magazines, books for library), legal fees (not applicable to one particular project), professional services (architects, engineers, CPAs) not billable to a job, donations, travel (including company vehicles not charged to jobs), and club and association dues

Depreciation. Expenditures on office equipment, calculators, computers, and any other equipment not billed to a job. A certain percentage of the cost is written off as depreciation each year and is part of the general overhead expense of running a business. A separate account should be kept for these expenses.

Figure 6.1 is a sample of the estimated home office cost for a general contractor for one year. The exempt and nonexempt employees are designations from the Fair Labor Standards Act. The most obvious difference between these employee designations is that employers are exempt from paying overtime to exempt employees. Commonly these are referred to as salaried employees. The federal statute that developed this differentiation also goes into detail about how the job function relates to the designation. Obviously, for smaller contractors the list would contain considerably fewer items, and for large contractors it could fill pages, but the idea is the same.

Non-Reimbursable Salaries		
Exempt Employees		
President	$100,000	
Vice President for Operations	$95,000	
Comptroller	$60,000	
Chief Estimator	$60,000	
Estimator	$29,000	
Director of Human Resources	$22,000	
Non-exempt Employees		
Secretaries (2)	$40,000	
Payroll Clerk	$15,000	
Accounts Payable Clerk	$15,000	
Total Office Labor Costs	**$436,000**	$436,000
Benefits @ 38%		$165,680
Office Rent (Gross Lease) 2000 sq. ft. @ $12.00		$24,000
Telephone		$3,600
Office Supplies		$1,200
Office Equipment		$11,500
Advertising		$5,000
Trade Journals		$200
Donations		$15,000
Legal Services		$2,000
Accounting Services		$3,600
Insurance on Office Equipment		$800
Club & Assoc. Dues		$1,000
Travel & Entertainment		$12,000
Cars (2) w/ Insurance		$9,000
Anticipated Office Expense for Year		$690,580

FIGURE 6.1. Estimated Home Office Costs for One Year.

From this, it should be obvious that the more work that can be handled by field operations, the smaller the amount that must be charged for general overhead and the better the chance of being the low bidder. As a contractor's operation grows, attention should be paid to how much and when additional staff should be added. The current staff may be able to handle the extra work if the additional workload is laid out carefully and through the use of selective spot overtime. Another consideration to adding new staff is the cost of supporting that person with computers, communications equipment, office space, and furniture.

Once the home office overhead has been estimated, it becomes necessary to estimate the sales for the year. If that amount is to rise over the coming year, the plan must state how to make that happen with the associated costs included in the budget. Will this growth come about by bidding for additional jobs, and will that require additional estimators? Will the growth come by expanding into new markets; and, if so, what are the costs of becoming known in these new markets? These are very important strategic issues that need to be addressed by the key people in the construction company. Once the sales for the year are estimated, a percentage can be developed. This percentage can then be applied to all work that is pursued. If this is not included in the estimate, these services are being provided free to all customers. Example 6-1 takes the home office cost estimate and shows how it would be allocated to specific projects.

EXAMPLE 6-1   HOME OFFICE COST ALLOCATION

Anticipated Home Office Costs for Fiscal Year—$690,000.

Anticipated Sales Volume for Fiscal Year—$9.5 million.

Home office cost allocation

$$= \frac{\text{Annual estimated home office costs}}{\text{Estimated annual revenue}}$$

Home office cost allocation

$$= \frac{\$690,580}{\$9,500,000} = 7.3\% \qquad \blacksquare$$

Some contractors do not allow for the category "General Overhead Expense" separately in their estimates; instead they figure a larger percentage for profit or group overhead and profit together. This, in effect, "buries" part of the expenses. From the estimator's viewpoint, it is desirable that all expenses be listed separately so that they can be analyzed periodically. In this manner, the amount allowed for profit is actually figured as profit—the amount left after all expenses are figured.

6–3 JOB OVERHEAD (GENERAL CONDITIONS, DIRECT OVERHEAD)

Also referred to as *general conditions, direct overhead,* or *indirect field costs,* job overhead comprises all costs that can be readily charged to a specific project but not to a specific item of work on that project. The list of job overhead items is placed on the first of the general estimate sheets under the heading of job overhead, general conditions, direct overhead, or indirect field costs. Most of these items are a function of the project duration; therefore, having a good estimate of the project duration is critical in developing a good job overhead estimate.

Itemizing each cost gives the estimator a basis for determining the amount of that expense and also provides for comparisons between projects. A percentage should not be added to the cost simply to cover overhead. It is important that each portion of the estimate be analyzed for accuracy to determine whether the estimator, in the future, should bid an item higher or lower.

Salaries.   Salaries include those paid to the project superintendent, assistant superintendent, timekeeper, material clerk, all foremen required, and security personnel if needed. Some companies and some cost-plus contracts also include the project manager and assistant project mangers in the project overhead. These costs must also include vehicle, mobile phone, travel, and job-related living expenses for these people.

The salaries of the various workers required are estimated per week or per month, and that amount is multiplied by the estimated time it is expected that each will be required on the project. The estimator must be neither overly optimistic nor pessimistic with regard to the time each person will be required to spend on the job. Figure 6.2 is a good example of how a bar chart schedule can be used to estimate the labor costs and then used during construction to control these costs.

Temporary Office.   The cost of providing a temporary job office for use by the contractor and architect during the construction of the project should also include office expenses such as electricity, gas, heat, water, telephone, and office equipment. Check the specification for special requirements pertaining to the office. A particular size may be required; the architect may require a temporary office, or other requirements may be included.

If the contractor owns the temporary office, a charge is still made against the project for depreciation and return on investment. If the temporary office is rented, the rental cost is charged to the project. Because the rental charges are generally based on a monthly fee, carefully estimate the number of months required. At $250 per month, three extra months amount to $750 from the profit of the project. Check whether the monthly fee includes setup and return of the office. If not, these costs must also be included.

Temporary Buildings, Barricades, Enclosures.   The cost of temporary buildings includes all tool sheds, workshops, and material storage spaces. The cost of building and maintaining barricades and providing signal lights in conjunction with the barricades must also be included. Necessary enclosures include fences, temporary doors and windows, ramps, platforms, and protection over equipment.

Temporary Utilities.   The costs of temporary water, light, power, and heat must also be included. For each of these items, the specifications must be read carefully to determine which contractor must arrange for the installation of the temporary utilities and who will pay for the actual amounts of each item used (power, fuel, water).

Water may have to be supplied to all subcontractors by the general contractor. This information is included in the project manual and must be checked for each project. The contractor may be able to tie into existing water mains. In this case, a plumber will have to be hired to make the connection. Other times, with permission of the municipality, the contractor may obtain water from a fire hydrant. Sometimes, the people who own the adjoining property will allow use of their water supply, usually for a fee. Water may be drawn from a nearby creek with a pump, or perhaps a water truck will have to be used. No matter where the water comes from, temporary water will be required on many projects, and the contractor must include its cost. The water source is one of the items the estimator should investigate at the site. Many water departments charge for the water used; the estimator will have to estimate the volume of water that will be required and price it accordingly.

Electrical requirements may include lighting and power for the project. These items are often covered in the electrical specifications, and the estimator must review it to be certain that all requirements of the project will be met. In small projects, it may be sufficient to tap existing power lines run them to a meter and string out extension lines through the project, from which lights will be added and power will be taken off. On large complexes, it may be necessary to install poles, transformers, and extensive wiring so that all power equipment being used to construct the project will be supplied. If the estimator finds that the temporary electricity required will not

PROJECT STAFF PLAN AND ESTIMATE																
Title	Monthly Gross Pay	Months On Project	Cost	Month												
				1	2	3	4	5	6	7	8	9	10	11	12
Project Manager	6,500	10	$65,000.00												
Assistant Project Manager	5,500	10	$55,000.00												
Civil Superintendent	5,000	2	$10,000.00												
Concrete Superintendent	4,000	5	$20,000.00												
Purchasing Agent	2,500	4	$10,000.00												
Clerical	1,500	8	$12,000.00												
Payroll	1,400	12	$16,800.00												
Total Project Staff Costs			$188,800.00												

FIGURE 6.2. Field Staff Plan and Estimate.

meet the needs of the project, the first call should be to the architect/engineer to discuss the situation. The architect may decide to issue an addendum that revises the specifications. If not, the estimator will have to include these costs in the general construction estimate. Typically, the temporary wiring and transformers are included in the scope of the electrical subcontractor's work. All power consumed must be paid for, and how this is handled should be in the project manual; often the cost is split on a percentage basis among the contractors, and it is necessary to make an allowance for this item also.

Heat is required if the project will be under construction in the winter. Much of the construction process requires maintenance of temperatures at a certain point. The required heat may be supplied by a portable heater using kerosene, liquefied petroleum gas, or natural gas for small projects. The total cost includes the costs of equipment, fuel, and required labor (remember that the kerosene type must be filled twice a day). On large projects, the heating system is sometimes put into use before the rest of the project is complete; however, costs to run and maintain the system and replace the filters before turning the project over to the owner must be included in the estimate.

Sanitary Facilities.   All projects must provide toilets for the workers. The most common type in use is the portable toilet, which is most often rented. The rent often includes the cleaning and servicing of the toilets on a regular basis. Large projects will require several portable toilets throughout the job. This item is also one of the first things that must be on the job site. Another important aspect with regard to portable toilets is waste hauling—how often the waste will be removed and whether weekend service is available. If the contractor or one of the subcontractors works on weekends, the portable toilets must be cleaned. This often will incur an additional charge that needs to be included in the estimate to ensure that there will be adequate funds in the project budget.

Drinking Water.   The cost of providing drinking water in the temporary office and throughout the project must be included. It is customary to provide cold or ice water throughout the summer. Keep in mind that the estimated cost must include the containers, cups, and someone to service them. If the construction is taking place in a hot and dry area, special electrolyte liquids may be required to maintain the health of the labor force.

Photographs.   Many project specifications require photographs at various stages of construction, and even if they are not required, it is strongly suggested that a digital camera and a video camera be kept at each job site. The superintendent should make use of them at all important phases of the project to record progress. In addition to the cost of the above items, the cost of processing and any required enlarging of the pictures should be considered. The use of webcams is becoming increasingly popular. These devices come with a wide variety of features and options. These cameras allow persons to view the project via the Internet. If a webcam is desired or required, the hardware and operating costs need to be included.

Surveys.   If a survey of the project location on the property is required, the estimator must include the cost for the survey in the estimate. A survey may be used to lay out the corners and grid lines of a building even if it is not required by the contract documents, which costs need to be included in the estimate. Check the project manual to see if a licensed surveyor is required, and then ask several local surveyors to submit a proposal.

Cleanup.   Throughout the construction's progress, rubbish will have to be removed from the project site. The estimator needs to estimate how many trips will be required and a cost per trip. In addition, a plan needs to be devised concerning where the rubbish can be dumped. As landfills become sparse, it is becoming more difficult and expensive to dump construction waste. Because some landfills will not accept construction waste, the debris may have to be hauled for long distances. In addition, recycling of construction waste and "green construction" is an area that is receiving lots of attention, which may require wastes to be separated by type. As the cost and difficulty of disposing of construction waste increases, so does the feasibility of some alternate methods of disposal.

Before acceptance of the project by the owner, the contractor will have to clean the floors, the job site, and, in many cases, even the windows. This is typically performed by a cleaning service and estimated by the square-foot.

Winter Construction.   When construction will run into or through the winter, several items of extra cost must be considered, including the cost of temporary enclosures, heating the enclosure, heating concrete and materials, and the cost of protecting equipment from the elements.

Protection of Property.   Miscellaneous items that should be contemplated include the possibility of damaging adjacent buildings, such as breaking windows, and the possible undermining of foundations or damages by workers or equipment. Protection of the adjacent property is critical. It is recommended that the contractor's insurance carrier be contacted prior to commencing construction, as the insurer can survey the adjacent structures for existing damages. This service is normally provided at no charge. If the insurance carrier does not provide this service, a consultant should be hired to take a photographic inventory.

All sidewalks and paved areas that are torn up or damaged during construction must be repaired. Many items of new construction require protection to avoid their damage during construction, including wood floors, carpeting, finished hardware, and wall finishes in heavily traveled areas. New work that is damaged will have to be repaired or replaced. Often, no one will admit to damaging an item, so the contractor must absorb this cost.

The supplementary general conditions should be checked carefully for other requirements that will add to the job overhead expense. Examples of these are job signs, billboards, building permits, testing of soil and concrete, and written progress reports. Section 6–6 is an indirect estimate checklist that is helpful in estimating this portion of the project.

6-4 SCHEDULING

A major underlying determinant of indirect costs associated with a project is how long it will take to complete the project. This length of time especially affects the estimator with regard to project overhead items, wages paid to supervisory and home office personnel, rental on trailers and toilet facilities, and any guards, traffic directors, and barricading required. It also affects the estimate in terms of how long equipment will be required on the project.

Traditionally, the estimator assumes an approximate project duration that is the basis for the estimated project overhead costs. There are many computer programs such as Microsoft Project, Suretrak (by Primavera), Primavera Contractor, and Primavera P6, that are useful tools in sequencing the construction process and assisting in developing the project duration. When competitively bidding projects, this process is time-consuming and expensive. The estimator and the appropriate management personnel must make a choice of how much time and cost should be expended in the estimating phase of the project. The job site overhead costs often impact whether a particular contractor is the low bidder. The speed at which the project is constructed has a heavy impact on the ultimate project cost. At minimum, a bar chart schedule should be developed to estimate a reasonable project duration. One approach is to bring together proposed members of the project team and have them develop this schedule. The advantage of this approach is that the persons who are actually going to direct the work in the field will have ownership and input in the construction schedule. This cooperative effort will reduce the conflict between the home office and the field and will add credibility to the schedule.

However, if the work is being negotiated and the decision focuses not upon which contractor to select but upon the economic feasibility of the project, then detailed network schedules may be appropriate. In addition to forcing everyone to think about the construction process, it shows the owner that the contractor is serious about planning the project and executing according to the plan.

The basics of scheduling the project can be broken into four steps:

1. List all activities required for the completion of the project.

2. Assign a duration to each of the activities listed in step 1. It is most important that *all* of the times be reasonably accurate. If the work is to be subcontracted, contact the subcontractor for her input.

3. Write each activity and its duration on a "post-it" note and have the construction team develop a network diagram that shows the sequence in which the activities will be performed. The most popular type of network diagram used today is the precedence diagram.

4. Perform a forward pass calculation (this is a schedule calculation) to determine the estimated project duration. If one of the commercially available scheduling programs is available, the post-it network will become the guide for entering the data.

It is beyond the scope of this textbook to show the complete workings of computerized scheduling. By using a small example, however, the basics can be explained. The small office shown in Figure 6.3 will be used. Using those simple drawings, the activity list in Figure 6.4 was developed.

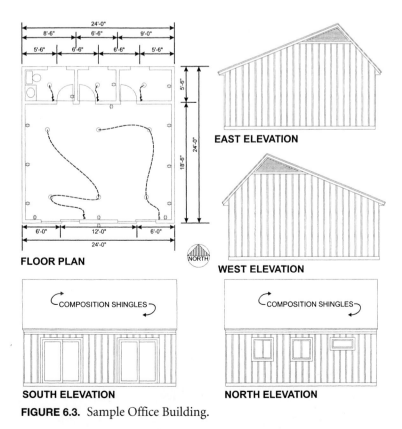

FIGURE 6.3. Sample Office Building.

Activity Id.	Description	Duration
10	Clear site	2
20	Scrape topsoil	2
30	Gravel fill	3
40	Plumbing rough-in	2
50	Form, concrete slab	2
60	Pour and finish concrete	1
70	Rough Carpentry	10
80	Electrical, rough-in	2
90	Insulation	2
100	Roofing	3
110	Plumbing, top out	1
120	Drywall	4
130	Interior trim	2
140	Exterior trim	5
150	Telephone, rough-in	1
160	Plumbing, finish	1
170	H.V.A.C., rough in	3
180	H.V.A.C., finish	3
190	Painting	4
200	Stain, exterior	3
210	Carpet	1
220	Windows	1
230	Glass doors	1
240	Wood Doors	1
250	Final grade	1
260	Seed	1
270	Electrical, finish	1

FIGURE 6.4. Activity List.

Once the activity list has been developed, it can be organized into a precedence diagram network schedule. A discussion of this methodology is clearly outside the scope of this book, and Figure 6.5 is included for reference only. There are many excellent scheduling books on the market that cover scheduling in great detail.

Once the project duration has been determined, it needs to be converted into calendar days. In the example in Figure 6.5, the project duration is 50 days. For a 5-day workweek, the duration is 10 weeks or 70 calendar days. If there are nonwork periods in that intervening 70 days, the calendar duration would be extended by the number of nonwork periods. Typical nonwork periods are holidays such as New Year's Day or Christmas.

6-5 CONTINGENCIES

On virtually every construction project, some items are left out or not foreseen when the estimates are prepared. In some cases, the items left out could not have been anticipated at the time of estimating. Should a contingency amount be included? That is, should a sum of money (or percentage) be added to the bid for items overlooked or left out? This money would provide a fund from which the items could be purchased.

If the money is allowed, it is not necessary to be quite so careful in the preparation of an estimate. But, if an accurate estimate is not made, an estimator never knows how much to allow for these forgotten items.

Contingencies are often an excuse for using poor estimating practices. When estimators use them, they are not truly estimating a project. Instead of adding this amount, the proper approach is to be as careful as possible in listing all items from the plans and the project manual. This listing should include everything the contractor is required to furnish, it should estimate labor costs carefully and accurately, and it should include the job overhead expense. To these items, add the desired profit. The most rational use for contingencies is for price escalation.

6-6 CHECKLIST

Undistributed Labor

- Job superintendent
- Assistant superintendent
- Engineers
    Job engineers
    Field engineers (surveyors)
    Expediter
    Cost engineer
    Scheduling engineer
- Timekeepers
- Material clerks
- Security personnel
- Project meetings
- Submittal coordination
- Secretaries

To Estimate Cost
Labor

Number of weeks on the project × Weekly rate

Expenses   If the company policy is for per diem, the cost would be the following:

The duration of each staff person on the project × Per diem

If the project location is remote, the cost of relocation should be included:

Persons to be relocated × Cost per person
(get estimated cost from moving company)

If the staff personnel do not relocate but commutes on a regular basis, a policy on the frequency of trips needs to be established and that policy needs to be integrated into the estimate.

Cost of travel for executives for job site visits must be included.

Temporary Buildings, Enclosures, and Facilities

- Temporary fences
- Temporary sheds
- Storerooms
- Storage and handling
- Temporary enclosures

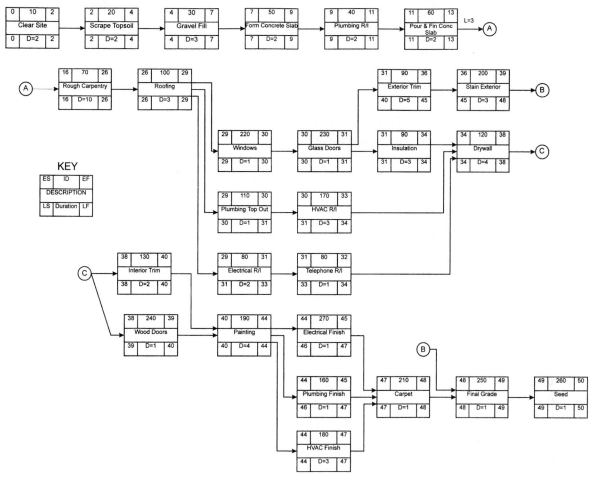

FIGURE 6.5. Sample Schedule.

- Ladders and stairs—used prior to permanent ones being installed
- Temporary partitions—used to separate new construction from existing facilities
- Temporary closures for doors and windows
- First aid
- Construction elevators, hoists, and cranes
- Noise control
- Dust control
- Water control
- Pest/rodent control

To Estimate Cost

Labor

Quantity required × Productivity rate

Material

Quantity required × Unit cost

Temporary Office

- Temporary job office
- Temporary office for architect/inspector

- Telephone
- Heat
- Lights and power
- Computers
- Stationery
- Project sign and associated signage

To Estimate Cost

Labor   Typically none, except for office set-up

Materials/Equipment

Duration required at job site × Monthly cost

Barricades and Signal Lights

- Building and maintaining barricades
- Cost of maintaining signal lights

To Estimate Cost

Labor

Quantity of materials × Productivity rate

Material

Quantity of materials × Unit price

Many contractors either rent barricades or contract for supply and maintenance. If so,

$$\text{Quantity item months} \times \text{Monthly rate}$$

Temporary Utilities

- Temporary toilets—Local codes, union rules, or accepted company ratios usually govern the quantity and types of toilets that must be supplied. Chemical, portable toilets will usually suffice. In the case of a remodeling project, the existing toilet facilities might be available for use; however, permission is typically needed.

To Estimate Cost

$$\text{Number of toilets} \times \text{Number of months required} \\ \times \text{Monthly rate}$$

- Temporary water—On most projects, the general contractor must provide water for all trades. This item can become costly if working in remote or extremely hot climates. If temporary water comes from a municipality, there may be some sewer requirements.

To Estimate Cost. Function of project location and environment

- Temporary light and power—This is determined by the power requirements of the project. If high-power electrical equipment is being used in the construction, the power drop from the pole may be anywhere from 2,300 volts to 480 volts. If this power is used, a transformer will be required to supply power for the office and power tools. In addition, there is typically a demand charge for utility installations that require more than 200 amps. If the power requirements are above this threshold, the kilowatt rate from the power company will have little meaning. Therefore, the best information comes from previous projects that had similar power requirements. In addition, there is the need to include temporary lights after the building is closed in and any associated power cords and bulbs.

    The estimator also needs to be aware of the anticipated start-up of the mechanical equipment. If this equipment is started early to improve the work environment, a substantial utility cost will be associated with keeping this equipment running during the remainder of the construction process.

To Estimate Cost. Determine the power requirements and get cost information from the utility supplier. If familiar with working in the area, a historical monthly allowance would be acceptable.

- Temporary heat—If the project runs through the winter, it may be necessary to rent heaters and provide needed fuel.

To Estimate Cost

$$\text{Number of heaters} \times \text{Number of months required} \\ \times \text{Monthly rental rate}$$

Repairs and Protection

- Repairs to streets and pavement—This covers the cost of repairing all streets damaged during construction.
- Damage to adjoining structures—There is always the chance of damaging adjacent structures, such as windows, foundations, and walls improperly shored. These items should be considered when estimating the project.
- Protecting new work from damage during construction—During construction, it becomes necessary to protect certain classes of work such as cut stone, marble, terra cotta, granite, and all types of floors and wood products.
- Repairing new work damaged during construction—Patching damaged plaster, replacing broken glass, and so forth are the responsibility of the contractor. General contractors should keep a close watch on these items, as they can be back-charged to a subcontractor if they broke or damaged an item.

Relocating Utilities

- Water lines
- Electric lines

To Estimate Cost. Identify items and solicit prices or get the cost from the agency that owns the lines.

Cleaning

- Removal of rubbish (typically weekly)
- Cleaning for final acceptance

To Estimate Cost Estimate the number of loads or frequency of loads. Get a price quote on a per haul basis.

Cleaning for final acceptance. Get a per-square-foot cost from a cleaning service.

Permits

To Estimate Cost. Contact local government agency to determine what is required and the cost. This should be done when going to or coming from the pre-bid site visit. In addition to building permits, permits may be required for work done in the public right of ways, such as a road cut. The contractor may have to submit other documents, such as a Storm Water Pollution Prevention Plan (SWPPP), to government agencies before construction can begin. The cost of preparing these documents and obtaining the necessary permits needs to be included in the project overhead.

Professional Services

- Surveys
- Photographs
- Testing

To Estimate Cost. The amount of surveying required is a function of the layout that will be performed by the field engineer and that which is to be done by a professional engineer/surveyor. It is best to contact a local firm and get their hourly billing rate for these services.

The photographic requirements are part of the specifications and company policy. With the advent of low-priced video cameras and digital cameras, this item is becoming less costly.

The testing requirements for materials are typically found in the specifications. This provides for the quantity of tests required. The cost of these tests can be determined by contacting a testing laboratory to determine its charge for specific types of tests. In addition, the estimator must know what the laboratory charges to collect samples. The cost of collecting the samples can exceed the cost of the test.

Labor Burdens and Taxes

- Cash equivalents and allowances
- Social security
- Medicare
- Federal unemployment taxes
- State unemployment taxes
- Workers' compensation insurance
- Benefits
- Sales tax

To Estimate Cost. Labor burden should be included in the labor costs used in the estimate. For example, the labor burden for the carpenter framing a wall should be included in the carpentry labor rate used when estimating the cost of the wall. The only labor burden that should be included in the overhead is the labor burden associated with the overhead labor. Historically, labor burden was difficult to track and bill to the individual cost codes or items. Labor burden was often charged to overhead and then allocated. With computer accounting systems, the charging of labor burden costs to the individual cost codes or items is effortless. Because the estimate becomes the basis of the budget used in the accounting system, the labor burden must be included in each item so that the estimate is consistent with the account system, making it easier to manage job costs.

Similarly, tax should be included in each class of items (for example, lumber, concrete, and so forth). When the material bills are paid, they will include the sales tax for the materials. Consistency between the accounting system and the estimating process must be maintained as part of good cost control procedures.

Bonds and Insurance

- Bid bond
- Performance and payment bonds
- Maintenance bonds
- Permit bonds
- Contractor's public liability insurance

To Estimate Cost. Bonds and insurance are functions of project requirements and the experience rate of the contractor. The exact cost of these items can be determined by contacting the contractor's insurance or bonding agent.

Miscellaneous Equipment and Services

- Pick-up truck(s)
- Flatbed truck(s)
- Pumps

To Estimate Cost. Determine the quantity of service vehicles and their approximate duration on the job site. This can be converted into a cost item by charging the vehicle to the project on a per month basis. If the trucks are rented, the cost would be the number of months per vehicle times the monthly rate. If a long-term lease agreement is used, there may be substantial penalties for returning the vehicle early. These costs could be charged to the project if the vehicle is not needed on another project.

WEB RESOURCES

www.primavera.com

office.microsoft.com/en-us/project/default.aspx

REVIEW QUESTIONS

1. What is overhead? Why must it be included in the cost of the project?

2. What is the difference between general and job overhead costs?

3. How are the items that the estimator will include in each type of overhead determined?

4. Why is the length of time it will take to complete the project so important in determining the overhead costs?

5. What competitive advantage is there for a company that has kept its general overhead low?

6. How may the weather, climate, and season during which the project is to be constructed affect the overhead costs?

7. Where would the cost of temporary utilities be put in the estimate?

8. How would the estimator determine who is responsible for what cost portion of supplying the temporary utilities?

9. How can a preliminary work schedule of the project help the estimator?

10. Define contingency amounts and how some estimators use them.

11. If a full set of contract documents is available, make a list of overhead expenses required for the project. If a set is not available, make another trip to the plan room and review a set of documents on file there.

12. Using the plans and outline specifications in Appendix A, prepare a bar chart schedule for the construction of the project.

CHAPTER SEVEN

LABOR

7–1 LABOR HOURS AND PRODUCTIVITY

The basic principles of estimating labor costs are discussed in this chapter; they form a basis for the labor costs, which will be illustrated in each chapter that covers quantity take-off. Estimating labor requires determining the number of labor hours to do a specific task and then applying a wage rate. A labor hour is defined as one worker working for one hour. Determining the labor hours requires knowing the quantity of work to be placed and the productivity rate for the specific crew that will perform the work. The crew is an aggregation of construction trades working on a specific task. The productivity rate is often expressed as a number of labor hours per unit of work, although it may also be expressed as the quantity of work performed by a crew during a standard eight-hour day. In this book, we will use the number of labor hours per unit of work.

The productivity rates can come from a number of sources, but the most reliable source is historical data. The advantage of historical data is that it reflects how a particular company's personnel perform the tasks. The historical productivity rate is determined by dividing the total number of labor hours to complete a task by the actual quantity of work performed, as shown in Formula 7-1.

$$\text{Productivity rate} = \frac{\text{Labor hours}}{\text{Quantity}} \quad \textbf{Formula 7-1}$$

Productivity Rate

For historical data to be useful, the accounting system should not only track the cost to complete a task, but also track the number of labor hours required and the quantity of work performed. For example, if a job requires the construction of a concrete slab, it is insufficient to know that the labor to construct the slab cost $5,523. The historical data must also tell us that it took 170 labor hours and that 10,000 square feet of slab was constructed. Without the labor hours or the quantity of work performed, it

is impossible to determine a productivity rate from historical data.

EXAMPLE 7-1 HISTORICAL PRODUCTIVITY RATE

Type of work—8" × 8" × 16" Concrete Masonry Units (CMUs)

Quantity of work—1,700 square feet

Labor cost—$6,987

Labor hours—170 labor hours

$$\text{Productivity rate} = \frac{170 \text{ labor hours}}{1,700 \text{ sf}}$$
$$= 0.1 \text{ labor hours per sf} \quad \blacksquare$$

Formula 7-2 is used to determine the estimated number of labor hours for a task using the productivity rate.

$$\text{Labor hours} = \text{Quantity takeoff} \times \text{Productivity rate}$$

Labor Hours   **Formula 7-2**

The productivity rate that is used, if derived from historical data, is for the average or standard conditions for the projects used in calculating the historical production rate. On many occasions, the project that is being bid deviates from these standard conditions. Therefore, the number of labor hours needs to be modified to take into consideration how the project that is being bid deviates from the standard condition. This is done by using a productivity factor. Formula 7-3 is the mathematical means by which the productivity factor is applied.

$$\text{Adjusted labor hours} = \text{Labor hours} \times \text{Productivity factor}$$

Adjusted Labor Hours   **Formula 7-3**

The productivity factor is a combination of several variables or conditions. This is perhaps one of the most complicated determinations. There are no hard-and-fast rules concerning productivity factors. The elements of experience and instinct are perhaps the best quality. When the conditions slow the progress of the work, a productivity factor greater than one is used, thus increasing the number of labor hours to complete a unit of work. When the conditions increase the speed at which the work progresses, a productivity factor less than one is used, thus decreasing the number of labor hours to complete a unit of work. Following are some variables that need to be considered when determining the productivity factor.

Availability and Productivity of Workers. When there is plenty of work available and workers are scarce, less-trained craft persons are accepted. These less-trained persons will require more time or labor hours to complete the required task, and a productivity factor greater than one is used. Conversely, when construction projects are scarce, workers may become motivated, and the contractor can be selective and hire only the most qualified workers. This will result in producing more work per labor hour, and a productivity factor less than one is used.

Climatic Conditions. Cold, hot, winds, rain, snow, and combinations of these all affect the amount of work that can be produced in an hour. Typically, any weather extremes will slow down the work pace and may require additional precautions that add labor hours to the project. The estimator must try to factor in each of these to determine the most cost-effective approach. Can the project be scheduled so that the concrete can be poured before the winter cold sets in? If not, extra time and materials will be necessary to make certain that the concrete does not freeze after it is poured. On the other hand, if the weather is too hot, precautions will need to be taken to ensure that the concrete does not set too quickly, requiring provisions to be in place to keep it damp. Adverse weather conditions, in and of themselves, do not warrant the use of a productivity factor greater than one. The anticipated weather condition must be worse than the average weather condition for the historical data. For example, if the historical data were for concrete poured during the winter months, the historical data already take into account the time it takes to protect concrete against the average winter conditions. However, if the historical data used were for concrete poured during the summer months or all year round, a productivity factor greater than one is used to account for the winter conditions. A productivity factor of less than one would be used when the anticipated weather conditions are better than the average weather conditions for the historical data.

Working Conditions. The job site working conditions can have a great effect on the rate of work. A project being built in the city with little working space, limited storage space, and difficult delivery situations typically has less work accomplished per labor hour just due to the difficulty of managing the resources. The same may be true of high-rise construction where workers may have to wait for the crane to deliver materials to them, have difficulty moving from floor to floor, and take extra time just to get from where they punch the time clock to where they will be working. Again, a productivity factor of greater than one is used when conditions are worse than average, and a productivity factor of less than one is used when conditions are better than average.

Projects that are far removed from the supply of workers and materials often have similar situations that the estimator must consider. How can material deliveries be made in a timely fashion? Where will the material be stored until needed? Will extra equipment and workers be required to transport the material from the storage area to where it will be installed? Will storage sheds be necessary for material that cannot be left out in the weather? If so, who will be responsible for receiving inventory and moving it to where it will be installed? Worker availability for remote jobs must also be considered. Are workers available, at what costs, and are any special incentives required? Again, a productivity factor of greater than one is used when conditions are worse than average, and a productivity factor of less than one is used when conditions are better than average.

Other Considerations. Workers seldom work a full 60 minutes during the hour. Studies of the actual amount of time worked per hour averaged 30 to 50 minutes. This is often referred to as system efficiency. Keep in mind that the time it takes to "start up" in the morning, coffee breaks, trips to the bathroom, a drink of water, discussing the big game or date last night, lunches that start a little early and may end a little late, and clean-up time all tend to shorten the work day. This list of variables is long, but these items must be considered. Again, the productivity factor is based upon the variance from average conditions, not from the ideal. A system efficiency lower than average will require a productivity factor greater than one, and a system efficiency greater than average will require a productivity factor less than one. Of most importance to the estimator are those items that can be done to make it "convenient" to work, such as placing restrooms, drinking water, and materials close to the work. This also involves providing adequate, well-maintained equipment; seeing that materials are delivered to the job just before they are needed; answering questions regarding the work to be performed in a timely manner; and anything else necessary to ensure that the work can proceed quickly.

When keeping historical records for the labor productivity, it is important that a record not only of the productivity rate be kept, but also under what conditions that productivity rate was achieved so that an appropriate productivity factor can be used.

EXAMPLE 7-2   LABOR HOURS

Type of work—8" × 8" × 16" Concrete Masonry Units—Decorative

Historical productivity rate—0.1 labor hours per sf

Productivity factor—1.1

Crew—three masons and two helpers

$$\text{Labor hours} = 1{,}000 \text{ sf} \times 0.10 \text{ labor hours per sf}$$
$$= 100 \text{ labor hours}$$

$$\text{Adjusted labor hours} = 100 \text{ labor hours} \times 1.1$$
$$= 110 \text{ labor hours}$$

Sixty percent (3/5) of the hours will be performed by masons and 40 percent (2/5) will be performed by their helpers.

$$\text{Mason labor hours} = 0.60 \times 110 \text{ labor hours}$$
$$= 66 \text{ labor hours}$$

$$\text{Mason helper labor hours} = 0.40 \times 110 \text{ labor hours}$$
$$= 44 \text{ labor hours} \quad \blacksquare$$

Another method of determining the productivity rate is cycle time analysis, which is used when the work is performed in a repeatable cycle. An example of a cycle is a truck hauling earthen materials from the borrow pit to the job site and returning to the borrow pit to make a second trip. Cycle time analysis is used extensively in excavation estimating. Cycle time analysis is performed by timing a number of cycles, ideally at least 30. Using the average cycle time (in minutes), the productivity rate is calculated using Formula 7-4. The average cycle time is determined by summing the cycle time and by dividing the sum by the number of observations. The productivity factor is the same as the productivity factor used in Formula 7-3. The crew size is the number of people in the crew and determines the number of labor hours per clock hour. The system efficiency takes into account that workers seldom work a full 60 minutes per hour. Typical system efficiencies range from 30 to 50 minutes per hour. The quantity per cycle is the number of units of work produced by one cycle.

Productivity rate

$$= \frac{\text{Average cycle time} \times \text{Productivity factor} \times \text{Crew size}}{\text{System efficiency} \times \text{Quantity per cycle}}$$

Productivity Rate   **Formula 7-4**

EXAMPLE 7-3   PRODUCTIVITY RATE USING CYCLE TIME

Type of work—Hauling materials from the borrow pit

Average cycle time—35 minutes

Truck capacity—17 tons

Crew—One driver

Productivity factor—0.95

System efficiency—45 minutes per hour

$$\text{Productivity rate} = \frac{35 \text{ minutes} \times 0.95 \times 1}{45 \text{ minutes per hour} \times 17 \text{ tons}}$$

Productivity rate = 0.0435 labor hours per ton   ■

When the work is performed linearly (such as paving or striping a road, placing a concrete curb using slip-forming machine, or grading a road), the rate of progress may be used to determine the productivity rate. The productivity rate is calculated using Formula 7-5. The quantity is the quantity of work to be performed; in the case of placing concrete curb, it would be the length (in feet) of the curb to be placed with the slip-forming machine. The rate of progress is the number of units of work that can be performed by the crew each minute when they are performing the work. In the case of placing a concrete curb, the rate of progress would be the number of feet of curb that is placed in one minute. The travel time is the time (in minutes) that the equipment is not working because it is being moved from one section of work to another section of work. In the case of placing the concrete curb, sections of curb will be left out where there is a tight radius or a driveway approach. The travel time would be the time it takes to move the equipment forward through the driveway approaches and sections where the curb is not being placed by the machine. The crew size is the number of people in the crew and determines the number of labor hours per clock hour. The system efficiency takes into account that workers seldom work a full 60 minutes per hour. Typical system efficiencies range from 30 to 50 minutes per hour.

Productivity rate

$$= \frac{\left(\dfrac{\text{Quantity}}{\text{Rate of progress}} + \text{Travel time}\right) \times \text{Crew size}}{\text{System efficiency} \times \text{Quantity}}$$

Productivity Rate   **Formula 7-5**

EXAMPLE 7-4   PRODUCTIVITY RATE USING RATE OF PROGRESS

Type of work—Slip forming concrete curb

Quantity—2,200′

Number of approaches—30 each @ 3 minutes each

Number of curves—5 each @ 5 minutes each

Rate of progress—3′ per minute

Crew—One operator and two helpers

Productivity factor—0.95

System efficiency—45 minutes per hour

$$\text{Travel time} = 30 \times 3 \text{ minutes} + 5 \times 5 \text{ minutes}$$
$$= 115 \text{ minutes}$$

$$\text{Productivity rate} = \frac{\left(\dfrac{2{,}200'}{3 \text{ feet per minute}} + 115 \text{ minutes}\right) \times 3}{45 \text{ minutes per hour} \times 2{,}200'}$$

Productivity rate = 0.0257 labor hours per foot   ■

When company data are not available, data from published sources can be used. Figure 7.1 shows the productivity and costs for concrete unit masonry from R.S. Means *Building Construction Cost Data*. From Figure 7.1, we see that the

04 22 Concrete Unit Masonry

04 22 10 - Concrete Masonry Units

04 22 10.14 Concrete Block, Back-Up		Crew	Daily Output	Labor-Hours	Unit	Material	2007 Bare Costs Labor	Equipment	Total	Total Incl O&P
1100	6" thick	D-8	430	.093	S.F.	1.98	3.19		5.17	7
1150	8" thick		395	.101		2.15	3.47		5.62	7.65
1200	10" thick		320	.125		2.87	4.29		7.16	9.65
1250	12" thick	D-9	300	.160		3.11	5.35		8.46	11.50
04 22 10.16 Concrete Block, Bond Beam										
0010	**CONCRETE BLOCK, BOND BEAM**									
0020	Not including grout or reinforcing									
0130	8" high, 8" thick	D-8	565	.071	L.F.	2.19	2.43		4.62	6.10
0150	12" thick	D-9	510	.094		3.05	3.14		6.19	8.15
0525	Lightweight, 6" thick	D-8	592	.068		2.20	2.32		4.52	5.95
0530	8" high, 8" thick	"	575	.070		2.19	2.39		4.58	6.05
0550	12" thick	D-9	520	.092		3.60	3.08		6.68	8.65
2000	Including grout and 2 #5 bars									
2100	Regular block, 8" high, 8" thick	D-8	300	.133	L.F.	4.14	4.57		8.71	11.50
2150	12" thick	D-9	250	.192		5.55	6.40		11.95	15.85
2500	Lightweight, 8" high, 8" thick	D-8	305	.131		4.60	4.50		9.10	11.90
2550	12" thick	D-9	255	.188		6.10	6.30		12.40	16.25
04 22 10.18 Concrete Block, Column										
0010	**CONCRETE BLOCK, COLUMN**									
0050	Including vertical reinforcing (4-#4 bars) and grout									
0160	1 piece unit, 16" x 16"	D-1	26	.615	V.L.F.	17.50	20.50		38	50.50
0170	2 piece units, 16" x 20"		24	.667		23	22		45	59.50
0180	20" x 20"		22	.727		33.50	24.50		58	74
0190	22" x 24"		18	.889		47.50	29.50		77	97
0200	20" x 32"		14	1.143		51.50	38		89.50	115
04 22 10.19 Concrete Block, Insulation Inserts										
0010	**CONCRETE BLOCK, INSULATION INSERTS**									
0100	Inserts, styrofoam, plant installed, add to block prices									
0200	8" x 16" units, 6" thick				S.F.	1.13			1.13	1.24
0250	8" thick					1.13			1.13	1.24
0300	10" thick					1.33			1.33	1.46
0350	12" thick					1.40			1.40	1.54
0500	8" x 8" units, 8" thick					.93			.93	1.02
0550	12" thick					1.13			1.13	1.24
04 22 10.23 Concrete Block, Decorative										
0010	**CONCRETE BLOCK, DECORATIVE**									
0020	Embossed, simulated brick face									
0100	8" x 16" units, 4" thick	D-8	400	.100	S.F.	3.02	3.43		6.45	8.55
0200	8" thick		340	.118		4.17	4.03		8.20	10.75
0250	12" thick		300	.133		5.50	4.57		10.07	13
0400	Embossed both sides									
0500	8" thick	D-8	300	.133	S.F.	4.68	4.57		9.25	12.10
0550	12" thick	"	275	.145	"	5.90	4.99		10.89	14.10
1000	Fluted high strength									
1100	8" x 16" x 4" thick, flutes 1 side,	D-8	345	.116	S.F.	3.59	3.98		7.57	10
1150	Flutes 2 sides		335	.119		4.36	4.09		8.45	11.05
1200	8" thick		300	.133		5.65	4.57		10.22	13.20
1250	For special colors, add					.35			.35	.39
1400	Deep grooved, smooth face									
1450	8" x 16" x 4" thick	D-8	345	.116	S.F.	2.34	3.98		6.32	8.65
1500	8" thick	"	300	.133		4.04	4.57		8.61	11.40
2000	Formblock, incl. inserts & reinforcing									

FIGURE 7.1. Published Productivity Rates.

productivity for an 8" × 8" × 16" decorative concrete block is 0.118 labor hours per sf. Care must be exercised when using the costs from national data sources, because they are less accurate than historical data.

7-2 UNIONS—WAGES AND RULES

The local labor situation must be surveyed carefully in advance of making the estimate. Local unions and their work rules should be given particular attention, since they may affect the contractor in a given community. The estimator will have to determine whether the local union is cooperative and whether the union mechanics tend to be militant in their approach to strike or would prefer to talk first and strike as a last resort. The estimator will also have to determine whether the unions can supply the skilled workers required for the construction. These items must be considered in determining how much work will be accomplished on any particular job in one hour.

While surveying the unions, the estimator will have to get information on the prevailing hourly wages, fringe benefits, and holidays; also the date that raises have been negotiated, when the present contract expires, and what the results and attitudes during past negotiations have been. If the project will run through the expiration date of the union contract, the estimator will have to include enough in the prices to cover all work done after the expiration of the union contract. Often this takes a bit of research into price trends throughout the country; at best it is risky, and many contractors refuse to bid just before the expiration of union contracts unless the terms of the project's contract provide for the adjustment of the contract amount for such increases.

7-3 OPEN SHOP

The construction unions have experienced a decline in membership over the years. The open shop contractor or subcontractor does not have to deal with restrictive union work rules, which gives the contractor greater flexibility and allows craft persons with multiple skills to stay on the project longer and perform a greater portion of the work. The downside of this arrangement is that the quality of the craft persons is not known when hired. The union carpenter has gone through a structured apprentice program to become a union carpenter. On the other hand, open shop contractors cannot go to the union hall looking for craft persons. Rather, they must directly hire all their craft personnel, which adds to higher turnover and training costs if craft labor is in short supply.

7-4 LABOR BURDEN

The wages paid to labor are known as the bare hourly wage rate or bare wage rate (in the case of salaried employees). In addition to the bare hourly wage rate, the contractor incurs a number of costs associated with employing the labor that needs to be included the labor rate. These costs are known as labor burden. The bare hourly wage rate plus the labor burden is known as the burdened hourly labor rate. The burdened labor rate is calculated by totaling all of the costs of the employee over the course of the year or the project and dividing these costs by the number of hours that are billable to the project during the year or the duration of the project. The following costs should be included when calculating the burdened labor rate.

Cash Equivalents and Allowances. Cash allowances are funds paid to the employee for the employee to provide his own tools or for the use of his personal vehicle. Cash allowances are not reimbursements to the employee for actual expenses, but an allowance given to the employee to defer the cost of providing his own tools and vehicle. In contrast, reimbursements are based on the actual cost of the tools or the actual mileage the employee uses his vehicle for business. Cash allowances are treated as taxable income to the employee, whereas reimbursed expenses are not.

Cash equivalents are funds paid to the employee in lieu of providing other benefits, such as insurance and vacation. When the employer is required to pay Davis-Bacon wages, the employer has the choice to provide benefits equal to the required amount, pay the cash equivalent of those benefits, or a combination of the two. Cash equivalents are treated as taxable income to the employee.

Payroll Taxes. Employers and employees are required to pay social security and Medicare taxes by the Federal Insurance Contribution Act (FICA). In 2009, both the employer and the employee were required to pay 6.2 percent social security tax on the first $106,800 of taxable wages for the employee. The 6.2 percent rate has been the same for many years; however, each year the amount of wages subject to the social security tax increases. The amount of wages subject to social security tax for the year is published in *Circular E: Employer's Tax Guide* (IRS Publication 15). Both the employer and the employee are required to pay 1.45 percent Medicare tax on all of the employee's taxable wages. Some benefits, such as health insurance costs paid by the employee, may be deducted from the employee's gross wages when determining the employee's taxable wages. Estimators should check with their accountant to see which benefits can be deducted from the employee's wages before calculating the employee's taxable wages and the social security and Medicare taxes. The social security and Medicare taxes paid by the employer represent a cost to the employer and should be included in the labor burden. The social security and Medicare taxes paid by the employee are deducted from her wages and are not a cost to the employer.

Unemployment Insurance. The Federal Unemployment Tax Act (FUTA) and State Unemployment Tax Act (SUTA) require employers to provide unemployment insurance for their employees. This is done by paying a FUTA tax and, where state programs exist, a SUTA tax. The FUTA and SUTA tax is paid entirely by the employer and is a labor burden cost.

In 2009, the FUTA tax rate was 6.2 percent on the first $7,000 of each employee's wages. Companies that pay into state unemployment insurance programs may reduce their FUTA tax liability by the amount they pay into a state program provided they pay their SUTA tax on time. The maximum they may reduce their FUTA liability by is 5.4 percent on the first $7,000 of each employee's wages, leaving them paying 0.8 percent on the first $7,000 of wages when they take the maximum credit.

The SUTA tax rate paid by an employer is based in part on the company's unemployment claims history. Companies that frequently lay off employees, which leads to more unemployment claims, pay higher SUTA rates than do companies that have a stable workforce and low claims. The maximum and minimum rates and the amount of wages the

SUTA tax is paid on varies from state to state. For example, in 2006, Texas companies paid SUTA tax on the first $9,000 of each employee's wages and the rates varied from 0.10 percent to 6.10 percent, whereas Utah companies paid on the first $26,700 of each employee's wages and the rates varied from 0.10 percent to 9.1 percent. A company should contact the state agency that is responsible for administrating the state unemployment insurance program to obtain their SUTA rate.

Workers' Compensation Insurance.   Workers' compensation insurance provides medical insurance benefits and may reimburse some of the lost wages to employees injured on the job or who contract an occupational illness. It may also provide some death benefits for employees killed on the job. By law employers must provide workers' compensation insurance for all employees, except the owners of the company. The workers' compensation insurance is paid entirely by the employer and is a labor burden cost. The premium for workers' compensation insurance is based upon the dollar value of the payroll, the type of work being performed, and the accident history of the company. The rates paid for high-risk tasks, such as roofing, are higher than low-risk tasks, such as office work. Workers' compensation rates are based upon the loss history for the local area. For medium and large companies, these rates are modified by an experience modifier based upon the company's claims for the last three years, not including the most recent year. For example, the rate for 2010 is based upon the losses for the years 2006 through 2008. Experience modifiers range from about 0.6 to 2.0; with numbers less than one representing a lower than expected claim history and numbers greater than one representing a higher than expected claim history. Because the workers' compensation rates for medium and large companies are based on actual losses, it is important that the company take measures to improve safety and reduce the costs of accidents.

General Liability Insurance.   General liability insurance provides the company with protection from lawsuits arising from negligence of company employees, including bodily injury, property damage or loss, and damage to one's reputation due to slander. The general liability insurance premium is a percentage of the company's payroll. The rates are different for different classes of employees and are highest for management personnel who are responsible for making the decisions. General liability insurance is paid by the employer and is a labor burden cost.

Insurance Benefits.   Companies often provide employees with health, dental, life, or disability insurance. The companies may pay all, part, or none of the premiums. The portion of the premiums paid by the employer represents a

cost to the employer and is part of the labor burden. The portion of the insurance premiums paid by the employees is deducted from their wages; therefore, these costs are not part of the labor burden.

Retirement Contributions.   Many employers provide employees with access to 401(k) retirement programs or other retirement programs. Often the employer contributes money to the employee's retirement program. The amount the employer pays may be based upon many things, including matching a percentage of the money the employee contributes to the plan. For example, an employer may contribute $0.50 for every $1.00 the employee contributes on 6 percent of an employee's wages. In this example, the employer would contribute a maximum of 3 percent of the employee's wages. When estimating retirement costs based upon matching employee contribution, the employer should take into account the amount the typical employee contributes to her retirement program. When the employer pays money into an employee's retirement plan, this money represents a labor burden cost to the employer. The money the employee contributes to the plan is taken out of the employee's wages and should not be included in the labor burden.

Union Payments.   When employees belong to a union, the employer is responsible for making payments to the union. These funds are used by the union to provide benefits (such as retirement and insurance) and training (such as apprentice programs) for the employees. The amount the company must pay to the union is found in the union contract. The union may also require the company to deduct union dues from the employee's paycheck. The payments made by the company are a labor burden cost, whereas the union dues deducted from the employee's paychecks are not.

Vacation, Holidays, and Sick Leave.   Vacation, holidays, and sick leave are incorporated into the labor burden by including the wages paid for vacation, holidays, and sick leave in the employee's wages before other burden costs are calculated. By doing this, the burdened cost associated with vacation, holidays, and sick leave is included in the total wages. When determining the burden hourly wage rate, the vacation, holiday, and sick leave hours are not included in the billable hours. This incorporates the vacation, holiday, and sick leave costs in the burden.

Burdened Hourly Wage Rate.   The cost of any other benefits paid by the employer should be included in the cost of the employee before determining the burdened hourly wage rate. The burdened hourly wage rate is calculated using Formula 7-6. The billable hours should only include hours

Quantity	Craft	Bare Hourly Wage Rate	Labor Burden	Burdened Hourly Wage Rate	Total Labor Cost/Hour
3	Mason	29.00	12.41	41.41	124.23
2	Helpers	18.00	8.84	26.84	53.68
5				Total	177.91
		Weighted Average Burdened Wage Rate			35.58

FIGURE 7.2. Weighted Average Burdened Wage Rate.

that are billable to a project and should not include vacation, holiday, and sick leave hours.

$$\text{Burdened hourly wage rate} = \frac{\text{Wages} + \text{Benefits}}{\text{Billable hours}}$$

Burdened Hourly Wage Rate **Formula 7-6**

EXAMPLE 7-5 BURDENED HOURLY WAGE RATE

Craft—Mason

Wage rate—$29.00 per hour

Hours worked—50 hours per week for 20 weeks and 40 hours per week for 29 weeks

Paid vacation, holidays, and sick leave—Three weeks at 40 hours per week

Overtime—Time-and-a-half for any hours over 40 per week

Gas allowance—$100 per month

Annual bonus—$500

Social security—6.2 percent on the first $106,800 of wages

Medicare—1.45 percent of all wages

FUTA—0.8 percent on the first $7,000 of wages

SUTA—4.5 percent on the first $18,000 of wages

Worker's compensation insurance—$7.25 per $100.00 of wages

General liability insurance—0.75 percent of wages

Health insurance (company's portion)—$300 per month per employee

Retirement—$0.75 per $1.00 contributed by the employee on 6 percent of the employee's wages

$$\begin{aligned} \text{Wages} =\ & (40 \text{ hours per week} \times 52 \text{ weeks} \times \$29.00/\text{hour}) \\ & + (10 \text{ hour per week} \times 20 \text{ weeks} \times \$29.00/\text{hour} \times 1.5) \\ & + \$100 \text{ per month} \times 12 \text{ months} + \$500 \end{aligned}$$

$$\text{Wages} = \$70{,}720$$

$$\text{Social security} = \$70{,}720 \times 0.062 = \$4{,}385$$

$$\text{Medicare} = \$70{,}720 \times 0.0145 = \$1{,}025$$

$$\text{FUTA} = \$7{,}000 \times 0.008 = \$56$$

$$\text{SUTA} = \$18{,}000 \times 0.045 = \$810$$

$$\text{Workers' compensation insurance} = \$70{,}720 \times \frac{\$7.25}{\$100} = \$5{,}127$$

$$\text{General liability insurance} = \$70{,}720 \times 0.0075 = \$530$$

$$\begin{aligned} \text{Health insurance} &= \$300/\text{month} \times 12 \text{ months} \\ &= \$3{,}600 \end{aligned}$$

The company contributes up to 4.5 percent (75 percent of 6 percent) of the employee's wages to the employee's retirement account. Assume that the employees take full advantage of this benefit.

$$\text{Retirement} = \$70{,}720 \times 0.045 = \$3{,}182$$

$$\begin{aligned} \text{Benefits} =\ & \$4{,}385 + \$1{,}025 + \$56 \\ & + \$810 + \$5{,}127 + \$530 \\ & + \$3{,}600 + \$3{,}182 \end{aligned}$$

$$\text{Benefits} = \$18{,}715$$

$$\begin{aligned} \text{Billable hours} =\ & 50 \text{ hours per week} \times 20 \text{ weeks} \\ & + 40 \text{ hours per week} \\ & \times 29 \text{ weeks} \end{aligned}$$

$$\text{Billable hours} = 2{,}160 \text{ hours}$$

$$\text{Burdened hourly wage rate} = \frac{\$70{,}720 + \$18{,}715}{2{,}160 \text{ hours}}$$

$$\text{Burdened hourly wage rate} = \$41.41 \text{ per hour} \quad\blacksquare$$

7-5 PRICING LABOR

To price labor, first the estimator must estimate the labor hours required to do the work. These labor hours can then be multiplied by the burdened wage rate to develop the labor costs. However, when the crew is made up of different crafts being paid different wage rates, a weighted average burdened wage rate must be determined. This is done by determining the total cost for the crew for an hour and by dividing that amount by the number of persons on the crew, as shown in Figure 7.2. In Figure 7.1, a D-8 crew was used in the placement of the decorative block. Figure 7.3 shows the makeup of the crew, which includes three bricklayers and two bricklayer helpers.

Once the average crew wage rate has been found, it can be multiplied by the number of labor hours to determine the labor costs. Formula 7-7 is used to determine the labor costs.

$$\begin{aligned} \text{Labor cost} =\ & \text{Adjusted labor hours} \\ & \times \text{Weighted average burdened wage rate} \end{aligned}$$

Labor Cost **Formula 7-7**

EXAMPLE 7-6 DETERMINING LABOR COST

From Example 7-2, there were 110 labor hours. The labor costs would be the following:

$$\text{Labor cost} = 110 \text{ labor hours} \times \$35.58/\text{hour} = \$3{,}913.80 \quad\blacksquare$$

Crews

Crew C-23A

Crew No.	Bare Costs Hr.	Daily	Incl. Subs O & P Hr.	Daily	Bare Costs	Incl. O&P
1 Labor Foreman (outside)	$30.75	$246.00	$47.90	$383.20	$32.39	$49.70
2 Laborers	28.75	460.00	44.75	716.00		
1 Equip. Oper. (crane)	39.80	318.40	60.00	480.00		
1 Equip. Oper. Oiler	33.90	271.20	51.10	408.80		
1 Crawler Crane, 100 Ton		1717.00		1888.70		
3 Conc. buckets, 8 C.Y.		518.40		570.24	55.88	61.47
40 L.H., Daily Totals		$3531.00		$4446.94	$88.28	$111.17

Crew C-24

Crew No.	Bare Costs Hr.	Daily	Incl. Subs O & P Hr.	Daily	Bare Costs	Incl. O&P
2 Skilled Worker Foremen	$40.00	$640.00	$62.25	$996.00	$38.17	$59.05
6 Skilled Workers	38.00	1824.00	59.15	2839.20		
1 Equip. Oper. (crane)	39.80	318.40	60.00	480.00		
1 Equip. Oper. Oiler	33.90	271.20	51.10	408.80		
1 Lattice Boom Crane, 150 Ton		1852.00		2037.20	23.15	25.47
80 L.H., Daily Totals		$4905.60		$6761.20	$61.32	$84.52

Crew C-25

Crew No.	Bare Costs Hr.	Daily	Incl. Subs O & P Hr.	Daily	Bare Costs	Incl. O&P
2 Rodmen (reinf.)	$41.30	$660.80	$67.75	$1084.00	$32.33	$53.67
2 Rodmen Helpers	23.35	373.60	39.60	633.60		
32 L.H., Daily Totals		$1034.40		$1717.60	$32.33	$53.67

Crew C-27

Crew No.	Bare Costs Hr.	Daily	Incl. Subs O & P Hr.	Daily	Bare Costs	Incl. O&P
2 Cement Finishers	$35.55	$568.80	$52.20	$835.20	$35.55	$52.20
1 Concrete Saw		118.40		130.24	7.40	8.14
16 L.H., Daily Totals		$687.20		$965.44	$42.95	$60.34

Crew C-28

Crew No.	Bare Costs Hr.	Daily	Incl. Subs O & P Hr.	Daily	Bare Costs	Incl. O&P
1 Cement Finisher	$35.55	$284.40	$52.20	$417.60	$35.55	$52.20
1 Portable Air Compressor, gas		14.75		16.23	1.84	2.03
8 L.H., Daily Totals		$299.15		$433.82	$37.39	$54.23

Crew D-1

Crew No.	Bare Costs Hr.	Daily	Incl. Subs O & P Hr.	Daily	Bare Costs	Incl. O&P
1 Bricklayer	$38.05	$304.40	$57.90	$463.20	$33.35	$50.75
1 Bricklayer Helper	28.65	229.20	43.60	348.80		
16 L.H., Daily Totals		$533.60		$812.00	$33.35	$50.75

Crew D-2

Crew No.	Bare Costs Hr.	Daily	Incl. Subs O & P Hr.	Daily	Bare Costs	Incl. O&P
3 Bricklayers	$38.05	$913.20	$57.90	$1389.60	$34.51	$52.63
2 Bricklayer Helpers	28.65	458.40	43.60	697.60		
.5 Carpenter	36.70	146.80	57.15	228.60		
44 L.H., Daily Totals		$1518.40		$2315.80	$34.51	$52.63

Crew D-3

Crew No.	Bare Costs Hr.	Daily	Incl. Subs O & P Hr.	Daily	Bare Costs	Incl. O&P
3 Bricklayers	$38.05	$913.20	$57.90	$1389.60	$34.40	$52.42
2 Bricklayer Helpers	28.65	458.40	43.60	697.60		
.25 Carpenter	36.70	73.40	57.15	114.30		
42 L.H., Daily Totals		$1445.00		$2201.50	$34.40	$52.42

Crew D-4

Crew No.	Bare Costs Hr.	Daily	Incl. Subs O & P Hr.	Daily	Bare Costs	Incl. O&P
1 Bricklayer	$38.05	$304.40	$57.90	$463.20	$33.05	$50.16
2 Bricklayer Helpers	28.65	458.40	43.60	697.60		
1 Equip. Oper. (light)	36.85	294.80	55.55	444.40		
1 Grout Pump, 50 C.F./hr		122.00		134.20	3.81	4.19
32 L.H., Daily Totals		$1179.60		$1739.40	$36.86	$54.36

Crew D-5

Crew No.	Bare Costs Hr.	Daily	Incl. Subs O & P Hr.	Daily	Bare Costs	Incl. O&P
1 Bricklayer	$38.05	$304.40	$57.90	$463.20	$38.05	$57.90
8 L.H., Daily Totals		$304.40		$463.20	$38.05	$57.90

Crew D-6

Crew No.	Bare Costs Hr.	Daily	Incl. Subs O & P Hr.	Daily	Bare Costs	Incl. O&P
3 Bricklayers	$38.05	$913.20	$57.90	$1389.60	$33.48	$51.01
3 Bricklayer Helpers	28.65	687.60	43.60	1046.40		
.25 Carpenter	36.70	73.40	57.15	114.30		
50 L.H., Daily Totals		$1674.20		$2550.30	$33.48	$51.01

Crew D-7

Crew No.	Bare Costs Hr.	Daily	Incl. Subs O & P Hr.	Daily	Bare Costs	Incl. O&P
1 Tile Layer	$35.50	$284.00	$52.10	$416.80	$31.38	$46.05
1 Tile Layer Helper	27.25	218.00	40.00	320.00		
16 L.H., Daily Totals		$502.00		$736.80	$31.38	$46.05

Crew D-8

Crew No.	Bare Costs Hr.	Daily	Incl. Subs O & P Hr.	Daily	Bare Costs	Incl. O&P
3 Bricklayers	$38.05	$913.20	$57.90	$1389.60	$34.29	$52.18
2 Bricklayer Helpers	28.65	458.40	43.60	697.60		
40 L.H., Daily Totals		$1371.60		$2087.20	$34.29	$52.18

Crew D-9

Crew No.	Bare Costs Hr.	Daily	Incl. Subs O & P Hr.	Daily	Bare Costs	Incl. O&P
3 Bricklayers	$38.05	$913.20	$57.90	$1389.60	$33.35	$50.75
3 Bricklayer Helpers	28.65	687.60	43.60	1046.40		
48 L.H., Daily Totals		$1600.80		$2436.00	$33.35	$50.75

Crew D-10

Crew No.	Bare Costs Hr.	Daily	Incl. Subs O & P Hr.	Daily	Bare Costs	Incl. O&P
1 Bricklayer Foreman	$40.05	$320.40	$60.95	$487.60	$36.64	$55.61
1 Bricklayer	38.05	304.40	57.90	463.20		
1 Bricklayer Helper	28.65	229.20	43.60	348.80		
1 Equip. Oper. (crane)	39.80	318.40	60.00	480.00		
1 S.P. Crane, 4x4, 12 Ton		575.00		632.50	17.97	19.77
32 L.H., Daily Totals		$1747.40		$2412.10	$54.61	$75.38

Crew D-11

Crew No.	Bare Costs Hr.	Daily	Incl. Subs O & P Hr.	Daily	Bare Costs	Incl. O&P
1 Bricklayer Foreman	$40.05	$320.40	$60.95	$487.60	$35.58	$54.15
1 Bricklayer	38.05	304.40	57.90	463.20		
1 Bricklayer Helper	28.65	229.20	43.60	348.80		
24 L.H., Daily Totals		$854.00		$1299.60	$35.58	$54.15

Crew D-12

Crew No.	Bare Costs Hr.	Daily	Incl. Subs O & P Hr.	Daily	Bare Costs	Incl. O&P
1 Bricklayer Foreman	$40.05	$320.40	$60.95	$487.60	$33.85	$51.51
1 Bricklayer	38.05	304.40	57.90	463.20		
2 Bricklayer Helpers	28.65	458.40	43.60	697.60		
32 L.H., Daily Totals		$1083.20		$1648.40	$33.85	$51.51

Crew D-13

Crew No.	Bare Costs Hr.	Daily	Incl. Subs O & P Hr.	Daily	Bare Costs	Incl. O&P
1 Bricklayer Foreman	$40.05	$320.40	$60.95	$487.60	$35.32	$53.87
1 Bricklayer	38.05	304.40	57.90	463.20		
2 Bricklayer Helpers	28.65	458.40	43.60	697.60		
1 Carpenter	36.70	293.60	57.15	457.20		
1 Equip. Oper. (crane)	39.80	318.40	60.00	480.00		
1 S.P. Crane, 4x4, 12 Ton		575.00		632.50	11.98	13.18
48 L.H., Daily Totals		$2270.20		$3218.10	$47.30	$67.04

Crew E-1

Crew No.	Bare Costs Hr.	Daily	Incl. Subs O & P Hr.	Daily	Bare Costs	Incl. O&P
1 Welder Foreman	$43.35	$346.80	$78.50	$628.00	$40.52	$69.65
1 Welder	41.35	330.80	74.90	599.20		
1 Equip. Oper. (light)	36.85	294.80	55.55	444.40		
1 Welder, gas engine, 300 amp		115.20		126.72	4.80	5.28
24 L.H., Daily Totals		$1087.60		$1798.32	$45.32	$74.93

FIGURE 7.3. Standard Crews.

WEB RESOURCES

www.irs.gov

www.constructionbook.com

www.rsmeans.com

REVIEW QUESTIONS

1. What unit of time is used to measure labor? What does it represent?

2. How do climatic conditions influence the amount of work actually completed in an hour?

3. Where would you use cycle time to estimate productivity?

4. Where would you use the rate of progress to estimate productivity?

5. What effect can upcoming labor union negotiations have on a bid?

6. Why do many contractors hesitate to bid just before the expiration of a union contract?

7. What effect could an extreme shortage of skilled workers have on the cost of a project?

8. How can working conditions on the job site affect worker productivity?

9. How may union work rules affect the pricing of labor?

10. What costs should be included in the labor burden?

11. How do allowances differ from reimbursements?

12. How are vacation, holidays, and sick leave included in the labor burden?

13. How can crews be used in the estimating of labor? How does this compare with using individual workers?

CHAPTER EIGHT

EQUIPMENT

8-1 GENERAL

One problem an estimator faces is the selection of equipment suitable to use for a given project. The equipment must pay for itself. Unless a piece of equipment will earn money for the contractor, it should not be used.

Because it is impossible for contractors to own all types and sizes of equipment, the selection of equipment will be primarily from that which they own. However, new equipment can be purchased if the cost can be justified. If the cost of the equipment can be charged off to one project or written off in combination with other proposed uses of the equipment, the equipment will pay for itself and should be purchased. For example, if a piece of equipment costing $15,000 will save $20,000 on a project, it should be purchased regardless of whether it will be used on future projects or whether it can be sold at the end of the current project.

Figuring the cost of equipment required for a project presents the same problems to estimators as figuring labor. It is necessary for the estimator to decide what equipment is required for each phase of the work and for what length of time it will have to be used.

If the equipment is to be used for a time and then will not be needed again for a few weeks, the estimator should ask the following: What will be done with it? Will it be returned to the main yard? Is there room to store it on the project? If rented, will it be returned so that the rental charge will be saved?

Equipment that is required throughout the project is included under equipment expenses, because it cannot be charged to any particular item of work. This equipment often includes the hoist towers and material-handling equipment such as lift trucks and long-reach forklifts.

As the estimator does a takeoff of each item, all equipment required should be listed so that the cost can be totaled in the appropriate column.

Equipment required for one project only or equipment that might be used infrequently may be purchased for the one project and sold when it is no longer needed. In addition to the operating costs, the difference between the purchase price and the selling price would then be charged to the project. Alternately, the equipment may be rented.

8-2 OPERATING COSTS

The costs of operating the construction equipment should be calculated on the basis of the working hour since the ownership or rental cost is also a cost per hour. Included are items such as fuel, grease, oil, electricity, miscellaneous supplies, and repairs. Operators' wages and mobilization costs are not included in equipment operation costs.

Costs for power equipment are usually based on the horsepower of the equipment. Generally, a gasoline engine will use between 0.06 and 0.07 gallons of gasoline per horsepower per hour when operating at full capacity. Fuel costs are calculated using Formula 8-1. When operating, the engine will probably operate at 55 to 80 percent of full capacity, or 55 to 80 percent of its available power will be utilized. This is known as power utilization and reduces fuel consumption. In addition, it will not be operated for the full hour. Typically, equipment is operated between 30 and 50 minutes per hour. This is known as the system efficiency or use factor and is expressed as a percentage of the hour that the equipment is operating. For example, 45 minutes per hour would be 75 percent (0.75) and 50 minutes per hour would be 83 percent (0.83).

$$\text{Fuel cost} = \text{hp rating} \times \text{Power utilization} \times \text{Use factor} \times \text{Consumption rate} \times \text{Fuel cost}$$

Fuel Cost per Hour **Formula 8-1**

EXAMPLE 8-1   FUEL COST

What is the estimated fuel cost of a 120-horsepower payloader? A job condition analysis indicates that the unit will operate about 45 minutes per hour (75 percent) at about 70 percent of its rated horsepower.

Assumptions:

Fuel cost—$1.10 per gallon

Consumption rate—0.06 gallons per hp per hour

Power utilization—70%

Use factor—75%

Fuel cost per machine hour = hp rating $\times$ Power utilization
$\times$ Use factor
$\times$ Consumption rate
$\times$ Fuel cost

Fuel cost per machine hour = 120 hp $\times$ 0.70 $\times$ 0.75
$\times$ 0.06 gal per hp per hour
$\times$ $1.10 gal

Fuel cost per machine hour = $4.16 ■

The diesel engine requires about 0.04 to 0.06 gallons of fuel per horsepower per hour when operating at full capacity. Because the equipment is usually operated at 55 to 80 percent of capacity and will not operate continuously each hour, the amount of fuel actually used will be less than the full per hour requirement. The full capacity at which the equipment works, the portion of each hour it will be operated, the horsepower, and the cost of fuel must all be determined from a job condition analysis.

Lubrication.   The amount of oil and grease required by any given piece of equipment varies with the type of equipment and job conditions. A piece of equipment usually has its oil changed and is greased every 100 to 150 hours. Under severe conditions, the equipment may need much more frequent servicing. Any oil and grease consumed between oil changes must also be included in the cost.

EXAMPLE 8-2   EQUIPMENT LUBRICATION

A piece of equipment has its oil changed and is greased every 120 hours. It requires six quarts of oil for the change. The time required for the oil change and greasing is estimated at 2.5 hours.

Assumptions:

Oil cost—$1.30 per quart

Oiler labor rate—$17.50 per hour

Lubrication cost = 6 qts. of oil $\times$ $1.30 per qt.
= $7.80

Labor cost = 2.5 hours
$\times$ $17.50 per labor hour
= $43.75

Total cost for oil change = $7.80 + $43.75 = $51.55

Cost per machine work hour = $51.55/120 hours
= $0.43 per hour ■

Tires.   The cost of tires can be quite high on an hourly basis. Because the cost of tires is part of the original cost, it is left in when figuring the cost of interest, but taken out for the cost of repairs and salvage values. The cost of tires, replacement, repair, and depreciation should be figured separately. The cost of the tires is depreciated over the useful life of the tires, and the cost of repairs is taken as a percentage of the depreciation, based on past experience.

EXAMPLE 8-3   TIRES

Four tires for a piece of equipment cost $5,000 and have a useful life of about 3,500 hours; the average cost for repairs to the tires is 15 percent of depreciation. What is the average cost of the tires per hour?

Tire depreciation = $5,000/3,500 hours = $1.43 per hour

Tire repair = 15% $\times$ $1.43 per hour = $0.21 per hour

Tire cost = $1.43 + $0.21
= $1.64 per machine work hour ■

8-3 DEPRECIATION

As soon as a piece of equipment is purchased, it begins to decrease (*depreciate*) in value. As the equipment is used on the projects, it begins to wear out, and in a given amount of time it will have become completely worn out or obsolete. If an allowance for depreciation is not included in the estimate, there will be no money set aside to purchase new equipment when the equipment is worn out. This is not profit, and the money for equipment should not be taken from profit.

On a yearly basis, for tax purposes, depreciation can be figured in a number of ways. But for practical purposes, the total depreciation for any piece of equipment will be 100 percent of the capital investment minus the scrap or salvage value, divided by the number of years it will be used. For estimating depreciation costs, assign the equipment a useful life expressed in years, hours, or units of production, whichever is the most appropriate for a given piece of equipment.

EXAMPLE 8-4   EXAMPLE DEPRECIATION

If a piece of equipment had an original cost of $67,500, an anticipated salvage value of $10,000, and an estimated life of five years, what would be the annual depreciation cost?

Depreciable cost = Original cost $-$ Salvage value

Depreciable cost = $67,500 $-$ 10,000 = $57,500

Depreciable cost per year = Depreciable cost/Useful life

Depreciable cost = $57,500/5 years
= $11,500 per year

If the piece of equipment should last 10,000 machine work hours, the hourly cost would be found as follows:

Depreciable cost per hour = Depreciable cost/Useful life (hours)

Depreciable cost per hour = $57,500/10,000 hours

Depreciable cost per hour = $5.75 per machine work hour ■

8-4 INTEREST

Interest rates must be checked by the estimator. The interest should be charged against the entire cost of the equipment, even though the contractor paid part of the cost in cash. Contractors should figure that the least they should get for the use of their money is the current rate of interest. Interest is paid on the unpaid balance. On this basis, the balance due begins at the cost price and decreases to virtually nothing when the last payment is made. Since the balance on which interest is being charged ranges from 100 to 0 percent, the average amount on which the interest is paid is 50 percent of the cost.

$$\text{Approximate interest cost} = \frac{C \times I \times L}{2} \quad \textbf{Formula 8-2}$$

Approximate Interest Costs

Where

C = Amount of loan

I = Interest rate

L = Life of loan

Because the estimator will want to have the interest costs in terms of cost per hour, the projected useful life in terms of working hours must be assumed. The formula used to figure the interest cost per hour would be

$$\text{Approximate interest cost} = \frac{C \times I \times L}{2 \times H} \quad \textbf{Formula 8-3}$$

Approximate Interest Costs per Hour

Where

H = Useful life of equipment (working hours)

Formula 8-3 is used to determine interest costs toward the total fixed cost per hour for a piece of equipment, as shown in the discussion of ownership costs (Section 8-5). Remember that 8 percent, when written as a decimal, is 0.08 since the percent is divided by 100.

Other fixed costs are figured in a manner similar to that used for figuring interest. The costs to be considered include insurance, taxes, storage, and repairs. Depreciation (Section 8-3) must also be considered. These items are taken as percentages of the cost of equipment, minus the cost of the tires, and are expressed as decimals in the formula. When the expenses are expressed in terms of percent per year, they must be multiplied by the number of years of useful life to determine accurate costs.

8-5 OWNERSHIP COSTS

To estimate the cost of using a piece of equipment owned by the contractor, the estimator must consider depreciation, major repairs, and overhaul as well as interest, insurance, taxes, and storage. These items are most often taken as a percentage of the initial cost to the owner. Also to be added later is the cost for fuel, oil, and tires. The cost to the owner should include all freight costs, sales taxes, and preparation charges.

EXAMPLE 8-5   COST OF OWNERSHIP

Estimate the cost of owning and operating a piece of equipment on a project with the following costs:

Assumptions:

Actual cost (delivered)—$47,600

Horsepower rating—150 hp

Cost of tires—$4,500

Salvage or scrap value—3 percent

Useful life—7 years or 14,000 hours

Total interest—8 percent per year

Length of loan—7 years

Total insurance, taxes, and storage—6 percent per year

Fuel cost—$1.10 per gallon

Consumption rate—0.06 gallons per hp per hour

Power utilization—62 percent

Use factor—70 percent

Lubrication—4 quarts oil at $1.25

Oiler labor—2 hours labor at $16.50

Lubrication schedule—every 150 hours

Life of tires—4,000 hours

Repair to tires—12 percent of depreciation

Repairs to equipment—65 percent over useful life

Fixed Cost (per hour):

$$\text{Approximate interest cost} = \frac{\$47,600 \times 0.08 \text{ per year} \times 7 \text{ years}}{2 \times 14,000 \text{ hours}}$$

Approximate interest cost = $0.95 per equip. work hour

Salvage value = $47,600 × 0.03 = $1,428

Depreciable cost = $47,600 − $1,428 = $46,172

Depreciable cost per work hour = $46,172/14,000 hours

Depreciable cost per work hour = $3.30 per equip. work hour

$$\text{Repairs} = (\$47,600 \times 0.65)/14,000 \text{ hours}$$
$$= \$2.21 \text{ per equip. work hour}$$

$$\text{Insurance, taxes, and storage} = (0.06 \text{ per year} \times 7 \text{ years}$$
$$\times \$47,600)/14,000 \text{ hours}$$

Insurance, taxes, and storage = $1.43 per equip. work hour

See Figure 8.1 for the total fixed cost.

Operating Cost (per hour):

$$\text{Tire depreciation} = (\$4,500/4,000 \text{ hours})$$
$$= \$1.13 \text{ per equip. work hour}$$

$$\text{Tire repair} = 0.12 \times \$1.13 \text{ per work hour}$$
$$= \$0.14 \text{ per equip. work hour}$$

$$\text{Fuel cost} = 150 \text{ hp} \times 0.06 \text{ gal per hp her hour}$$
$$\times 0.62 \times 0.7 \times \$1.10 \text{ per gal}$$

Fuel cost = $4.30 per equip. work hour

$$\text{Lubrication cost} = 4 \text{ qts. of oil} \times \$1.50 \text{ per qt.}$$
$$= \$6.00$$

Recap of Fixed Costs	
Item	**$ / Work Hour**
Approximate Interest Cost	0.95
Depreciable cost	3.30
Repairs	2.21
Insurance Taxes & Storage	1.43
Total Fixed Cost	7.89

FIGURE 8.1. Fixed Costs.

Recap of Operating Costs	
Item	**$ / Work Hour**
Tire Depreciation	1.13
Tire Repair	0.14
Fuel Cost	4.30
Lubrication Cost	0.26
Total Operating Cost	5.83

FIGURE 8.2. Operating Costs.

$$\text{Lubrication labor} = 2 \text{ work hours}$$
$$\times \$16.50 \text{ per work hour} = \$33.00$$

$$\text{Lubrication cost per hour} = (\$6.00 + \$33.00)/150 \text{ hours}$$

$$\text{Lubrication cost per hour} = \$0.26 \text{ per equip. work hour}$$

$$\text{Total ownership cost} = \text{fixed cost} + \text{operating cost}$$

$$\text{Total ownership cost} = \$5.83 \text{ per hour} + \$7.89 \text{ per hour}$$

$$\text{Total ownership cost} = \$13.72 \text{ per equip. work hour}$$

See Figure 8.2 for the total operating cost. ∎

8-6 RENTAL COSTS

If a project is a long distance from the contractor's home base or if the construction involves the use of equipment that the contractor does not own and will not likely use after the completion of this one project, the estimator should seriously consider renting the equipment. In considering the rental of equipment, the estimator must investigate the available rental agencies for the type and condition of equipment available, the costs, and the services the rental firm provides. The estimator must be certain that all terms of rental are understood, especially those concerning the repair of the equipment.

Contractors tend to buy equipment even when it is more reasonable to rent. Many rental firms have newer equipment than a contractor might purchase. They also may have a better maintenance program. Estimators should check the rental firms carefully, especially when doing work in a given locale for the first time. The price of the rental is important, but the emphasis should be on the equipment's condition and service. If no reputable rental agency is available, the contractor may be forced to purchase the required equipment.

Equipment is generally rented for a short time, and lease agreements are arranged when that time extends to one year or more. Rental rates are usually quoted by the month, week, or day. These costs must be broken down into costs per hour or per unit of work so that they may be accurately included in the estimate and checked during construction. The rental charge will be based on a day of eight hours (or less). If the equipment is to be used more than eight hours per day, a proportional charge will be added. To this must be added the other costs of operating the equipment; usually these include mobilization, repairs (except ordinary wear and tear), and day-to-day maintenance, as well as the costs of fuel, insurance, taxes, and cleaning.

8-7 MISCELLANEOUS TOOLS

Examples of miscellaneous tools are wheelbarrows, shovels, picks, crowbars, hammers, hoses, buckets, and ropes. The mechanics that work on the projects have their own small tools, but the contractor will still need a supply of miscellaneous tools and equipment. The estimator should list the equipment required and estimate its cost. The life of this type of equipment and tools is generally taken as an average of one year. Loss of miscellaneous tools and equipment due to disappearance (theft) is common, and all attempts must be made to keep it under control.

8-8 COST ACCOUNTING

The costs for equipment cannot come from thin air; the estimator must rely heavily on equipment expense data for future bids. Especially in heavy construction, the cost accounting is important since the contractor has a great deal of money invested, and the equipment costs become a large percentage of the costs of the project. It is important that equipment costs be constantly analyzed and kept under control. See *Construction Accounting and Financial Management* by Steven J. Peterson for information on how to set up cost accounting for equipment.

Small, miscellaneous equipment and tools are not subjected to this cost control analysis and are generally charged to each project on a flat-rate basis. The procedure for determining equipment expenses varies from contractor to contractor, but the important point is that the expenses must be determined. Generally the equipment expense is broken down into a charge per hour or a charge per unit of work. Field reports of equipment time must include only the time during which the equipment is in use. When excessive idle time occurs, estimators must check to see whether it can be attributed to bad weather, poor working conditions, or management problems on the project. Management problems sometimes include poor field supervision, poor equipment maintenance, poor equipment selection, and an excessive amount of equipment on the project.

A report on quantities of work performed is required if a cost per unit of work is desired. Generally the work is measured on a weekly basis; sometimes the work completed is estimated as a percentage of the total work to be performed. This type of report must be stated in work units that are compatible with the estimate.

8-9 MOBILIZATION

The estimate must also include the cost of transporting all equipment required for the project to the job and then back again when the work is completed. Obviously, this cost will vary with the distance, type and amount of equipment, method of transportation used, and the amount of dismantling required for the various pieces of equipment. Mobilization costs must also be considered for rental equipment since it must be brought to the job site. The cost of erecting some types of equipment, such as hoists, scaffolding, or cranes, must also be included, as well as the costs of loading the equipment at the contractor's yard and unloading it at the job site.

8–10 CHECKLIST

Equipment listings are given in each chapter and are considered in relation to the work required on the project. Equipment that may be required throughout the project includes the following:

Lifting cranes	Hoisting engines
Hoisting towers	Heaters
Lift trucks	Scaffolding

REVIEW QUESTIONS

1. What are the advantages to a small contractor of renting equipment instead of owning?

2. What is depreciation on equipment?

3. What operating costs must be considered?

4. Why should interest be included in the equipment costs if the contractor paid cash for the equipment?

5. What is the approximate interest cost if the cost of equipment is $75,000, the interest rate 9.5 percent, and the life of the loan is six years?

6. Why must mobilization be included in the cost of equipment?

7. Why is it important that reports from the field pertaining to equipment be kept?

8. If there is excessive idle time for equipment on the job, to what factors may this be attributed?

CHAPTER NINE

EXCAVATION

9–1 GENERAL

Calculating the quantities of earth that must be excavated is considered to be one of the most difficult aspects of the estimator's task. Calculating the excavation for the project often involves a great deal of work. The number of cubic yards to excavate is sometimes easy enough to compute, but calculating the cost for this portion of the work is difficult because of the various hidden items that may affect the cost. These include such variables as the type of soil, the required slope of the bank in the excavated area, whether bracing or sheet piling will be required, and whether groundwater will be encountered and pumping will be required.

9–2 SPECIFICATIONS

The estimator must carefully check the specifications to see exactly what is included in the excavation. Several questions demand answers: What is the extent of work covered? What happens to the excess excavated material? Can it be left on the site or must it be removed? If the excess must be removed, how far must it be hauled? All of these questions and more must be considered. Who does the clearing and grubbing? Who removes trees? Must the topsoil be stockpiled for future use? Where? Who is responsible for any trenching required for the electrical and mechanical trades?

If the owner is using separate contracts, it is important that the estimator understand exactly what work each contractor is performing. On the other hand, if the general contractor is the sole contractor, that person becomes responsible for addressing all of the coordination issues.

9–3 SOIL

One of the first items the estimator must consider is the type of soil that will be encountered at the site. The estimator may begin by investigating the soil borings shown on the drawings or included in the specifications. When soil borings are provided, the contract documents often absolve the architect/engineer and the owner of any responsibility for their correctness. The estimator must be certain to check any notes on the drawings and the specifications in this regard. Because of such notes and because the specifications for some projects provide no soil information, it is a common practice for the estimator to investigate the soil conditions when visiting the site. Bringing a long-handled shovel or a post-hole digger will allow the estimator to personally check the soil and then record all observations in the project notebook. Example 9-1 details the required accuracy.

9–4 CALCULATING EXCAVATION

Excavation is measured by the cubic yard for the quantity takeoff (27 cf = 1 cy). Before excavation, when the soil is in an undisturbed condition, it weighs about 100 pounds per cf; rock weighs about 150 pounds per cf.

The site plan is the key drawing for determining earthwork requirements and is typically scaled in feet and decimals of a foot. There is usually no reason to change to units of feet and inches; however, at times they must be changed to decimals. Remember that when estimating quantities, the computations need not be worked out to an exact answer.

EXAMPLE 9-1   REQUIRED ACCURACY

Given the following dimensions, determine the quantity to be excavated.

$$\text{Length} = 52.83\,\text{ft}$$

$$\text{Width} = 75.75\,\text{ft}$$

$$\text{Depth} = 6.33\,\text{ft}$$

$$\text{Volume (cf)} = L' \times W' \times D'$$

$$\text{Volume (cf)} = 52.83' \times 75.75' \times 6.33' = 25{,}331.853\,\text{cf}$$

Use 25,332 cf or (23,332 cf/27 cf per cy) 938 cy   ■

Percentage of Swell & Shrinkage		
Material	Swell	Shrinkage
Sand and Gravel	10 to 18%	85 to 100%
Loam	15 to 25%	90 to 100%
Dense Clay	20 to 35%	90 to 100%
Solid Rock	40 to 70%	130%

FIGURE 9.1. Swell and Shrinkage Factors. (Solid rock when compacted is less dense than its bank condition.)

Swell and Compaction. Material in its natural state is referred to as bank materials and is measured in bank cubic yards (bcy). When bank materials are excavated, the earth and rocks are disturbed and begin to swell. This expansion causes the soil to assume a larger volume; this expansion represents the amount of *swell* and is generally expressed as a percentage gained above the original volume. Uncompacted excavated materials are referred to as loose materials and are measured in loose cubic yards (lcy). When loose materials are placed and compacted (as fill) on a project, it will be compressed into a smaller volume than when it was loose, and with the exception of solid rock it will occupy less volume than in its bank condition. This reduction in volume is referred to as *shrinkage*. Shrinkage is expressed as a percentage of the undisturbed original or bank volume. Materials that have been placed and compacted are referred to as compacted materials and are measured in compacted cubic yards (ccy). Bank, loose, and compacted cubic yards are used to designate which volume we are talking about. Figure 9.1 is a table of common swell and shrinkage factors for various types of soils. When possible, tests should be performed to determine the actual swell and shrinkage for the material.

EXAMPLE 9-2   DETERMINING SWELL AND HAUL

If 1,000 bank cubic yards (in place at natural density) of dense clay (30 percent swell) needs to be hauled away, how many loose cubic yards would have to be hauled away by truck?

Cubic yards of haul = In-place quantity × (1 + Swell percentage)

Cubic yards of haul = 1,000 bcy × (1 + 0.3) = 1,300 lcy

If 7 cy dump trucks will be used to haul this material away, how many loads would be required?

Loads = lcy of haul/lcy per load

Loads = 1,300 lcy/7 lcy per load = 186 loads ■

EXAMPLE 9-3   DETERMINING SHRINKAGE AND HAUL

If 500 compacted cubic yards in-place of sand/gravel is required, how many loads would be required? The material has a swell of 15 percent and shrinkage of 95 percent.

Required bcy = Required in-place cy/Shrinkage

Required bcy = 500 in-place ccy/0.95 = 526 bcy

The volume of material to be transported is calculated as follows:

Loose cubic yards = Bank cubic yards × (1 + Swell)

Loose cubic yards = 526 bcy × (1 + 0.15)

Loose cubic yards = 605 lcy

If the same 7 cy dump truck is used, the following would be the number of loads required.

Loads = 605 lcy/7 lcy per load = 86 loads ■

9-5 EQUIPMENT

Selecting and using suitable equipment is of prime importance. The methods available vary considerably depending on the project size and the equipment owned by the contractor. Hand digging should be kept to a bare minimum, but almost every job requires some handwork. Equipment used includes trenching machines, bulldozers, power shovels, scrapers, front-end loaders, backhoes, and clamshells. Each piece of equipment has its use; and as the estimator does the takeoff, the appropriate equipment for each phase of the excavation must be selected. If material must be hauled some distance, either as the excavated material hauled out or the fill material hauled in, equipment such as trucks or tractor-pulled wagons may be required.

The front-end loader is frequently used for excavating basements and can load directly into the trucks to haul the excavated material away. A bulldozer and a front-end loader are often used in shallow excavations, provided the soil excavated is spread out near the excavation area. If the equipment must travel over 100 feet in one direction, it will probably be more economical to select other types of equipment.

The backhoe is used for digging trenches for strip footings and utilities and for excavating individual pier footings, manholes, catch basins, and septic tanks. The excavated material is placed alongside the excavation. For large projects, a trenching machine may be economically used for footing and utility trenches.

A power shovel is used in large excavations as an economical method of excavating and loading the trucks quickly and efficiently. On large-sized grading projects, tractor-hauled and self-propelled scrapers are used for the cutting and filling requirements.

9-6 EARTHWORK—NEW SITE GRADES AND ROUGH GRADING

Virtually every project requires a certain amount of earthwork. It generally requires cutting and filling to reshape the grade. *Cutting* consists of bringing the ground to a lower level by removing earth. *Filling* is bringing soil in to build the land to a higher elevation.

The estimator with little or no surveying or engineering knowledge can still handle smaller, uncomplicated projects

ESTIMATE WORK SHEET

Project:	Little Office Building
Location	Littleville, Tx
Architect	U.R. Architects
Items	Rough Grading

CUT & FILL WORK SHEET

Estimate No.	1234
Sheet No.	1 of 1
Date	11/11/20XX
By	LHF	Checked	JBC

Grid	Fill									Cut								
	Fill At Intersections					Points	Average	Area	Total	Cut At Intersections					Points	Average	Area	Total
	1	2	3	4	5					1	2	3	4	5				
1																		
2																		
3																		
4																		
5																		
6																		
7																		
8																		
9																		
10																		
11																		
12																		
13																		
14																		
15																		
16																		
17																		
18																		
19																		
20																		
21																		
22																		
23																		
24																		
25																		
26																		
27																		
28																		
29																		
30																		
31																		
32																		
33																		
34																		
35																		
36																		
37																		
38																		
39																		
40																		
41																		
42																		

TOTAL FILL - Cubic feet	TOTAL CUT - Cubic feet
Cubic Yards	Cubic Yards
Shrinkage Factor	Swell Factor
Required Cubic Yards of Fill	Cubic Yards of Cut to Haul
Net Cubic Yards to Purchase	

FIGURE 9.2. Cut and Fill Workup Sheet.

but should obtain help from someone more experienced if a complex project is being bid.

Cut and fill sheets similar to the one found in Figure 9.2 are helpful in performing the quantification of rough grading. A spreadsheet version of Figure 9.2 (Cut-Fill.xls) is provided on the companion disk. Regardless of the type of form used, it is essential that the cut and fill quantities be kept separate to allow the estimator to see whether the available cut material can be used for fill. In addition, the estimator needs these quantities to estimate the amount of effort required to convert the cut material into the fill material. For example, if the cut material is to be used for the fill material, it must be hauled to where it will be used and then compacted. This requires decisions concerning what type of hauling equipment will be used and the need for compaction equipment and their associated operators. In addition, because of shrinkage, one cubic yard of cut is not equal to one cubic yard of fill.

The primary drawing for site excavation is the site plan. This drawing typically shows contour lines and spot elevations, and locates all site improvements. Contour lines connect points of equal elevation. Typically, contour lines are in one-foot increments and are based upon some benchmark, which is a permanent point of known elevation. Most commonly, the existing elevations are shown with dashed contour lines while the proposed new elevations are denoted with solid lines. Water flows down and perpendicular to contour lines. In Figure 9.3, the water currently flows from a ridge at elevation 105′ to the top right-hand corner and the bottom left-hand corner of the site. The revised contour lines change this flow and create a level area at elevation 104′.

Spot elevations detail an exact elevation of a point or object on the site. Because the contour lines are typically shown in one-foot increments, the elevation of any point between those lines must be estimated. Through the use of spot elevations, the designer increases the accuracy of

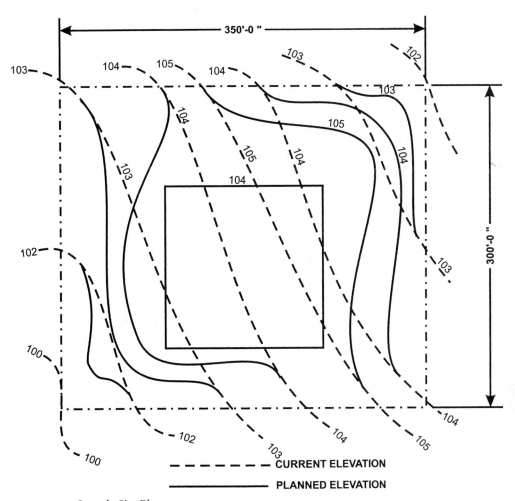

FIGURE 9.3. Sample Site Plan.

the site drawings. For example, the top of a grate elevation on a catch basin may be denoted on the drawing as elevation 104.3′. Because that elevation is critical for the drainage of the parking lot, it is specified as an exact dimension.

Cross-Section Method

The cross-section method entails dividing the site into a grid and then determining the cut or fill for each of the grids. The size of the grid should be a function of the site, the required changes, and the required level of accuracy. If the changes in elevation are substantial, the grid should be small. The smaller the grid, the more accurate the quantity takeoff. In Figure 9.4, the site was divided into a 50-foot grid in both directions. Each line on the grid should be given a number or letter designation. If the horizontal lines are numeric, the vertical lines would be alphabetic. The opposite is also true. By using this type of labeling convention, points on the site plan can be easily found and referenced. In addition to this numbering system, it is also helpful to number each resulting grid square.

The next step is to determine the approximate current and planned elevation for each grid line intersection. Once these are noted, the cut and fill elevation changes can also be noted. Figure 9.5 shows the labeling convention that should be used for this process.

Because contour lines rarely cross the grid intersections, it is necessary to estimate the current and proposed elevations at each of the grid intersection points. If the proposed elevation is greater than the current elevation, fill will be required. Conversely, if the planned elevation is less than the current elevation, cutting will be needed. Figure 9.6 shows the previous site plan with all the elevations and the cut and fill requirements. Once all of the grids have been laid out with existing and proposed elevations and cut or fill, examine them to see which grids contain both cut and fill. This is done by checking the corners of the individual grid boxes. In Figure 9.6, these are grids 3, 4, 10, 11, 12, 17, 18, 19, 25, 27, 32, 34, 39, and 41. These grids require special consideration. In these grids, some of the materials will be converted from cut to fill, or vice versa. The quantity of cut and fill needs to be separated out within these grids so that the cost of converting materials from cut to fill or fill to cut

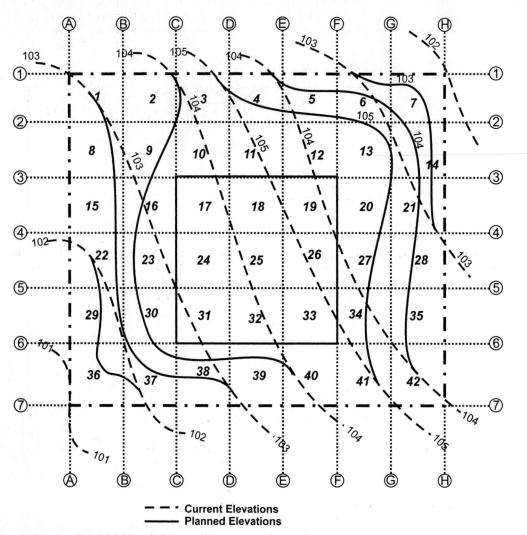

FIGURE 9.4. Site Plan Divided into 50′ Square Grid.

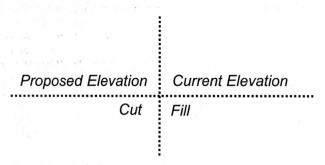

FIGURE 9.5. Labeling Convention.

can be estimated for that specific grid, as well as dealing with shrinkage.

For the squares that contain only cut or fill, the changes in elevation are averaged and then multiplied by the grid

area to determine the required volume of cut or fill. Those quantities are then entered in the appropriate column on the cut and fill worksheets.

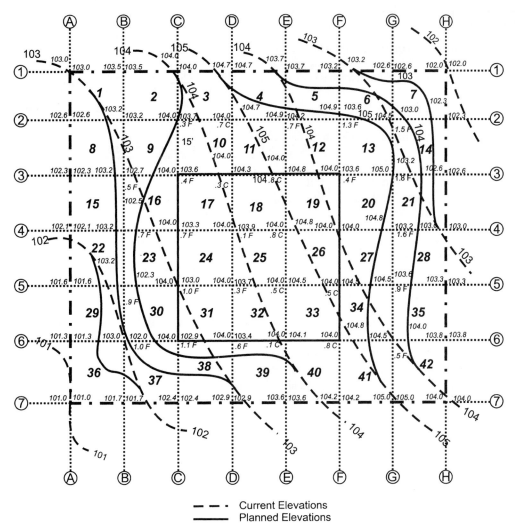

FIGURE 9.6. Grid with Elevations.

EXAMPLE 9-4 FILL VOLUME

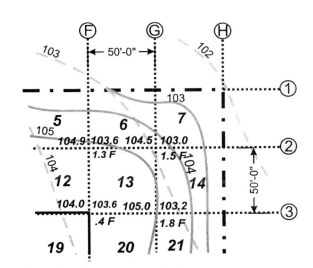

FIGURE 9.7. Excerpt of Grid 13.

Using grid 13 (Figure 9.7) from Figure 9.6 as an example, determine the fill quantity. From Figure 9.7, the information in Figure 9.8 is known about grid 13.

$$\text{ccf of fill} = \frac{\text{Sum of fill at intersections}}{\text{Number of intersections}} \times \text{Area}$$

$$\text{ccf of fill} = \frac{1.3' + 1.5' + 0.4' + 1.8'}{4} \times 2{,}500 \text{ sf}$$

$$= 3{,}125 \text{ ccf of fill}$$

That amount of fill is then entered in the fill column of the cut and fill worksheet (Figure 9.16).

Point	Planned Elevation	Existing Elevation	Fill (ft.)
F2	104.9'	103.6'	1.3
G2	104.5'	103.0'	1.5
F3	104.0'	103.6'	0.4
G3	105.0'	103.2'	1.8

FIGURE 9.8. Data for Grid 13. ■

EXAMPLE 9-5 CUT VOLUME

The volume of cut is determined in exactly the same fashion for cuts as fills. The information in Figure 9.9 is known from using grid 40; as an example (Figure 9.10), the following information is known.

$$\text{bcf of cut} = \frac{0.1' + 0.8' + 0' + 0'}{4} \times 2,500 \text{ sf} = 563 \text{ bcf cut}$$

That amount of cut is then entered in the cut column on the cut and fill worksheet (Figure 9.16).

Point	Planned Elevation	Existing Elevation	Cut (ft.)
E6	104.0'	104.1'	0.1
F6	104.0'	104.8'	0.8
E7	103.6'	103.6'	0.0
F7	104.2'	104.2'	0.0

FIGURE 9.9. Data for Grid 40.

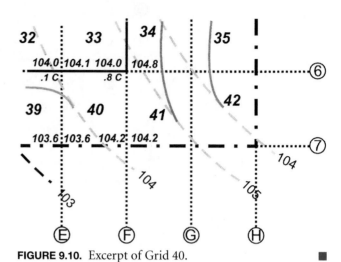

FIGURE 9.10. Excerpt of Grid 40. ■

When a specific grid contains both cut and fill, that grid needs to be divided into grids that contain only cut, only fill, or no change. These dividing lines occur along theoretical lines that have neither cut nor fill. These lines of no change in elevation are found by locating the grid sides that contain both cut and fill. Theoretically, as one moves down the side of the grid, there is a transition point where there is neither cut nor fill. These transition points, when connected, develop a line that traverses the grid and divides it into cut and fill areas and, in some instances, areas of no change.

EXAMPLE 9-6 CUT AND FILL IN THE SAME GRID

Grid 10 (Figure 9.11) from Figure 9.6 is an example of a square that contains both cut and fill. Along line 2, somewhere between lines C and D, there is a point where there is no change in elevation. This point is found first by determining the total change in elevation and by dividing that amount by the distance between the points; second, determine the change in elevation per foot of run.

Total change in elevation (C–D)

$$= 0.3' + 0.7' = 1.0' \text{ change in elevation}$$

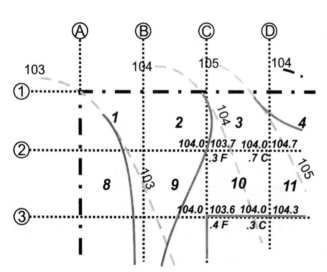

FIGURE 9.11. Grid 10.

Change in elevation per foot of run (C–D)

$$= 1.0'/50' = 0.02' \text{ per foot of run}$$

Because the elevation change is 0.02 foot per foot of run, the estimator can determine how many feet must be moved along that line until there has been a 0.3-foot change in elevation.

Distance from C2 = 0.3'/0.02' per foot of run = 15'

This means that as one moves from point C2 toward point D2 at 15 feet past point C2, there is the theoretical point of no change in elevation, or the transition point between the cut and the fill. Because the same thing occurs along line 3 between points C3 and D3, the same calculations are required.

Total change in elevation (C–D)

$$= 0.4' + 0.3' = 0.7' \text{ change in elevation}$$

Change in elevation per foot of run (C–D)

$$= 0.7'/50' = 0.014' \text{ per foot of run}$$

From this calculation, the distance from point C3 to the point of no change in elevation can be found.

Distance from C3 = 0.4'/0.014' per foot of run = 29'

Given this information, grid 10 can be divided into two distinct grids: one for cut and one for fill. Figure 9.12 details how the grid would be divided.

The next step is to determine the area of the cut and fill portions. A number of methods are available. Perhaps the most simple is to divide the areas into rectangles and/or triangles.

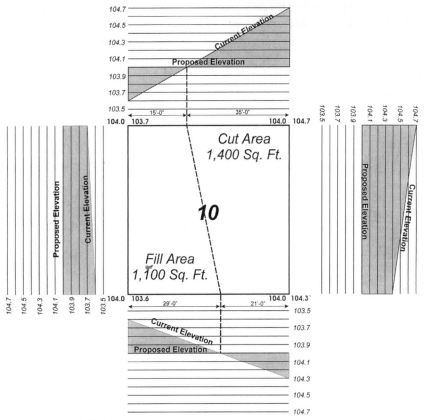

FIGURE 9.12. Grid 10 Layout.

Fill area$_{\text{Rectangle}}$ = 15' × 50' = 750 sf

Fill area$_{\text{Triangle}}$ = 0.5 × 14' × 50' = 350 sf

Total fill area = 750 sf + 350 sf = 1,100 sf

$$\text{Fill} = \frac{0.3' + 0.4' + 0' + 0'}{4} \times 1,100 \text{ sf} = 193 \text{ ccf of fill}$$

The area of the cut equals the area of the grid less the area of the fill.

Cut area = 2,500 sf − 1,100 sf = 1,400 sf

$$\text{Cut} = \frac{0.3' + 0.7' + 0' + 0'}{4} \times 1,400 \text{ sf} = 350 \text{ bcf of cut}$$

These cuts and fills are entered into the cut and fill columns on the cut and fill worksheet (Figure 9.16).

EXAMPLE 9-7   CUT AND FILL

Occasionally when the grid is divided, a portion of the grid will be neither cut nor fill. Grid 3 is an example of such an occurrence. Figure 9.13 is an excerpt from the site plan. In that grid, the change from fill to cut occurs on line 2 between C and D.

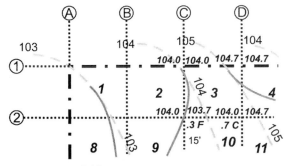

FIGURE 9.13. Grid 3.

Total change in elevation (C–D)

= 0.3' + 0.7' = 1.0' change in elevation

Change in elevation per foot of run (C–D)

= 1.0'/50' = 0.02' per foot of run

From this calculation, the distance from point C2 to the point of no change in elevation can be found.

Distance from C2 = 0.3'/0.02' per foot of run = 15'

Fill area = 0.5 × 50' × 15' = 375 sf

Cut area = 0.5 × 50' × 35' = 875 sf

$$\text{Fill} = \frac{0.3' + 0' + 0'}{3} \times 375 \text{ sf} = 38 \text{ ccf of fill}$$

$$\text{Cut} = \frac{0.7' + 0' + 0'}{3} \times 875 \text{ sf} = 204 \text{ bcf of cut}$$

Figure 9.14 shows the dimensions and proportions between cut, fill, and the unchanged area of grid 3. The remaining 1,250 sf theoretically have no cut or fill. Figure 9.15 is the entire site plan with the areas of no cut and fill shown. Figure 9.16 is the completed cut and fill worksheet for the entire plot.

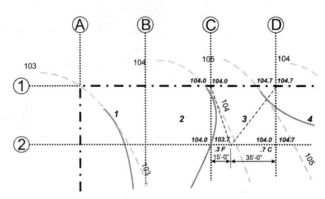

FIGURE 9.14. Cut and Fill Area.

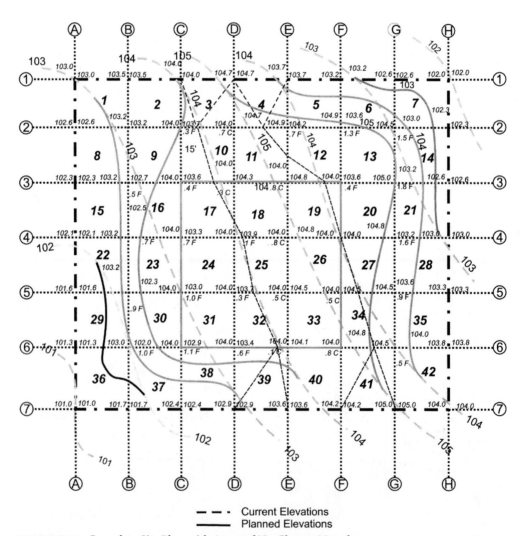

Current Elevations
Planned Elevations

FIGURE 9.15. Complete Site Plan with Areas of No Change Noted.

ESTIMATE WORK SHEET

Project:	Little Office Building		Estimate No.	1234
Location	Littleville, TX		Sheet No.	1 of 1
Architect	U. R. Architects	CUT & FILL	Date	1/1/20XX
Items	Rough Grading		By LHF	Checked JBC

Grid	Fill									Cut									
	Fill At Intersections					Points	Average	Area	Total	Cut At Intersections					Points	Average	Area	Total	
	1	2	3	4	5					1	2	3	4	5					
1						0	0		0						0	0		0	
2	0	0	0	0.3		4	0.075	2500	188						0	0		0	
3	0	0	0.3			3	0.1	375	38	0	0	0.7			3	0.233333	875	204	
4	0	0	0.7			3	0.233333	625	146	0	0	0.7			3	0.233333	625	146	
5	0.7	1.3	0	0		4	0.5	2500	1,250						0	0		0	
6	1.3	1.5	0	0		4	0.7	2500	1,750						0	0		0	
7	1.5	0	0	0		4	0.375	2500	938						0	0		0	
8	0	0.5	0	0		4	0.125	2500	313						0	0		0	
9	0.5	0.4	0	0.3		4	0.3	2500	750						0	0		0	
10	0.3	0	0.4	0		4	0.175	1100	193	0	0.7	0	0.3		4	0.25	1400	350	
11	0	0	0.7			3	0.233333	287.5	67	0.7	0	0	0.8	0.3	5	0.3	2212.5	664	
12	0	0.7	1.3	0.4	0	5	0.48	2054.5	986	0	0	0.8			3	0.266667	445.5	119	
13	1.3	1.5	0.4	1.8		4	1.25	2500	3,125						0	0		0	
14	1.5	0	1.8	0		4	0.825	2500	2,063						0	0		0	
15	0	0.7	0	0.5		4	0.3	2500	750						0	0		0	
16	0.5	0.4	0.7	0.3		4	0.475	2500	1,188						0	0		0	
17	0.4	0	0	0.1	0.3	5	0.1	2101	210	0	0.3	0			3	0.1	399	40	
18	0	0	0.1			3	0.033333	36	1	0	0	0.8	0.8	0.3	5	0.32	2464	788	
19	0	0.4	0			3	0.133333	425	57	0.8	0	0	0.8		4	0.4	2075	830	
20	0	1.6	0.4	1.8		4	0.95	2500	2,375						0	0		0	
21	1.6	0	1.8	0		4	0.85	2500	2,125						0	0		0	
22	0	0.9	0	0.7		4	0.4	2500	1,000						0	0		0	
23	0.7	0.3	0.9	1		4	0.725	2500	1,813						0	0		0	
24	1	0.3	0.7	0		4	0.5	2500	1,250						0	0		0	
25	0.1	0	0	0.3		4	0.1	625	63	0	0.8	0	0.5		4	0.325	1875	609	
26						0	0		0	0.8	0.5	0.5	0		4	0.45	2500	1,125	
27	0	1.6	0.9	0		4	0.625	2050	1,281	0	0.5	0			3	0.166667	450	75	
28	0.9	0	1.6	0		4	0.625	2500	1,563						0	0		0	
29	0	0.9	0	1		4	0.475	2500	1,188						0	0		0	
30	0.9	1	1	1.1		4	1	2500	2,500						0	0		0	
31	1.1	0.6	1	0.3		4	0.75	2500	1,875						0	0		0	
32	0.3	0	0.6	0		4	0.225	1550	349	0	0.5	0.1	0		4	0.15	950	143	
33						0	0		0	0.5	0.5	0.1	0.8		4	0.475	2500	1,188	
34	0	0.9	0	0.5		4	0.35	1275	446	0.5	0	0.8	0		4	0.325	1225	398	
35	0.9	0	0.5	0		4	0.35	2500	875						0	0		0	
36	0	0	0	1		4	0.25	2500	625						0	0		0	
37	1	1.1	0	0		4	0.525	2500	1,313						0	0		0	
38	1.1	0.6	0	0		4	0.425	2500	1,063						0	0		0	
39	0.6	0	0			3	0.2	1075	215	0.1	0	0			3	0.033333	175	6	
40						0	0		0	0.1	0.8	0	0		4	0.225	2500	563	
41	0.5	0	0			3	0.166667	475	79	0.8	0	0			3	0.266667	775	207	
42	0.5	0	0	0		4	0.125	2500	313						0	0		0	
						0			0						0	0		0	

TOTAL FILL - Compacted Cubic Feet		36,317	TOTAL CUT - Bank Cubic Feet		7,453
Compacted Cubic Yards		1,345	Bank Cubic Yards		276
Shrinkage Factor	0.95		Swell Factor	0.25	
Required Bank Cubic Yards of Fill		1,416	Loose Cubic Yards of Cut to Haul		345
Net Bank Cubic Yards to Purchase		1,140			

FIGURE 9.16. Completed Cut and Fill Worksheet.

In the previous examples, it was assumed that the finish grade was the point at which the earthwork took place; however, this is typically not true. In Figure 9.17, the planned contour lines on the parking lot represent the top of the asphalt. Therefore, the rough grading will be at an elevation different from the one shown on the site plan. In this scenario, the elevation for the rough grading needs to be reduced by the thickness of the asphalt and base material.

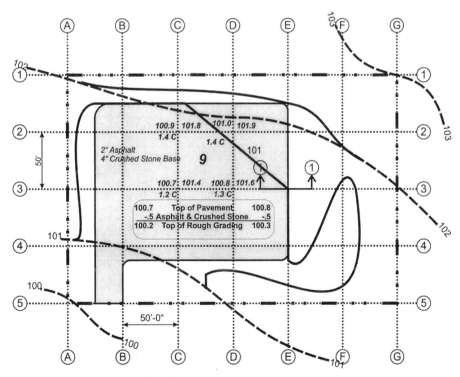

FIGURE 9.17. Parking Lot Site Plan.

EXAMPLE 9-8   CUT AND FILL WITH PAVING

Using Figures 9.17 and 9.18, determine cut for grid 9. In Figure 9.18, the top of the rough grade is 0.50 foot below the top of pavement. The cuts are the differences between the existing elevation and the top of the rough grade, and are shown in Figure 9.19.

$$\text{Cut} = \frac{1.4' + 1.4' + 1.2' + 1.3'}{4} \times 2{,}500 \text{ sf}$$
$$= 3{,}313 \text{ bcf or } 123 \text{ bcy}$$

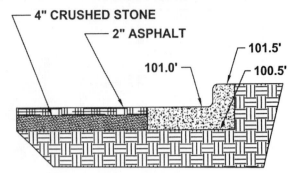

FIGURE 9.18. Cross-Section Through Pavement.

Point	Top of Pavement	Top of Rough Grade	Existing Elevation	Cut (ft.)
C2	100.9'	100.4'	101.8'	1.4'
D2	101.0'	100.5'	101.9'	1.4'
C3	100.7'	100.2	101.4'	1.2'
D3	100.8'	100.3	101.6'	1.3'

FIGURE 9.19. Data for Grid 9.

Average End Area

The average end area method of quantifying cut and fill is often used when dealing with long narrow tracts, such as for roads. In this method, the site is divided into stations. This labeling convention comes from plane surveying using 100-foot measuring tapes. The first numbers are the number of tapes, and the last numbers are the number of feet on the partial tape. In Figure 9.20, station 00+00 is the beginning. Station 00+75 is 75 feet from the beginning station, and station 01+75 is 175 feet from the beginning station. The positioning of station lines is a function of the contour and requires accuracy. The closer the station lines, the greater the accuracy.

The first step in determining the volume using the average end area method is to draw a profile at the station lines. Next, the cut and fill area for each of the profiles is calculated, and finally, the cut or fill area of two adjacent profiles is averaged and multiplied by the distance between the two stations to determine the cut and fill quantity between the stations.

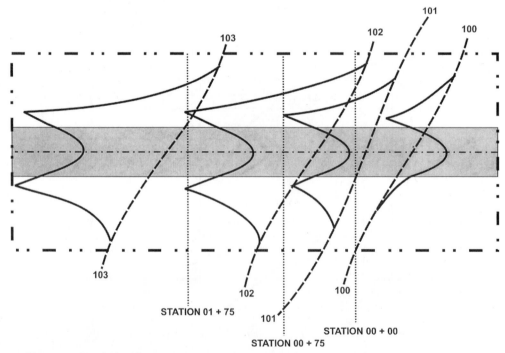

FIGURE 9.20. Road Site Plan.

EXAMPLE 9-9   CUT AND FILL BETWEEN STATION 00+00 AND STATION 00+75

Determine the volume of cut and fill between stations 00+00 and 00+75. The first step is to profile stations 00+00 and 00+75.

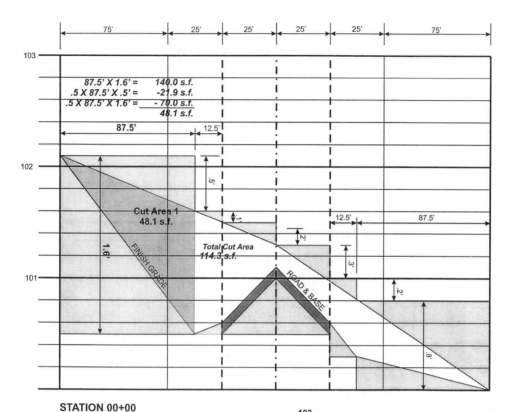

$$87.5' \times 1.6' = \quad 140.0 \text{ s.f.}$$
$$.5 \times 87.5' \times .5' = \quad -21.9 \text{ s.f.}$$
$$.5 \times 87.5' \times 1.6' = \quad - 70.0 \text{ s.f.}$$
$$\overline{\quad 48.1 \text{ s.f.}}$$

Cut Area 1
48.1 s.f.

Total Cut Area
114.3 s.f.

STATION 00+00

FIGURE 9.21. Example Profiles.

Figure 9.21 shows the profile for all stations. A quick observation of these profiles shows that they contain only cut. The next step is to determine the cut area for each of the profiles. Because these profiles are drawn to scale, albeit a different vertical and horizontal scale, they can be calculated easily by breaking up these profiles into rectangles and triangles to find the area. Alternately, the area of cuts and fills can be measured using estimating software such as On-Screen Takeoff. Figure 9.22 is an example of how the cut area for station 00+00 was calculated.

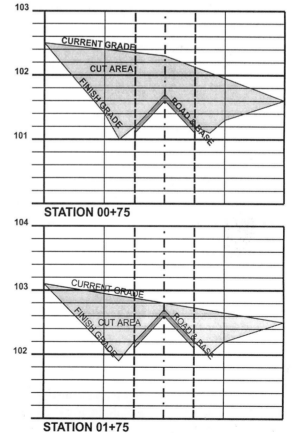

STATION 00+75

STATION 01+75

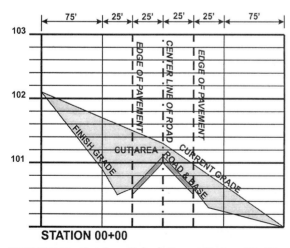

STATION 00+00

FIGURE 9.22. Cut Area Calculation at Station 00+00.

Figure 9.23 shows that the cut area at station 00+00 is 114.3 sf and 232.4 sf at station 00+75. These amounts are averaged and then multiplied by 75 feet, the distance between the two stations to find that the estimated cut is 482 bcy.

Cut between station 00+00 and 00+75 is calculated as follows:

$$\text{Cut} = \frac{114.3\,\text{sf} + 232.4\,\text{sf}}{2} \times 75' = 13,001\,\text{bc} \quad \text{or} \quad 482\,\text{bcy}$$

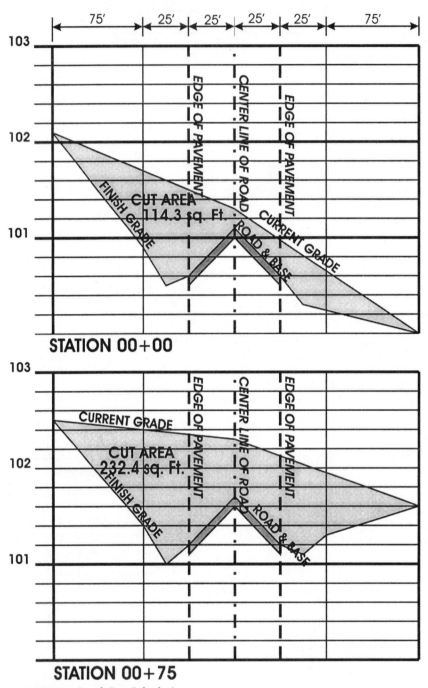

FIGURE 9.23. Road Cut Calculation.

9-7 PERIMETER AND AREA

Throughout the estimate, some basic information is used repeatedly. The perimeter of a building is one such basic dimension that must be calculated. The perimeter is the distance around the building; it is the total length around the building expressed in linear feet.

EXAMPLE 9-10   PERIMETER CALCULATIONS

Find the perimeter of the building shown in Figure 9.24, which is excerpted from Appendix A. Starting in the upper left corner of the building and proceeding clockwise,

$$\text{Perimeter} = 85' + 25'$$
$$+ 15' + 35' + 30' + 10' + 30' + 10' + 40' + 60'$$
$$\text{Perimeter} = 340'$$

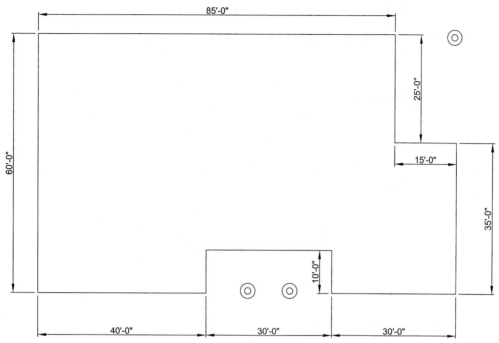

FIGURE 9.24. Sample Perimeter and Area Calculation.

EXAMPLE 9-11   BUILDING AREA

Find the area of the building shown in Figure 9.24. The area of the building will be calculated by finding the area of the $100' \times 60'$ rectangle and by subtracting the recesses.

Basic area $100' \times 60'$	6,000	sf
Bottom recess $10' \times 30'$	−300	sf
Top recess $15' \times 25'$	−375	sf
Net building area	5,325	sf

9–8   TOPSOIL REMOVAL

The removal of topsoil to a designated area where it is to be stockpiled for finished grading and future use is included in many specifications. Thus, the estimator must determine the depth of the topsoil, where it will be stockpiled, and what equipment should be used to strip the topsoil and move it to the stockpile area. Topsoil is generally removed from all building, walk, roadway, and parking areas. The volume of topsoil is figured in cubic yards. A clearance around the entire basic plan must also be left to allow for the slope required for the general excavation; usually about 5 feet is allowed on each side of a building and 1 to 2 feet for walks, roadways, and parking areas.

EXAMPLE 9-12   TOPSOIL REMOVAL

In Figure 9.25, the "footprint" of the building has been enlarged by 5 feet to compensate for accuracy and slope. Assuming that the topsoil to be removed is 9 inches thick, determine the quantity of topsoil to be removed and stockpiled.

Quantity of topsoil to be removed (bcf)
$$= 110' \times 70' \times 0.75' = 5,775 \text{ bcf}$$
Quantity of topsoil to be removed (bcy)
$$= 5,775 \text{ bcf}/27 \text{ cf per cy} = 214 \text{ bcy}$$

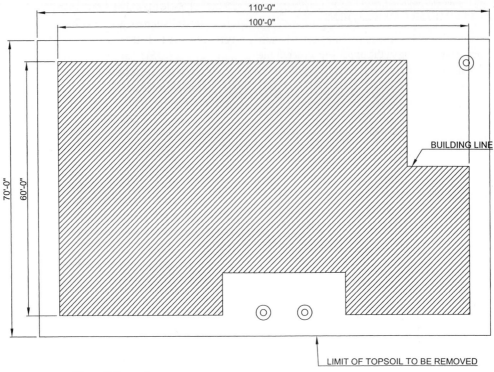

FIGURE 9.25. Topsoil Quantity.

EXAMPLE 9-13   EQUIPMENT AND LABOR COST

Equipment selection for the removal of topsoil will probably be limited to either a bulldozer or a front-end loader. Assume that a 1-cy bucket front-end loader is selected (see Figures 9.26 and 9.27) and its production rate is estimated to be an average of 24 bcy per hour. Mobilization time is estimated at 2.5 hours, the operating cost per hour for the equipment is estimated at $11.35, and the cost for an operator is $17.75 per hour. Estimate the number of hours and the cost to strip the topsoil.

First, the total hours required to complete the topsoil removal (Example 9-12) must be calculated. Divide the total cubic yards to be excavated by the rate of work done per hour, and add the mobilization time; the answer is the total hours for this phase of work.

$$\text{Hours} = \frac{214 \text{ bcy}}{24 \text{ bcy per hour}} + 2.5 \text{ hours} = 11.4 \text{ hours}$$

The total number of hours is then multiplied by the cost of operating the equipment per hour, plus the cost of the crew for the period of time.

Equipment cost = $11.35 per hour $\times$ 11.4 hours = $129

Lobor cost = $17.75 per hour $\times$ 11.4 hours = $202

Total cost = $129 + $202 = $331

| Soil | Dozer | | | | Tractor shovel | | Front end loader | | Backhoe | |
| | 50 ' haul | | 100' haul | | No haul | | 50' haul | 100' haul | No haul | |
	50 hp	120 hp	50 hp	120 hp	1 c.y.	2.25 c.y.	1 c.y.	2.25 c.y..	.5 c.y..	1 c.y..
Medium	40	100	30	75	40	70	24	30	25	55
Soft, sand	45	110	35	85	45	90	30	40	25	60
Heavy soil or stiff clay	15-20	40	10-15	30-35	15-20	35	10	12	10	15

FIGURE 9.26. Equipment Capacity (cy per Hour).

Load and haul		
Truck size	**Haul**	**c.y.**
6 c.y.	1 mile	12-16
6 c.y.	2 miles	8-12
12 c.y.	1 mile	18-22
12 c.y.	2 miles	12-14

FIGURE 9.27. Truck Haul (cy per Hour).

9–9 GENERAL EXCAVATION

Included under general (mass) excavation is the removal of all types of soil that can be handled in fairly large quantities, such as excavations required for a basement, mat footing, or a cut for a highway or parking area. Power equipment such as power shovels, front-end loaders, bulldozers, hydraulic excavators, and graders are typically used in this type of project.

When calculating the amount of excavation to be done for a project, the estimator must be certain that the dimensions used are the measurements of the outside face of the footings and not those of the outside of the building. The footings usually project beyond the wall. Also, an extra 6 inches to 1 foot is added to all sides of the footing to allow the workers to install and remove forms. The estimator must also allow for the sloping of the banks to prevent a cave in. The amount of slope required must be determined by the estimator who considers the depth of excavation, type of soil, and possible water conditions. Some commonly used slopes—referred to as the angle of repose (Figure 9.28)—are given in Figure 9.29. The minimum angle of repose or slope is set by OSHA standards, which govern construction safety. If job conditions will not allow the sloping of soil, the estimator will have to consider using sheet piling (Figure 9.30) or some type of bracing to shore up the bank. Any building or column footing projection is a separate calculation, and the result is added to the amount of excavation for the main portion of the building. Allow about 1 foot all around for working space.

The actual depth of cut is the distance from the top of grade to the bottom of the fill material used under the concrete floor slab. If topsoil has been stripped, the average depth of topsoil is deducted from the depth of cut. If the fill material such as gravel is not used under the concrete floor, the depth is then measured to the bottom of the floor slab. Because the footings usually extend below the fill material, a certain amount of excavation will be required to bring the excavation down to the proper elevation before footings can be placed. This would also be included under the heading of "general excavation," but would be kept separate from topsoil.

Before estimators can select equipment, they will have to determine what must be done with the excess excavation—whether it can be placed elsewhere on the site or whether it must be hauled away. If it must be hauled away, they should decide how far. The answers to these questions will help determine the types and amount of equipment required for the most economical completion of this phase of the work.

To determine the amount of general excavation, it is necessary to determine the following:

1. The size of building (building dimensions).
2. The distance the footing will project beyond the wall.
3. The amount of working space required between the edge of the footing and the beginning of excavation.
4. The elevation of the existing land, by checking the existing contour lines on the plot (site) plan.
5. The type of soil that will be encountered. This is determined by first checking the soil borings (on the drawings), but must also be checked during the site investigation (Section 4-6). Almost every specification clearly states that the soil borings are for the contractor's information, but they are not guaranteed.
6. Whether the excavation will be sloped or shored. Slope angles (angles of repose) are given in Figure 9.29.
7. The depth of the excavation. This is done by determining the bottom elevation of the cut to be made. Then take the existing elevation, deduct any topsoil removed, and subtract the bottom elevation of the cut to determine the depth of the general excavation.

When sloping sides are used for mass excavations, the volume of the earth that is removed is found by developing the average cut length in both dimensions and by multiplying them by the depth of the cut. The average length of the cut can be found as shown in Figure 9.31, or the top of cut and bottom of cut dimensions can be averaged. Either method will result in the same answer.

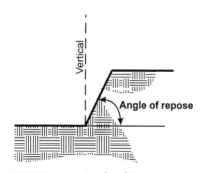

FIGURE 9.28. Angle of Repose.

Material	Angle		
	Wet	Moist	Dry
Gravel	15-25	20-30	24-40
Clay	15-25	25-40	40-60
Sand	20-35	35-50	25-40

FIGURE 9.29. Earthwork Slopes.

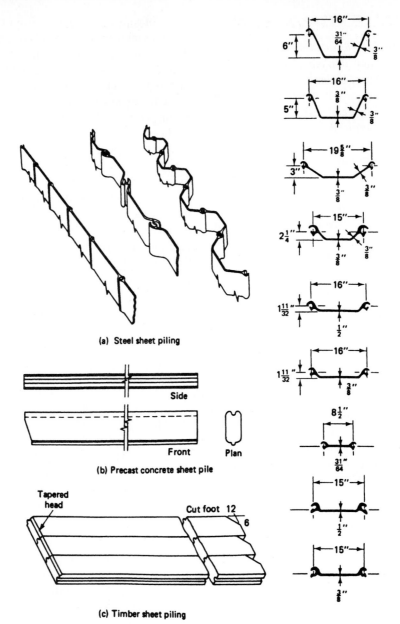

(a) Steel sheet piling

(b) Precast concrete sheet pile

Side

Front

Plan

Tapered head

Cut foot 12

6

(c) Timber sheet piling

FIGURE 9.30. Sheet Pilings.

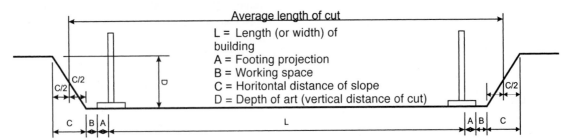

Average length of cut

L = Length (or width) of building
A = Footing projection
B = Working space
C = Horitontal distance of slope
D = Depth of art (vertical distance of cut)

FIGURE 9.31. Typical Excavation.

EXAMPLE 9-14   BASEMENT EXCAVATION

Determine the amount of general excavation required for the basement portion of the building shown in Figure 9.32. This building is an excerpt of the one found in Appendix A.

To get information about this building, look at the building and wall sections. From drawing S8.1 details 3 and 11 found in Appendix A, the following sketches can be drawn (Figures 9.33 and 9.34).

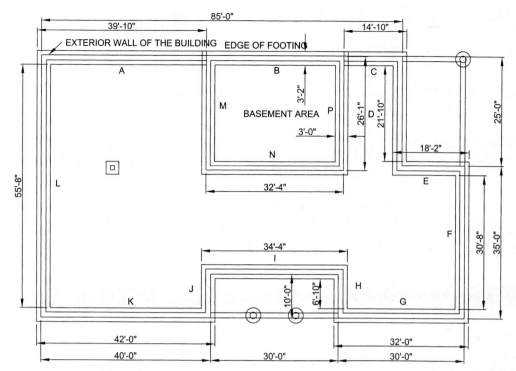

FIGURE 9.32. Building Plan.

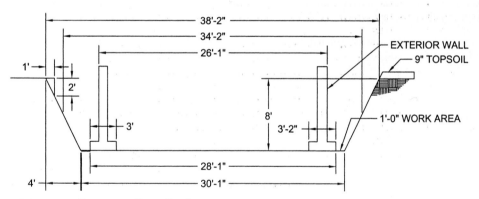

FIGURE 9.33. Basement Cross-Section.

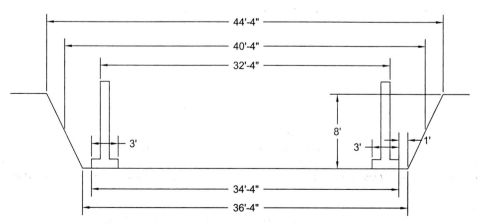

FIGURE 9.34. Basement Cross-Section.

1. From the building plan, the exterior dimensions of the basement are 26'1" by 32'4".

2. From the wall section, the footing projects out 1 foot from the foundation wall.

3. The workspace between the edge of the footing and the beginning of the excavation will be 1 foot in this example.

4. The elevation of the existing land, by checking the existing contour lines on the plot (site) plan, is found and noted. In this example, the expected depth of the cut is 8 feet after a deduction for the topsoil that would have already been removed.

5. Check the soil borings. For this example, a slope of 2:1 will be used, which means for every 2 feet of vertical depth an additional 1 foot of horizontal width is needed. Since the alternative is shoring or sheet piling on this project, the sloped excavation will be used.

6. The bottom elevation of the general excavation cut will be at the bottom of the gravel. Since this elevation is rarely given, it may have to be calculated. Generally, the drawings will give the elevation of the basement slab or bottom of the footing; the depth of cut is calculated from these.

Average Width of Cut:
$$2'0'' + 1'0'' + 28'1'' + 1'0'' + 2'0'' = 34'1''$$

Average Length of Cut:
$$2'0'' + 1'0'' + 34'4'' + 1'0'' + 2'0'' = 40'4''$$

General Excavation:

$$\text{General excavation (bcf)} = 34'1'' \times 40'4'' \times 8'$$

$$\text{General excavation (bcf)} = 34.083' \times 40.33' \times 8' = 10{,}997 \text{ bcf}$$

$$\text{General excavation (bcy)} = 10{,}997 \text{ bcf}/27 \text{ cf per cy} = 407 \text{ bcy}$$

Required equipment: Backhoe with 1-cy bucket
Mobilization: 2 hours
Rate of work for backhoe: 55 bcy per hour
Equipment cost: $16.75 per hour
Operator cost: $19.75 per hour

$$\text{Equipment hours} = 407 \text{ bcy}/55 \text{ bcy per hour} = 7.4 \text{ hours}$$

$$\text{Total hours} = 7.4 \text{ equipment hours}$$
$$+ 2 \text{ hours for mobilization} = 9.4 \text{ hours}$$

$$\text{Equipment cost} = 9.4 \text{ hours} \times \$16.75 \text{ per hour} = \$157$$

$$\text{Labor cost} = 9.4 \text{ hours} \times \$19.75 \text{ per hour} = \$186$$

$$\text{Total cost} = \$157 + \$186 = \$343$$

If this were to be hauled off the site, assuming a 30 percent swell factor, it would take 76 loads using a 7-cy truck.

$$\text{Required haul (lcy)} = 407 \text{ bcy} \times 1.3 = 529 \text{ lcy}$$

$$\text{Required loads} = 529 \text{ lcy}/7 \text{ lcy per load} = 76 \text{ loads} \quad \blacksquare$$

EXAMPLE 9-15   CONTINUOUS FOOTING EXCAVATION

Determine the amount of general excavation required for the continuous footings of the building shown in Figure 9.32. This building is an excerpt of the one found in Appendix A. Figure 9.35 is a sketch of the continuous footing with dimensions. In this example the slope is 1.5:1, which means that for every 1.5 feet of vertical rise, there is 1 foot of horizontal run.

The simplest way to approximate the amount of cut is to multiply the average cut width times the perimeter of the building times the depth. From Example 9-10, the building perimeter is 340 feet. However, 32'4" of that perimeter was included in the basement wall (refer to Figure 9.32). Therefore, the linear distance of continuous footing is 307'8".

$$\text{General excavation (bcf)} = 7'10'' \times 307'8'' \times 4'$$

$$\text{General excavation (bcf)} = 7.833' \times 307.67' \times 4' = 9{,}640 \text{ bcf}$$

$$\text{General excavation (bcy)} = 9{,}640 \text{ bcf}/27 \text{ cf per cy} = 357 \text{ bcy}$$

Required equipment: Backhoe with 0.5-cy bucket
Mobilization: 2 hours
Rate of work for backhoe: 25 bcy per hour
Equipment cost: $16.25 per hour
Operator cost: $19.75 per hour

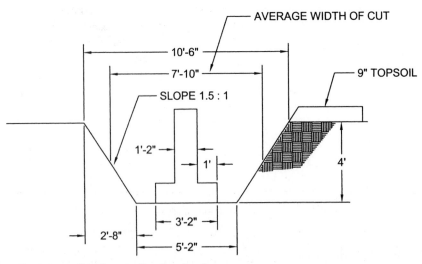

FIGURE 9.35. Continuous Footing Section.

Equipment hours = 357 bcy/25 bcy per hour = 14 hours

Total hours = 14 equipment hours + 2 hours for mobilization
= 16 hours

Equipment cost = 16 hours × $16.25 per hour = $260

Labor cost = 16 hours × $19.75 per hour = $316

Total cost = $260 + $316 = $576   ■

EXAMPLE 9-16   SPREAD FOOTING EXCAVATION

There is one spread footing shown in Figure 9.32. The details of this spread footing can be found on sheet S8.1 detail 16 in Appendix A. From those details, given a slope of 1.5:1, Figure 9.36 can be sketched.

Since the spread footing is square, the general excavation can be found by squaring the average cut width and by multiplying that by the depth.

General excavation (bcf) = 8′4″ × 8′4″ × 5′0″

General excavation (bcf) = 8.33′ × 8.33′ × 5.0′ = 347 bcf

General excavation (bcy) = 347 bcf/27 cf per cy = 13 bcy

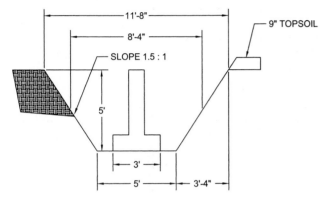

FIGURE 9.36. Sketch of Spread Footing.   ■

9–10 SPECIAL EXCAVATION

Usually the special excavations are the portions of the work that require hand excavation, but any excavation that requires special equipment to be used for a particular portion other than general (mass) excavation may be included.

Portions of work most often included under this heading are footing holes, small trenches, and the trench-out areas below the general excavation for wall and column footings, if required. On a large project, a backhoe may be brought in to perform this work, but a certain amount of hand labor is required on almost every project.

The various types of excavation must be kept separately on the estimate; and if there is more than one type of special or general excavation involved, each should be considered separately and then grouped together under the heading "special excavation" or "general excavation."

In calculating the special excavation, the estimator must calculate the cubic yards, select the method of excavation, and determine the cost.

9–11 BACKFILLING

Once the foundation of the building has been constructed, one of the next steps in construction is the backfilling required around the building. *Backfilling* is the putting back of the excess soil that was removed from around the building during the general excavation. After the topsoil and general and special excavations have been estimated, it is customary to calculate the amount of backfill.

The material may be transported by wheelbarrows, scrapers, front-end loaders with scoops or buckets, bulldozers, and perhaps trucks if the soil must be transported a long distance. The selection of equipment will depend on the type of soil, weather conditions, and distance the material must be moved. If tamping or compaction is required, special equipment will be needed, and the rate of work per hour will be considerably lower than if no tamping or compaction is required.

One method for calculating the amount of backfill to be moved is to determine the total volume of the building within the area of the excavation. This would be the total volume of the basement area, figured from the underside of the fill material, and would include the volume of all footings, piers, and foundation walls. This volume is deducted from the volume of excavation that had been previously calculated. The volume of backfill required is the result of this subtraction. The figures should not include the data for topsoil, which should be calculated separately.

A second method for calculating backfill is to compute the actual volume of backfill required. The estimator usually makes a sketch of the actual backfill dimensions and finds the required amount of backfill.

The following examples illustrate how to calculate backfill quantities.

EXAMPLE 9-17   BACKFILLING THE BASEMENT WALLS

Using the sketches in Figures 9.37, 9.38, and 9.39, the following volume calculations can be performed.

Building volume (cf) = 32′4″ × 26′1″ × 8′

Building volume (cf) = 32.33′ × 26.083′ × 8′ = 6,746 cf

Footing volume (cf) = 1′ × 1′ × 26′1″ × 2

Footing volume (cf) = 1′ × 1′ × 26.083′ × 2 = 52 cf

Footing volume (cf) = 1′ × 1′ × 32′4″ × 2

Footing volume (cf) = 1′ × 1′ × 32.33′ × 2 = 65 cf

Total Footing volume (cf) = 52 cf + 65 cf = 117 cf

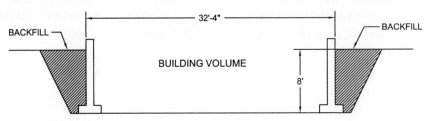

FIGURE 9.37. Backfill Section.

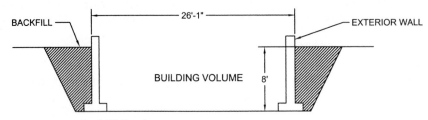

FIGURE 9.38. Backfill Section.

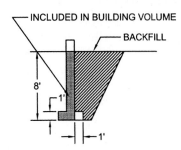

FIGURE 9.39. Footing Backfill.

From Example 9-14, the basement excavation is 10,997 bcf.

Backfill (bcf) = General excavation − Building volume
− Footing volume

Backfill (bcf) = 10,997 bcf − 6,746 cf − 117 cf = 4,134 ccf

Backfill (bcy) = 4,134 ccf/27 cf per cy = 153 ccy

Equipment: Dozer (120 hp) at $16.85 per hour
Dozer work rate: 100 ccy per hour (Figure 9.26)
Operator: $18.30 per hour
Laborer: $11.50 per hour
Mobilization: 1.5 hours

Backfill time (hours) = 153 ccf/100 ccy per hour = 1.5 hours

Total hours = 1.5 hours + 1.5 mobilization hours
= 3 hours

Operator cost ($) = 3 hours × $18.30 per hour = $54.90

Laborer cost ($) = 3 hours × $11.50 per hour = $34.50

Equipment cost ($) = 3 hours × $16.85 per hour = $50.55

Total cost = $54.90 + $34.50 + $50.55
= $139.95 ∎

EXAMPLE 9-18 BACKFILLING THE FOUNDATION WALLS

There are two ways in which the quantity of backfill can be determined. Both will yield virtually the same answer. The first is to subtract the area of the footing from the area backfill and multiply that number by the length of the footing. Figure 9.40 is a sketch of the backfill requirements. The excavation for the footing was discussed in Example 9-15.

Volume of footing (cf) = 3′2″ × 1′ × 307′8″

Volume of footing (cf) = 3.167′ × 1′ × 307.67′
= 974 cf

Volume in foundation wall (cf) = 3′ × 1′2″ × 307′8″

Volume in foundation wall (cf) = 3′ × 1.167′ × 307.67′
= 1,077 cf

Volume in foundation and footing (cf) = 974 cf + 1,077 cf
= 2,051 cf

Volume of backfill (ccf) = 9,640 bcf − 2,051 cf
= 7,589 ccf

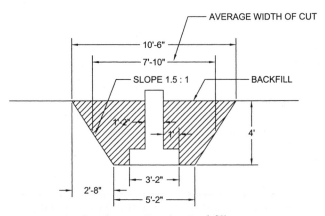

FIGURE 9.40. Continuous Footing Backfill.

Volume of backfill (ccy) = 7,589 ccf/27 cf per cy
= 281 ccy

Alternately, the area of backfill can be figured as shown in Figure 9.41 by figuring the area and by multiplying that amount by the length.

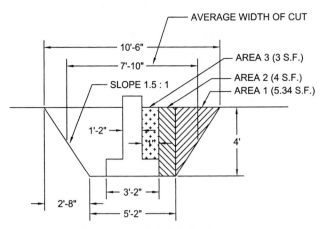

FIGURE 9.41. Alternate Backfill Method.

Fill area (sf) = (5.34 sf + 4 sf + 3 sf) × 2 = 24.68 sf

Volume of backfill (ccf) = 24.68 sf × 307.67′ = 7,593 bcf

Volume of backfill (ccf) = 7,593 ccf/27 cf per cy = 281 ccy ■

9–12 EXCESS AND BORROW

Once the earthwork has been calculated in terms of excavation, backfill, and grading, the estimator must compare the total amounts of cut and fill required and determine whether there will be an *excess* of materials that must be discarded, or whether there is a shortage of materials and some must be brought in (*borrow*). Topsoil is not included in the comparison at this time; topsoil must be compared separately because it is much more expensive than other fill that might be required, and the excess topsoil is easily sold. Shrinkage and swell must also be taken into account.

The specifications must be checked for what must be done with the excess material. Some specifications state that it may be placed in a particular location on the site, but many times they direct the removal of excess materials from the site. If the material is to be hauled away, the first thing the estimator must know is how many cubic yards are required; then the estimator must find a place to haul the material. Remember that soil swells; if the estimator calculated a haul of 100 bcy and the swell is estimated at 15 percent, 115 lcy must be hauled away. Finding where the excess material can be hauled is not always a simple matter, because it is desirable to keep the distance as short as possible from the site. If the haul distance is far and/or there is a restriction on the number of available dump trucks, the haul time may dictate the excavation equipment cost rather than the actual amount of earth to be removed. If the backhoe or the front-end loader has to wait for the truck to load, the equipment and operator costs are incurred for the wait time even though no productive work is being performed. The estimator should check into this when visiting the site. If material must be brought in, the estimator must first calculate the amount required and then set out to find a supply of material as close to the job site as is practical. Check the specifications for any special requirements pertaining to the type of

soil that may be used. Keep in mind that the material being brought in is loose and will be compacted on the job. If it is calculated that 100 ccy are required, the contractor will have to haul in at least 110 to 140 lcy of soil—even more if it is clay or loam.

The next step is to select equipment for the work to be done, which will depend on the amount of material, type of soil, and the distance it must be hauled.

9–13 SPREADING TOPSOIL, FINISH GRADE

Many specifications call for topsoil to be placed over the rough grade to make it a *finish grade*. The topsoil that was stockpiled may be used for this purpose, but if there is no topsoil on the site, it will have to be purchased and hauled in—a rather costly proposition. The estimator must be certain to check the specification requirements and the soil at the site to see whether the existing soil can be used as topsoil. On most projects, the equipment needed for finish grading will consist of a bulldozer or a scraper. The quantity of topsoil is usually calculated by multiplying the area used for rough grading by the depth of topsoil required.

After the topsoil has been spread throughout the site in certain areas, such as in the courtyards, around the buildings, or along the parking areas, it is necessary to hand rake the topsoil to a finish grade. The volume of soil to be moved must be calculated and entered on the estimate.

9–14 LANDSCAPING

Most specifications require at least some landscaping work. If the work is seeding and fertilizing, the general contractor may do it or subcontract it out. The estimator should check the specifications for the type of fertilizer required. The specifications may also state the number of pounds per square (1 square = 100 sf) that must be spread. The seed type will be included in the specifications, which will also state the number of pounds to be spread per square. Who is responsible for the growth of grass? If the contractor is, be certain that it will receive adequate water and that the soil will not erode before the grass begins to grow. Often the seeded area is covered with straw or special cloth, which helps keep in moisture and reduce erosion. The estimator will have to calculate the area to be fertilized and seeded, determine what equipment will be used, arrange covering and water, and then arrange the removal of the covering if it is straw.

Sodding is often required when the owner wants an "instant lawn." The specifications will state the type of sod required, and the estimator must determine the area to be covered. The general contractor or a subcontractor may handle this phase of work. The estimator, in calculating labor costs, must consider how close to the actual area being sodded the truck bringing the sod can get, or whether the sod will have to be transferred from the truck to wheelbarrows or lift trucks and brought in.

The trees and shrubs required for the project will generally be handled by a subcontractor who specializes in landscaping. The estimator should note the number, size, and species required. The landscape contractor may submit a proposal to perform all the landscape work; however, it is up to the estimator to make a preliminary decision about how it will be handled.

In the estimate, there must also be an allowance for the maintenance of the landscaping for whatever period is required in the specifications. Many specifications also require a guarantee period during which the contractor must replace landscaping that fails to grow or dies.

9–15 PUMPING (DEWATERING)

Almost all specifications state that the contractor provides all pumping and dewatering required. The estimator should examine any soil investigation reports made on the property for the possibility of a high water table. Also to be considered is the time of year when the soil investigation was made, as the water table varies throughout the year depending on the particular area in which the project is located.

Water can present a problem to the contractor in almost any location. Even if the groundwater table presents no problem, the possibility exists of rainfall affecting the construction and requiring pumps for its removal.

Some projects require constant pumping to keep the excavation dry, whereas others require a small pump simply to remove excess groundwater. The variation in costs is extreme, and the estimator must rely on past experience in a given area as a guide. If the area is unfamiliar, the problem should be discussed with people who are familiar with the locale. Who should the estimator ask? If the contacts in a given area are limited, the following may be good sources of information: the local building department or any retired construction engineers, superintendents, surveyors (who are the most readily available source of information)—generally, these people will be most honest in their appraisal of a situation and will be delighted to be of help. Other sources might be local sales representatives, the architect/engineer, and any local subcontractor, such as an excavator.

The problem of water will vary depending on the season of the year, location, type of work, general topography, and weather. The superintendent should be certain that during construction the general slope of the land is away from the excavation so that, in case of sudden rainfall, it is not "washed out."

9–16 ROCK EXCAVATION

The excavation of rock differs from the excavation of ordinary soils. Rock is generally classified as soft, medium, or hard, depending on the difficulty of drilling it. Almost all rock excavation requires drilling holes into the rock and then blasting the rock into smaller pieces so that it may be handled on the job. Types of drills include rotary and core

jackhammers. The drill bits may be detachable, carbide, or have diamond-cutting edges, solid or hollow. Among the types of explosives used are blasting powder and dynamite. When a contractor has had no experience in blasting, a specialist is called to perform the required work. The estimator working for a contractor who has had blasting experience will make use of the cost information from past projects.

Factors that affect rock excavation include the type of rock encountered, the amount of rock to be excavated, whether bracing is required, the manner of loosening the rock, equipment required, length of haul, delays, and special safety requirements. Also, the estimator must decide whether it is to be an open cut or a tunnel in which water might be encountered. Once loosened, a cubic yard of rock will require as much as 70 percent more space than it occupied initially, depending on the type of rock encountered.

The cost of blasting will be affected by the type of rock, equipment, explosive, and depth of drilling. Only the most experienced personnel should be used, and all precautions should be taken. Mats are often used to control the possibility of flying debris.

9–17 SUBCONTRACTORS

Contractors who specialize in excavation are available in most areas. Specialized subcontractors have certain advantages over many general contractors. They own a large variety of equipment, are familiar with the soil encountered throughout a given area, and know where fill can be obtained and excess cut can be hauled. The subcontractor may bid the project as a lump sum or by the cubic yard. Either way the estimator must still prepare a complete estimate—first, to check the subcontractor's price to be certain that it is neither too high nor too low, and second, because the estimator will need the quantities to arrive at a bid price. All subcontractors' bids must be checked, regardless.

The estimator should always discuss with subcontractors exactly what will be included in their proposals and put this in writing so that both parties agree on what is being bid. It is customary for the general contractor to perform all hand excavation and sometimes the trenching. Selection of a subcontractor is most often based on cost, but also to be considered are the equipment owned by the subcontractor and a reputation for speed and dependability.

If the subcontractor does not meet the construction schedule, it will probably cost far more than the couple of hundred dollars that may have been saved on the initial cost.

9–18 EXCAVATION CHECKLIST

Clearing site:

removing trees and stumps

clearing underbrush

removing old materials from premises

removing fences and rails

removing boulders

wrecking old buildings

Foundations:

underpinning existing buildings

disconnecting existing utilities

clearing shrubbery

Excavation (including backfilling):

basement

footings

foundation walls

sheet piling

pumping

manholes

catch basins

backfilling

tamping (compacting)

blasting

grading (rough and fine)

utility trenches

grading and seeding lawns

trees

shrubbery

topsoil removal

topsoil brought in

9–19 PILES

Piles used to support loads are referred to as *bearing piles*. They may be wood, steel "H" sections, poured-in-place concrete with metal casing removed, poured-in-place with metal casing left in place, wood and concrete, or precast concrete.

The wide variations in design, conditions of use, types of soil, and depth to be driven make accurate cost details difficult to determine unless the details of each particular job are known. The various types and shapes available in piles also add to the difficulty, particularly in concrete piles. It is suggested that estimators approach subcontractors who specialize in this type of work until they gain the necessary experience. Figure 9.42 gives approximate requirements for labor when placing bearing piles.

Type and Size of Pile	Approximate No. of Feet Driven Per Hour
Wood	80-130
Steel	
30-100	50-100
100-200	30-70
Concrete	
100-200	70-100
200-400	10-40

FIGURE 9.42. Driving Bearing Piles.

Sheet piling may be wood, steel, or concrete. They are used when the excavation adjoins a property line or when the soil is not self-supporting. Sheet piling is taken off by the surface area to be braced or sheet-installed.

Wood sheet piling is purchased by board feet, steel piling by 100 pounds (cwt), and concrete piling by the piece. Depending on the type used, the quantity of materials required must also be determined.

Sheet piling may be placed in the excavation and braced to hold it erect, or it may be driven into the soil. No one way is less expensive. Often, after the work is done, the piles are removed *(pulled)*. Pulling may require the employment of a pile extractor for difficult jobs, or hand tools for the simple jobs. Figure 9.43 gives approximate requirements for labor when placing sheet piling.

Estimating.   The first step is to determine exactly what type and shape pile is required; next, the number of piles required; and third, decide the depth that the piles will have to be driven. With this information, the estimator can determine material quantities and their cost.

If the contractor's crew will drive the piles, the next determination should be the type of equipment required, how many hours the work will take, and cost per hour. Then the crew size, along with the hours and cost per hour required, is calculated. To arrive at this, an estimate of the linear footage placed per hour (including all cutting off, moving around, etc.) must be made so that the total number of hours required for the completion of the pile driving is determined. In dealing with sheet piling, if the sheets are to be pulled at the end of construction, the estimator must also include a figure for this. Costs must also comprise mobilization of workers, equipment, and material for pile driving, and demobilization when the work has been completed.

Type of Sheet Pile	Place, Drive By Hand	Place, Drive By Power Hammer	Brace	Pulling
Wood	4.0 – 9.0	3.3 – 8.0	2.0 – 3.0	3.0 – 5.0
Steel	3.0 – 9.0	3.0 – 7.5	2.0 – 3.0	3.0 – 5.0
Concrete	-	4.0 – 10.0	2.0 – 4.5	2.0 – 8.0

FIGURE 9.43. Sheet Piles, Approximate Hours Required.

Check the soil borings for the kind of soil recorded and the approximate depth the piles must be driven. The accuracy of this information is quite important.

Specifications.   Check the type of piles required and under what conditions sheet piling is required. Any requirements regarding soil conditions, equipment, or other special items should be noted. Many specifications require that only firms experienced in driving piles be allowed to do so on the job.

9–20 PILE CHECKLIST

Type:

wood

wood and concrete

H-piles

casing and poured in place

poured in place

precast

wood sheets

steel sheets

precast sheets

bracing

Equipment:

pile driver

compressors

derricks

cranes

9–21 ASPHALT PAVING

The asphalt paving required on the project is generally subcontracted to someone specializing in paving. The general contractor's estimator will make an estimate to check the subcontractor's price.

Asphalt paving will most commonly be hot-mix and is generally classified by traffic (heavy, medium, or light) and use (walks, courts, streets, driveways, etc.).

The estimator will be concerned with subgrade preparation, subdrains, soil sterilization, insulation course, subbase course, base courses, prime and tack coats, and the asphalt paving required. Not all items are required on any given project, so the estimator should determine which items will be required, the material and equipment necessary for each portion of the work, and the requisite thickness and amount of compaction.

Specifications.   Check the requirements for compaction, thickness of layers, total thicknesses, and materials required for each portion of the work. The drawings will also have to be checked for some of these items. The drawings will show the location of most of the work to be completed, but the specifications should also be checked. The specifications and drawings will list different requirements for the various uses. (These are called traffic requirements.)

Estimate.   The number of square feet (or square yards) of surface area to be covered is determined, and the thickness (compacted) of each course and the type of materials required are noted. Base courses and the asphalt paving are often taken off by the ton, as this is the unit in which these materials must be bought. The type of asphalt and aggregate size required must also be noted. Two layers of asphalt paving are required on some projects: A coarse base mix may be used with a fine topping mix. Equipment required may include a steel-wheel roller, trailers to transport equipment, dump trucks, paving machines, and various small tools.

To estimate the tons of material required per 1,000 sf of surface area, refer to Figure 9.44. Different requirements will be listed for the various uses (walk, driveway, etc.), and the different spaces must be kept separately.

In many climates, the asphalt paving has a cutoff date in cold weather, and the paving that is not placed when the mixing plants shut down will not be laid until the start-up time in the spring. The plants may be shut down for as long as four months or more, depending on the locale.

Compacted Thickness	Asphalt[1] Paving	Granular[2] Material	Subgrade[3] Material
1"	6.5	5.25	4.6
2"	13.0	10.5	9.2
3"	19.5	15.75	13.8
4"	26.0	21.0	18.4
5"	32.5	26.25	23.0
6"	39.0	31.5	27.6
8"	52.0	42.0	36.8
10"	65.0	52.5	46.0
12"	78.0	63.0	55.2

Per 1,000 s.f. of surface area, figures include 10 percent waste.

1.   *Asphalt paving, 140 – 150 pounds per c.f.*
2.   *Granular material, 110- 120 pounds per c.f.*
3.   *Subgrade material, 95 – 105 pounds per c.f.*

FIGURE 9.44. Approximate Asphalt Paving Materials Tonnage.

EXAMPLE 9-19   ASPHALT TAKEOFF

Determine the quantity of asphalt and subgrade material required for the parking lot in the commercial building found in Appendix A. Figure 9.45 is an excerpt of the parking lot while Figure 9.46 is a cross-section. Assume that the asphalt required is 3 inches thick and the subgrade is 6 inches thick.

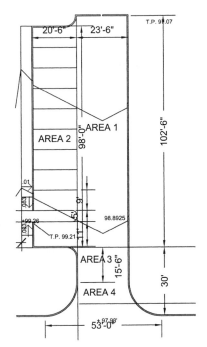

FIGURE 9.45. Parking Lot.

Area 1 = 23.5′ × 102.5′ = 2,409 sf

Area 2 = 20.5′ × 98.0′ = 2,009 sf

Area 3 = 15.5′ × 23.5′ = 364 sf

Area 4 = [(53.0′ + 23.5′) × 14.5′)]/2
= 555 sf

Total area = 5,337 sf

Asphalt = 19.5 tons per 1,000 sf 3″ thick
× 5.3 thousand sf

Asphalt = 104 tons

Subgrade material = 27.6 tons per 1,000 sf 6″ thick
× 5.3 thousand sf

Subgrade material = 246 tons

Because of the special equipment required, the installation is virtually always subcontracted.

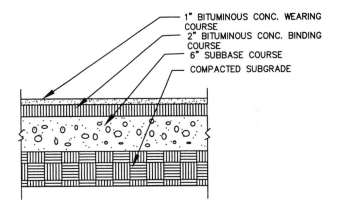

FIGURE 9.46. Parking Lot Section. ■

WEB RESOURCES

REVIEW QUESTIONS

1. What type of information about the excavation can the estimator learn from the specifications?

2. How does the type of soil to be excavated affect the estimate?

3. What is the unit of measure in excavation?

4. How will the type of soil, shape of the excavation, and amount of work to be done affect the equipment selection?

5. What is the difference between a bank cubic yard, a loose cubic yard, and a compacted cubic yard? Why are they important in estimating excavation?

6. What is meant by cut and fill?

7. What does the estimator have to consider if there is a substantial amount of cut on the job? What if there is a substantial amount of fill?

8. How can the estimator get an estimate of the depth of topsoil on the project?

9. What type of excavation is considered to be general excavation?

10. How does general excavation differ from special excavation?

11. What is excess and borrow, and how are each considered in the estimate?

12. How will the possibility of a high water table or underground stream affect the bid?

13. What are piles, and under what conditions might they be required on the project?

14. Calculate the general excavation for the building shown in Appendix C.

15. Calculate the topsoil to be moved, put back around the building, and hauled away for Appendix C.

16. Calculate the crushed gravel and asphalt required for the parking lot for Appendix C.

CHAPTER TEN

CONCRETE

10–1 CONCRETE WORK

The concrete for a project may be either ready mixed or mixed on the job. Most of the concrete used on commercial and residential work is ready mixed and delivered to the job by the ready-mix company. Quality control, proper gradation, water, and design mixes are easily obtained by the ready-mix producers. When ready-mix is used, the estimator must determine the amount of concrete required and the type and amount of cement, aggregates, and admixtures. These are discussed with the supplier, who then gives a proposal for supplying the specific concrete.

Concrete for large projects or those in remote locations will typically be mixed at the job site and will require a field batch plant. To successfully estimate a project of this magnitude, the estimator must have a thorough understanding of the design of concrete mixes. The basic materials required for concrete are cement, aggregates (fine and coarse), and water. Various admixtures, which modify the properties of the concrete, may also be required. If field batching is desired, the estimator will have to compute the amounts of each material required, evaluate the local availability, and determine if these ingredients will be used or if materials will be shipped to the site from somewhere else. If the materials are shipped in, they are bulky, heavy, and typically transported by rail. If the site is not adjacent to a rail line, provisions will need to be made to move the materials from the rail siding to the project batch plant. No attempt will be made here to show the design of concrete mixes. It is suggested that the estimator who is unfamiliar with mix design and control consult with a professional civil engineer.

10–2 ESTIMATING CONCRETE

Concrete is estimated by the cubic yard (cy) or by the cubic foot (cf) and then converted into cubic yards. Concrete quantities are measured in cubic yards as it is the pricing unit of the ready-mix companies, and most tables and charts available relate to the cubic yard.

Roof and floor slabs, slabs on grade, pavements, and sidewalks are most commonly measured and taken off in length, width, and thickness and converted to cubic feet and cubic yards (27 cf = 1 cy). Often, irregularly shaped projects are broken down into smaller areas for more accurate and convenient manipulation.

When estimating footings, columns, beams, and girders, their volume is determined by taking the linear footage of each item times its cross-sectional area. The cubic footage of the various items may then be tabulated and converted to cubic yards.

When estimating footings for buildings with irregular shapes and jogs, the estimator must be careful to include the corners only once. It is a good practice for the estimator to highlight on the plans which portions of the footings have been figured. When taking measurements, keep in mind that the footings extend out from the foundation wall; therefore, the footing length is greater than the wall length.

In estimating quantities, the estimator makes no deductions for holes smaller than 2 sf or for the space that reinforcing bars or other miscellaneous accessories take up. Waste ranges from 5 percent for footings, columns, and beams to 8 percent for slabs.

The procedure that should be used to estimate the concrete on a project is as follows:

1. Review the specifications to determine the requirements for each area in which concrete is used separately (such as footings, floor slabs, and walkways) and list the following:
    (a) Type of concrete
    (b) Strength of concrete
    (c) Color of concrete
    (d) Any special curing or testing
2. Review the drawings to be certain that all concrete items shown on the drawings are covered in the specifications. If not, a call will have to be made to the architect-engineer so that an addendum can be issued.

3. List each of the concrete items required on the project.

4. Determine the quantities required from the working drawings. Footing sizes are checked on the wall sections and foundation plans. Watch for different size footings under different walls.

Concrete slab information will most commonly be found on wall sections, floor plans, and structural details. Exterior walks and driveways will most likely be identified on the plot (site) plan and in sections and details.

EXAMPLE 10-1   CONTINUOUS FOOTINGS

The objective of this example is to determine the quantity of concrete in the 3'2"-wide continuous footings. From drawings S2.1 and S8.1 of the small commercial building found in Appendix A and excerpted in Figure 10.1, it can be discerned that there are two different sizes of continuous footings. The continuous footings on the perimeter of the building are 3'2" wide, and the ones found on the interior of the building are 3'0" wide. The following steps should be taken:

1. Determine the linear feet of footing for each width.

2. Determine the cross-sectional area for each of the differing sizes.

3. Determine the volume and convert into cubic foot.

Determining the linear feet of continuous footing typically requires some minor calculations. The dimensions listed on the drawings typically reference the exterior face of the building or the centerline of the structural framing. Neither of these dimensions is appropriate for finding the linear feet of footing. In addition, the overlap that would occur in the corners by taking the measurement along the exterior face of the footing needs to be compensated. Therefore, the best approach is to determine the dimensions along the exterior edge and dimension so that there is no overlap. Figure 10.1 shows the exterior dimensions and footing dimensions. The overlap in the footings for walls A and B and walls B and C is because the footing in wall B is below the footings in walls A and C. Figure 10.2 is a sketch derived from drawing S8.1 detail 13 (Appendix A).

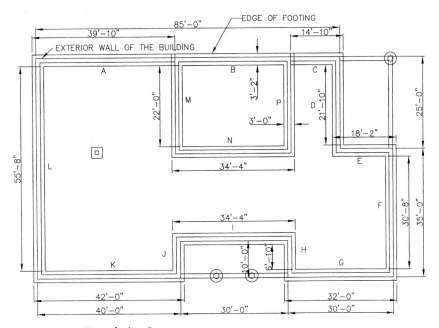

FIGURE 10.1. Foundation Layout.

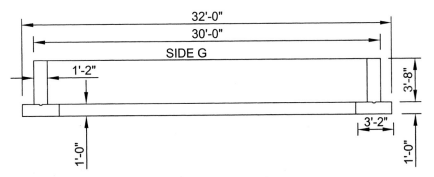

FIGURE 10.2. Footing and Foundation Wall Detail.

This sketch shows how the footing dimensions for side G were determined. Figure 10.3 is a tabulation of all dimensions for the 3'2"-wide footings. One of the advantages of takeoff software is that the estimator can measure the length of footings directly off the plans, marking the footing and keep a running total as she goes.

$$\text{Cross-sectional area (sf)} = \text{Width (ft)} \times \text{Height (ft)}$$

$$\text{Cross-sectional area (sf)} = 3'2'' \times 1'0''$$

$$\text{Cross-sectional area (sf)} = 3.167' \times 1.0' = 3.167 \text{ sf}$$

$$\text{Volume of concrete (cf)} = \text{Cross-sectional area (sf)} \times \text{Length (ft)}$$

$$\text{Volume of concrete (cf)} = 3.167 \text{ sf} \times 337.3' = 1,068 \text{ cf}$$

$$\text{Volume of concrete (cy)} = \text{cubic feet}/27$$

$$\text{Volume of concrete (cy)} = 1,068/27 = 39.6 \text{ cy}$$

Add 5 percent for waste and round off

$$\text{Volume of concrete (cy)} = 39.6 \text{ cy} \times 1.05 = 42 \text{ cy}$$

Side	Length
A	39'-10"
B	34'-4"
C	14'-10"
D	21'-10"
E	18'-2"
F	30'-8"
G	32'-0"
H	6'-10"
I	34'-4"
J	6'-10"
K	42'-0"
L	55'-8"
TOTAL	337'-4"

FIGURE 10.3. Linear Feet of 3'2"-Wide Continuous Footing. ■

EXAMPLE 10-2   SPREAD FOOTING

Drawing S8.1 detail 16 in Appendix A (Figure 10.4) of the small commercial building details the one spread footing.

The concrete contained in this footing is found in virtually the same fashion, as was the continuous footing.

$$\text{Volume of concrete (cf)} = \text{Length (ft)} \times \text{Width (ft)} \times \text{Height (ft)}$$

$$\text{Volume of concrete (cf)} = 3'0'' \times 3'0'' \times 1'0'' = 9 \text{ cf}$$

$$\text{Volume of concrete (cy)} = 9 \text{ cf}/27 \text{ cf per cy} = 0.33 \text{ cy}$$

Add 5 percent for waste and round off

$$\text{Volume of concrete (cy)} = 0.33 \text{ cy} \times 1.05 = 0.35 \text{ cy}$$

Use 1 cy

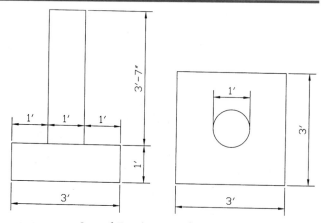

FIGURE 10.4. Spread Footing Detail. ■

Drilled Piers.   When dealing with drilled piers, some information about the soil is needed. If the soils are loose or the water table is high, it may be necessary to case the piers. The casing prevents the sides from caving in and water from seeping in. The casing, if steel, is typically removed as the concrete is placed; however, heavy-duty cardboard tubes can be used as casing. If this type of material is used, it is left in place and will eventually deteriorate. In clay soils, which are cohesive and impermeable, casings are rarely required.

To quantify a drilled pier, as shown in Figure 10.5, the shaft diameter, bell diameter, and angle of the bell must be known. The volume of the drilled piers can be calculated, or tables similar to the ones found in Figure 10.6 could be used to find the volume. If the volume is to be manually calculated, the bell diameter and angle can be used to find the height of the bell by using Formula 10-1 or 10-2. Then the volume of the bell is found using Formula 10-3 or 10-4, which is added to the volume of the shaft (Formula 10-5) to determine the volume for a specific pier.

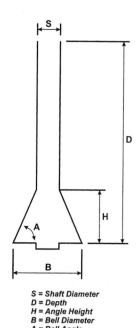

S = Shaft Diameter
D = Depth
H = Angle Height
B = Bell Diameter
A = Bell Angle

FIGURE 10.5. Cross-Section of Drilled Pier.

VOLUME OF BELLS IN DRILLED PIERS										
BELL ANGLE 50										
SHAFT DIAMETER (Inches)		BELL DIAMETER (Inches)									
		18	24	30	36	42	48	54	60	66	72
12	h (feet)	0.30	0.60	0.89	1.19	1.49	1.79	2.09	2.38	2.68	2.98
	C.F.	0.37	1.09	2.28	4.06	6.53	9.83	14.06	19.34	25.80	33.54
18	h (feet)		0.30	0.60	0.89	1.19	1.49	1.79	2.09	2.38	2.68
	C.F.		0.72	1.91	3.69	6.16	9.46	13.69	18.97	25.43	33.17
24	h (feet)			0.30	0.60	0.89	1.19	1.49	1.79	2.09	2.38
	C.F.			1.19	2.96	5.44	8.74	12.97	18.25	24.71	32.45
30	h (feet)				0.30	0.60	0.89	1.19	1.49	1.79	2.09
	C.F.				1.77	4.25	7.55	11.78	17.06	23.52	31.26
36	h (feet)					0.30	0.60	0.89	1.19	1.49	1.79
	C.F.					2.48	5.77	10.00	15.29	21.74	29.48
42	h (feet)						0.30	0.60	0.89	1.19	1.49
	C.F.						3.30	7.53	12.81	19.27	27.01

VOLUME OF BELLS IN DRILLED PIERS										
BELL ANGLE 60										
SHAFT DIAMETER (Inches)		BELL DIAMETER (Inches)									
		18	24	30	36	42	48	54	60	66	72
12	h (feet)	0.43	0.87	1.30	1.73	2.17	2.60	3.03	3.46	3.90	4.33
	C.F.	0.54	1.59	3.32	5.89	9.49	14.28	20.43	28.11	37.49	48.75
18	h (feet)		0.43	0.87	1.30	1.73	2.17	2.60	3.03	3.46	3.90
	C.F.		1.05	2.78	5.36	8.96	13.75	19.90	27.58	36.96	48.21
24	h (feet)			0.43	0.87	1.30	1.73	2.17	2.60	3.03	3.46
	C.F.			1.73	4.31	7.91	12.70	18.85	26.53	35.91	47.16
30	h (feet)				0.43	0.87	1.30	1.73	2.17	2.60	3.03
	C.F.				2.58	6.18	10.97	17.12	24.80	34.18	45.43
36	h (feet)					0.43	0.87	1.30	1.73	2.17	2.60
	C.F.					3.60	8.39	14.54	22.22	31.60	42.85
42	h (feet)						0.43	0.87	1.30	1.73	2.17
	C.F.						4.79	10.94	18.62	28.00	39.25

FIGURE 10.6. Volume in Bells 50-Degree Angle (top) and 60-Degree Angle (bottom).

EXAMPLE 10–3   DRILLED PIER

From the small commercial building example found in Appendix A, there are three identical drilled piers with the following dimensions:

Shaft diameter = 18 inches
Bell diameter = 42 inches
Angle of bell = 60 degrees

$$H = \frac{\sin(A) \times \dfrac{B-5}{2}}{\sin(90-A)} \qquad \text{Formula 10-1}$$

Height of Bell on Reamed Footing

$$H = \tan(A) \times \left(\frac{B-5}{2}\right) \qquad \text{Formula 10-2}$$

Height of Bell on Reamed Footing

From the given dimensions the bell height is

$$H = \frac{\sin(60°) \times \left(\dfrac{3.5' - 1.5'}{2}\right)}{\sin(90° - 60°)}$$

$$H = 1.732''$$

or

$$H = \tan(60°) \times \left(\frac{3.5' - 1.5'}{2}\right)$$

$$H = 1.732'$$

Once the bell height (H) is found, Formula 10-3 or 10-4 can be used to determine the volume of the belled portion of the footing.

$$\text{Volume of bell} = \left(\frac{\pi S^2 \times H}{4}\right) + \left(2\pi \left(\left(\frac{2S + B}{6}\right)\right.\right.$$
$$\left.\left. \times \left(\frac{(B - S) \times H}{4}\right)\right)\right) \qquad \text{Formula 10-3}$$

Volume of Bell

$$\text{Volume of bell} = \frac{\pi H}{12}(S^2 + B^2 + SB) \qquad \text{Formula 10-4}$$

Volume of Bell

The bell volume is as follows:

$$\text{Volume of bell} = \left(\frac{\pi 1.5'^2 \times 1.732'}{4}\right) + \left(2\pi \left(\left(\frac{(2 \times 1.5') + 3.5'}{6}\right)\right.\right.$$
$$\left.\left. \times \left(\frac{3.5' - 1.5') \times 1.732'}{4}\right)\right)\right)$$

Volume of bell = 8.96 cf

or

$$\text{Volume of bell} = \frac{\pi H}{12}(S^2 + B^2 + SB)$$

$$\text{Volume of bell} = \frac{\pi \times 1.732'}{12}(1.5'^2 + 3.5'^2 + 1.5' \times 3.5')$$

Volume of bell = 8.96 cf

		ESTIMATE WORK SHEET												

Project: *Little Office Building*
Location: *Littleville, Tx*
Architect: *U.R. Architects*
Items: *Foundation Concrete*

FOOTING CONCRETE

Estimate No. 1234
Sheet No. 1 of 1
Date 11/11/20xx
By LHF Checked JBC

Cost Code	Description	Length Ft.	Length In.	Witdth Ft.	Witdth In.	Height Ft.	Height In.			Length Ft.	Width Ft.	Height Ft.	Volume C.F.	Quantity	Unit
	3-2" Wide Footing (2,800 psi)														
	Side A	39	10	3	2	1	0			39.83	3.17	1.00	126.14		
	Side B	34	4	3	2	1	0			34.33	3.17	1.00	108.72		
	Side C	14	10	3	2	1	0			14.83	3.17	1.00	46.97		
	Side D	21	10	3	2	1	0			21.83	3.17	1.00	69.14		
	Side E	18	2	3	2	1	0			18.17	3.17	1.00	57.53		
	Side F	30	8	3	2	1	0			30.67	3.17	1.00	97.11		
	Side G	32	0	3	2	1	0			32.00	3.17	1.00	101.33		
	Side H	6	10	3	2	1	0			6.83	3.17	1.00	21.64		
	Side I	34	4	3	2	1	0			34.33	3.17	1.00	108.72		
	Side J	6	10	3	2	1	0			6.83	3.17	1.00	21.64		
	Side K	42	0	3	2	1	0			42.00	3.17	1.00	133.00		
	Side L	55	8	3	2	1	0			55.67	3.17	1.00	176.28		
										337.33			1,068.22		
	Add 5 % for Waste												1,121.63		
	Total Concrete In 3'-2" Wide Footings													41.5	C.Y.
	3'-0" Wide Footing (2,800 psi)														
	Side M	22	0	3	0	1	0			22.00	3.00	1.00	66.00		
	Side N	34	4	3	0	1	0			34.33	3.00	1.00	103.00		
	Side O	22	0	3	0	1	0			22.00	3.00	1.00	66.00		
										78.33			235.00		
	Add 5 % for Waste												246.75		
	Total Concrete In 3'-2" Wide Footings													9.1	C.Y.
	Spread Footing								Quantity						
	3' Square Column Footings	3	0	3	0	1	0		1	3.00	3.00	1.00	9.00		
	Add 5 % for Waste												9.45		
	Total Concrete In 3' Square Sp. Ftg													1.0	C.Y.
	Use 1 C.Y.														

FIGURE 10.7. Concrete Footing Takeoff.

The remaining element of the drilled piers is to determine the volume of the shaft, which is done by subtracting the height of the bell from the pier depth. However, the shaft typically extends 6 inches through the bell requiring an extra 6 inches to be added to the shaft length.

$$\text{Length of the shaft} = 16.0' - 1.732' + 0.5'$$

$$\text{Length of the shaft} = 14.8'$$

This length of the shaft is then multiplied by its cross-sectional area to determine its volume.

$$\text{Shaft volume} = \left(\frac{\pi S^2}{4}\right) \times (\text{Length of shaft}) \qquad \textbf{Formula 10-5}$$

Shaft Volume

$$\text{Shaft volume} = \left(\frac{\pi 1.5'^2}{4}\right) \times (14.8')$$

$$\text{Shaft volume} = 26.15 \text{ cf}$$

The volume of the shaft and pier can then be added together and multiplied times the number of piers to determine the total volume.

$$\text{Volume of concrete in piers} = \text{Count} \times (\text{Volume in bell} + \text{Volume in shaft})$$

$$\text{Volume of concrete in piers (cf)} = 3 \times (8.96 \text{ cf} + 26.15 \text{ cf})$$

$$\text{Volume of concrete (cy)} = 105 \text{ cf}/27 \text{ cf per cy} = 3.89 \text{ cy}$$

$$\text{Add 10 percent for waste}$$

$$\text{Volume of concrete (cy)} = 3.89 \text{ cy} \times 1.10 = 4.3 \text{ cy}$$

$$\text{Use 5 cy}$$

The bell volume can be verified using the 60-degree table in Figure 10.6. From this table, the bell height is 1.73 feet and the volume is 8.96 cf per bell, which match the previous calculations. Figure 10.7 is the workup sheet for the spread and continuous footings for the building in Appendix A. ∎

EXAMPLE 10-4   FORMED PIERS

Spread footings typically have projecting formed piers that support the building structure. The small commercial building has one formed pier. This pier, as shown in Figure 10.4, is 1 foot in diameter and 3 feet, 7 inches tall (sheet S8.1 details 1, 7, and 16 in Appendix A). The volume of the formed pier is found by multiplying the cross-sectional area by its height.

$$\text{Volume in formed piers} = \text{Count} \times \text{Cross−sectional area} \times \text{Height}$$

$$\text{Cross-sectional area} = \pi r^2 = \pi \times 0.5' \times 0.5' = 0.786 \text{ sf}$$

$$\text{Volume in formed piers} = 0.786 \text{ sf} \times 3'7'' = 2.8 \text{ cf}$$

$$\text{Use 1 cy}$$ ∎

EXAMPLE 10-5   FOUNDATION WALLS

The concrete for the foundation walls is done in substantially the same manner as the spread footings. In the small commercial building example, the building perimeter foundation walls are 1'2" thick and the interior walls are 1'0" thick. Figures 10.8 and the table in Figure 10.9 detail the linear feet of 1'2"-thick foundation walls. In addition, side B is 8'4" tall as compared with 3'8" high for the remaining 1'2"-thick walls. This is why, side B is not included in Figure 10.9.

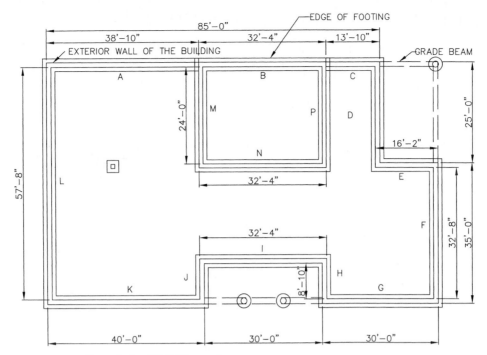

FIGURE 10.8. Foundation Wall Dimensions.

Side	Foundation Wall Length
A	38'-10"
C	13'-10"
D	23'-10"
E	16'-2"
F	32'-8"
G	30'-0"
H	8'-10"
I	32'-4"
J	8'-10"
K	40'-0"
L	57'-8"
TOTAL	303'-0"

FIGURE 10.9. Exterior Foundation Walls 3'8" Tall.

Foundation wall concrete (cf)
$$= \text{Linear feet (ft)} \times \text{Height (ft)} \times \text{Thickness (ft)}$$
3'8"-high wall

Quantity of foundation wall concrete (cf) $= 303' \times 3'8'' \times 1'2''$

Quantity of foundation wall concrete (cf) $= 303' \times 3.67' \times 1.17'$

Quantity of foundation wall concrete (cf) $= 1,301$ cf

8'4"-high wall

Quantity of concrete in foundation wall (cf)
$$= 32'4'' \times 8'4'' \times 1'2''$$

Quantity of concrete in tall foundation wall (cf)
$$= 32.33' \times 8.33' \times 1.17'$$

Quantity of concrete in tall foundation wall (cf) $= 315$ cf

Quantity of concrete in exterior foundation walls
$$= (315 \text{ cf} + 1,301 \text{ cf})/27 \text{ cf per cy}$$

Quantity of concrete in exterior foundation walls $= 59.9$ cy

Add 5 percent for waste and round off

Quantity of concrete in exterior foundation walls
$$= 59.9 \text{ cy} \times 1.05$$

Quantity of concrete in exterior foundation walls $= 62.9$ cy

Use 63 cy ∎

EXAMPLE 10-6   GRADE BEAMS

Grade beams are located in the front and right rear corner of the building (refer to Figures 10.10 and 10.11). These grade beams are required to tie the drilled piers to the remainder of the building foundation. The volume of concrete in the grade beam is found by multiplying the cross-sectional area of the grade beams by their length. From Figure 10.10, it can be discerned that the grade beams have different cross-sectional areas.

38'10" of 1'2"-wide grade beams (cross-sectional area $= 1.75$ sf)

30'0" of 1'6"-wide grade beams (cross-sectional area $= 2.25$ sf)

Volume of concrete in 1'2"-wide grade beams
$$= 38.833' \times 1.75 \text{ sf} = 68 \text{ cf}$$

Volume of concrete in 1'6"-wide grade beams
$$= 30.0' \times 2.25 \text{ sf} = 68 \text{ cf}$$

Total concrete in grade beams (cf) $= 68 \text{ cf} + 68 \text{ cf} = 136 \text{ cf}$

Total concrete in grade beams (cy) $= 136 \text{ cf}/27 \text{ cf per cy}$

Total concrete in grade beams (cf) $= 5.04$ cy

Add 8 percent for waste and round off

Total concrete in grade beams (cf) $= 5.04 \text{ cy} \times 1.08 = 5.44 \text{ cy}$

Use 6 cy

Figure 10.12 is a workup sheet for the concrete in the foundation walls and grade beams.

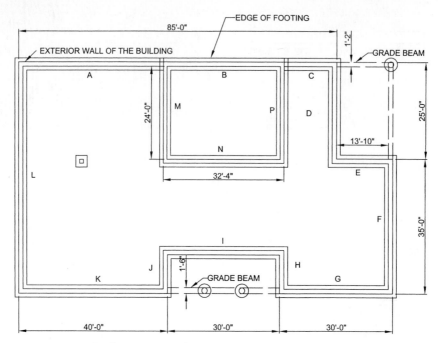

FIGURE 10.10. Grade Beam Location.

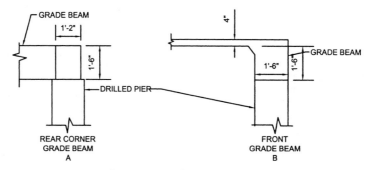

FIGURE 10.11. Grade Beam Details.

ESTIMATE WORK SHEET

Project:	Little Office Building									Estimate No.		1234
Location	Littleville, Tx									Sheet No.		1 of 1
Architect	U.R. Architects		Foundation Wall & Grade Beams							Date		11/11/20xx
Items	Foundation Concrete									By	LHF	Checked	JBC

Cost Code	Description	Dimensions								Length Ft.	Width Ft.	Height Ft.	Volume C.F.	Quantity	Unit
		Length		Width		Thickness									
		Ft.	In.	Ft.	In.	Ft.	In.								
	3'-8" Highe Foundation Walls														
	1'-2" Wide (2,800 psi)														
	Side A	38	10	1	2	3	8			38.83	1.17	3.67	166.12		
	Side C	13	10	1	2	3	8			13.83	1.17	3.67	59.18		
	Side D	23	10	1	2	3	8			23.83	1.17	3.67	101.95		
	Side E	16	2	1	2	3	8			16.17	1.17	3.67	69.16		
	Side F	32	8	1	2	3	8			32.67	1.17	3.67	139.74		
	Side G	30	0	1	2	3	8			30.00	1.17	3.67	128.33		
	Side H	8	10	1	2	3	8			8.83	1.17	3.67	37.79		
	Side I	32	4	1	2	3	8			32.33	1.17	3.67	138.31		
	Side J	8	10	1	2	3	8			8.83	1.17	3.67	37.79		
	Side K	40	0	1	2	3	8			40.00	1.17	3.67	171.11		
	Side L	57	8	1	2	3	8			57.67	1.17	3.67	246.69		
	8'-4" High Founfation Wall														
	1'-2" Wide (2,800 psi)														
	Side B	32	4	1	2	8	4			32.33	1.17	8.33	314.35		
	8'-4" High Foundation Wall														
	1'-0" Wide (2,800 psi)														
	Side M	24	0	1	0	8	4			24.00	1.00	8.33	200.00		
	Side N	32	4	1	0	8	4			32.33	1.00	8.33	269.44		
	Side P	24	0	1	0	8	4			24.00	1.00	8.33	200.00		
	Total Foundation Walls												2,279.96		
	Add 5% for Waste												88.67	89	C.Y.
	Grade Beams														
	1'-2" X 1'-6" Rear of Building	25	0	1	2	1	6			25.00	1.17	1.50	43.75		
	1'-2" X 1'-6" Rear of Building	13	10	1	2	1	6			13.83	1.17	1.50	24.21		
	1'-6" X 1'-6" Front of Building	30	0	1	6	1	6			30.00	1.50	1.50	67.50		
	Total Grade Beams												135.46		
	Add 8% fo Waste												5.42	6	C.Y.
	1 Only Formed Pier	1	0	1	0	4	0			1.00	1.00	4.00	4.00		
	Add 5 % for Waste												0.16	1	C.Y.

FIGURE 10.12. Foundation Wall and Grade Beam Takeoff.

EXAMPLE 10-7   CONCRETE SLABS

The volume of a reinforced slab is found by taking the square footage and by multiplying it by the depth of the slab. From the drawings in Appendix A and Figures 10.13, 10.14, and 10.15, there are four unique types of slabs. First, there is the 745 sf (25'3" × 31'4" less 46 sf for stair opening) of 2-inch-thick topping.

$$\text{Quantity of concrete (cf)} = 745 \text{ sf} \times 0.17' = 127 \text{ cf}$$

$$\text{Quantity of concrete (cy)} = 127 \text{ cf}/27 \text{ cf per cy} = 4.7 \text{ cy}$$

$$\text{Add 5\% waste} - \text{Use 5 cy}$$

The second slab is the 4-inch-thick slab that is in the front of the building. This slab is on grade and is 300 sf (30 feet × 10 feet) and 4 inches thick. Because the edges of this slab are thickened and it is on grade, a larger waste factor will be used to compensate for this situation.

$$\text{Quantity of concrete (cf)} = 300 \text{ sf} \times 4''$$

$$\text{Quantity of concrete (cf)} = 300 \text{ sf} \times 0.333' = 100 \text{ cf}$$

$$\text{Quantity of concrete (cy)} = 100 \text{ cf}/27 \text{ cf per cy} = 3.7 \text{ cy}$$

$$\text{Add 8 percent waste} - \text{Use 4 cy}$$

Third is the slab on grade for the remaining portion of the building that is at grade. This area is found by determining the area for the whole slab and by subtracting the 2-inch topping and 4-inch slab by the front entrance, as shown in Figure 10.13.

$$\text{Quantity of concrete (cf)} = 4,427 \text{ sf} \times 0.5' = 2,214 \text{ cf}$$

$$\text{Quantity of concrete (cy)} = 2,214 \text{ cf}/27 \text{ cf per cy} = 82 \text{ cy}$$

$$\text{Add 8 percent waste} - \text{Use 89 cy}$$

The other remaining slab is the floor for the portion of the building that is below grade. That takeoff is similar to the three previous examples and can be found in Figure 10.16 with all the slab takeoffs.

Description	Dimensions	Square Feet
Whole Building Area	99'-4" X 59'-4"	5,893
Covered Patio	25' X 15'	-375
2" Topping		-791
4" Slab On Grade (Front)		-300
Total		4,427

FIGURE 10.13. Slab Areas.

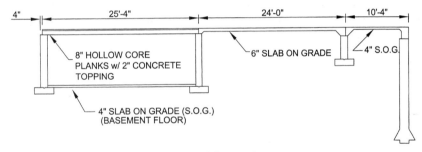

FIGURE 10.14. Foundation Wall and Slab Detail.

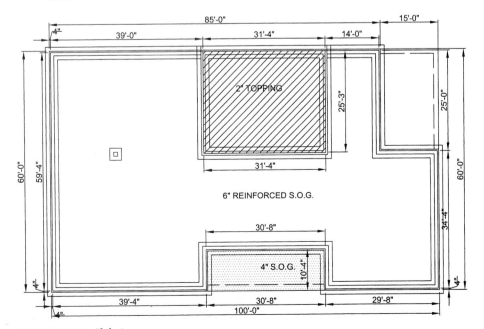

FIGURE 10.15. Slab Area.

ESTIMATE WORK SHEET

Project:	*Little Office Building*										Estimate No.		1234	
Location	*Littleville, Tx*					Slabs					Sheet No.		*1 of 1*	
Architect	*U.R. Architects*										Date		*11/11/20xx*	
Items	*Foundation Concrete*										By	*LHF*	Checked	*JBC*

Cost Code	Description	Dimensions								Length Ft.	Width Ft.	Height Ft.	Volume C.F.	Quantity	Unit
		Length		Witdth		Thickness									
		Ft.	In.	Ft.	In.	Ft.	In.								
	All Concrete is 2,800 psi														
	2" Thick Topping Over (2,800 psi)														
	Pre-Cast Hollow Core Planks	25	3	31	4	0	2			25.25	31.33	0.17	131.86		
	Add 5 % for Waste												138.45	5	C.Y.
	4" Thick Slab On Grade By The														
	Front Enterance	30	0	10	0	0	4			30.00	10.00	0.33	100.00		
	Add 8 % For Waste & Thickening												108.00	4	C.Y.
	6" Thick Slab On Grade (Whole Area)	99	4	59	4	0	6			99.33	59.33	0.50	2,946.89		
	Less Front Entrance	-30	0	10	0	0	6			-30.00	10.00	0.50	-150.00		
	Less 2" Topping	-25	3	31	4	0	6			-24.75	31.33	0.50	-387.75		
	Less Covered Patio	-25	0	15	0	0	6			-25.00	15.00	0.50	-187.50		
	Total Volume												2,221.64		
	Add 8% for Waste & Thickening												2,399.37	89	C.Y.
	4" Slab on Grade														
	(below Finish Floor)	22	0	28	4	0	4			22.00	28.33	0.33	207.78		
	Add 8% for Waste & Thickening												224.40	9	C.Y.

FIGURE 10.16. Slab Takeoff.

Type of Placement	Productivity Rate (Labor Hours/c.y)
Continuous Footing—Direct Chute	0.4
Spread Footing—Direct Chute	0.873
Drilled Piers	0.320
Formed Piers	0.873
Foundation Walls—Direct Chute	0.5
Grade Beams	0.4
Slab—Direct Chute	0.32

FIGURE 10.17. Labor Productivity Rates—Placing Concrete.

Labor Costs. The labor costs are found by multiplying the quantity takeoff by the appropriate productivity rate. Figure 10.17 shows various productivity rates for placing concrete in various situations.

EXAMPLE 10-8   LABOR COSTS

Find the labor cost for placing concrete in the 3'2"-wide continuous footings. From the concrete takeoff in Figure 10.7, there are 42 cy of concrete. Using that quantity and the labor productivity information from Figure 10.17, the following calculations can be performed.

Labor hours = Quantity (cy)
    × Productivity rate (labor hours/cy)

Labor hours = 42 cy × 0.4 labor hours/cy

Labor hours = 17 hours

Labor costs ($) = Hours × Wage rate

Labor costs ($) = 17 hours × $13.50/hour

Labor costs ($) = $229.50

Figure 10.18 is the priced-out estimate for all the concrete in the foundation of the small commercial building.

10-3   REINFORCING

The reinforcing used in concrete may be reinforcing bars, welded wire mesh (WWF), or a combination of the two. Reinforcing bars are listed (noted) by the bar number, which corresponds to the bar diameter in eighths of an inch. For example, a No. 7 bar (deformed) has a 7/8-inch nominal diameter. The No. 2 bar is a plain round bar, but all the rest are deformed round bars. The bar numbers, diameters, areas, and weights are given in Figure 10.19.

Reinforcing bars are taken off by linear feet. The takeoff (workup) sheet should be set up to include the number of the bars, pieces, lengths, and bends. Because reinforcing bars are usually priced by the hundredweight (100 pounds, cwt), the weight of reinforcing required must be calculated. The steel can be bought from the mill or main warehouses, and the required bars will be cut, bundled, and tied. Bars will also be bent to job requirements at these central points. Bars purchased at smaller local warehouses are generally bought in 20-foot lengths and cut and bent in the field. This process is usually more expensive and involves more waste. When time permits, the reinforcing bars should be ordered from the mill or main warehouse and be shop fabricated. Often the fabricators will provide the required shop drawings. Check the specifications to determine the type of steel required and whether it is plain, coated with zinc, painted with epoxy paint (typically green), or galvanized. Zinc coating and galvanizing can increase the material cost by as much as 150 percent and often delays delivery by many weeks. Rebar painted with epoxy paint increases the cost by about 20 percent and has the advantage that damage to the paint can be touched up in the field.

Allowance for splicing (lapping) the bars (Figure 10.20) must also be included (lap splicing costs may range from 5 to 15 percent, depending on the size of the bar and yield strength of steel used). Waste may range from less than 1 percent for precut and preformed bars to 10 percent when the bars are cut and bent on the job site.

ESTIMATE SUMMARY SHEET

Project: *Little Office Building*
Location: *Littleville, Tx*
Architect: *U.R. Architects, Inc.*
Items: *Foundation Concrete*

Estimate No. ____ 1234
Sheet No. ____ 1 of 1
Date ____ 11/11/20xx
By ____ LHF ____ Checked ____ JBC

Cost Code	Description	Q.T.O.	Waste Factor	Purch. Quan.	Unit	Crew	Prod. Rate	Wage Rate	Labor Hours	Unit Cost Labor	Unit Cost Material	Unit Cost Equipment	Labor	Material	Equipment	Total
	Continuous Footings 2800 psi															$0
	3'-2" Wide	40	5.00%	42	C.Y.		0.4	$13.50	15.8	$5.40	$52.50		$224	$2,181	$0	$2,405
	3'-0" Wide	9	5.00%	9	C.Y.		0.4	$13.50	3.5	$5.40	$52.50		$49	$480	$0	$529
	Spread Footing 2800 psi	1	0.00%	1	C.Y.		0.4	$13.50	0.4	$5.40	$52.50		$5	$53	$0	$58
	Drilled Piers - 2800 psi	4	10.00%	4	C.Y.		0.32	$13.50	1.2	$4.32	$52.50		$18	$225	$0	$243
	Foundation Walls - 2800 psi	84	5.00%	89	C.Y.		0.5	$13.50	42.2	$6.75	$52.50		$598	$4,655	$0	$5,253
	Grade Beams - 2800 psi	5	8.00%	5	C.Y.		0.4	$13.50	2.0	$5.40	$52.50		$29	$284	$0	$314
	Formed Pier - 2800 psi	1	0.00%	1	C.Y.		0.873	$13.50	0.9	$11.79	$52.50		$12	$53	$0	$64
	2" Topping - 2800 psi	5	5.00%	5	C.Y.		0.32	$13.50	1.6	$4.32	$52.50		$23	$283	$0	$306
	Slabs on Grade - 2800 psi															$0
	4" Thick at Front Entrance	4	8.00%	4	C.Y.		0.32	$13.50	1.3	$4.32	$52.50		$19	$227	$0	$245
	6" Slab (main level)	82	8.00%	89	C.Y.		0.32	$13.50	26.3	$4.32	$52.50		$384	$4,665	$0	$5,049
	4" Basement floor	8	8.00%	8	C.Y.		0.32	$13.50	2.5	$4.32	$52.50		$36	$436	$0	$472
																$0
																$0
																$0
																$0
																$0
																$0
																$0
																$0
																$0
																$0
																$0
																$0
																$0
																$0
																$0
										97.77				$1,399	$13,541	$0	$14,940

FIGURE 10.18. Cast-In-Place Concrete Estimate Summary.

Bar Designation	Nominal Diameter	Cross-Sectional Area (sq. in.)	Weight Pounds / Foot
2	¼"	.015	.167
3	3/8"	.11	.376
4	½"	.20	.668
5	5/8"	.31	1.043
6	¾"	.44	1.502
7	7/8"	.60	2.044
8	1"	.79	2.670
9	1.13"	1.00	3.400
10	1.27"	1.27	4.303
11	1.41"	1.56	5.313

FIGURE 10.19. Weights and Areas of Reinforcing Bars.

	Splice required when specified as a number of bar diameters	
Bar Size	24 d	30 d
3	1'-0"	1'-0"
4	1'-0"	1'-3"
5	1'-3"	1'-7"
6	1'-6"	1'-11"
7	1'-9"	2'-3"
8	2'-0"	2'-6"
9	2'-4"	2'-10"
10	2'-7"	3'-3"
11	2'-10"	3'-7"

FIGURE 10.20. Splice Requirements (Minimum splice is 1'0").

Mesh reinforcing may be a welded wire mesh or expanded metal. Welded wire mesh is an economical reinforcing for floor and driveways and is commonly used as temperature reinforcing and beam and column wrapping. It is usually furnished in flat sheets or by the roll. The rolls are 5 feet wide and 150 feet long. Wire mesh is designated on the drawings by wire spacing and wire gauge in the following manner: 6 × 6¹⁰/₁₀. This designation shows that the longitudinal and transverse wires are spaced 6 inches on center, while both wires are No. 10 gauge. Another example, 4 × 8⁸/₁₂, means that the longitudinal wires are spaced 4 inches on center, and the transverse wires are spaced 8 inches on center; the longitudinal wire is No. 8 gauge while the transverse is No. 12 gauge. The takeoff must be broken up into the various sizes required and the number of square feet required of each type. It is commonly specified that wire mesh have a lap of one square, and this allowance must be included: For example, a 6-inch lap requires 10 percent extra mesh, whereas a 4-inch lap requires almost 6.7 percent extra. The mesh may be either plain or galvanized; this information is included in the specifications. Galvanized mesh may require special ordering and delivery times of two to three weeks.

The 1/4-inch rib lath is designed primarily as reinforcement for concrete floor and roof construction. The 1/4-inch ribs are spaced 6 inches on center, and this reinforcement is available only in coated copper alloy steel, 0.60 and 0.75 pounds per square foot. The width available is 24 inches with lengths of 96, 120, and 144 inches, packed six sheets to a bundle (96, 120, and 144 sf, respectively). Allow for any required lapping and 3 percent for waste (for rectangular spaces).

The reinforcing steel must be elevated into the concrete to some specified distance. This can be accomplished by using concrete bricks, bar chairs, spacers, or bolsters, or it may be suspended with wires. The supports may be plastic, galvanized or zinc-coated steel, steel with plastic-coated legs, and other materials. If the finished concrete will be exposed to view and the supports are touching the portion to be exposed, consideration should be given to using noncorrosive supports. Steel, even zinc or galvanized steel, has a tendency to rust, and this rust may show through on the exposed finish. A wide selection of accessories and supports are available.

When the reinforcing must be fabricated into a special shape—perhaps round, spiral, or rectangular—it is usually cheaper to have this done at a fabricating shop, which has the equipment for bending these shapes. Thus, there would be a charge in addition to the base cost of the reinforcing, but the process provides speed and economy in most cases.

Corrugated steel subfloor systems are also used for reinforcing concrete. When corrugated steel floor deck material is used as reinforcing for the concrete, it also acts as a form for the concrete that is to be poured on top of it. The system may simply be corrugated deck with concrete or may be as elaborate as supplying in-floor distribution of electricity, hot air, and telephone requirements. The more elaborate the system, the more coordination that is required between the trades.

Steel deck subfloors are taken off by the square footage required (also discussed in Chapter 11). Available in a variety of heights and widths, the type used will depend on the span and loading requirements of the job. Finishes include galvanized, galvanized with primer on the underside, and phosphate treated on upper surfaces with primer on the underside.

Wire mesh is sometimes specified for use as temperature steel, in conjunction with the steel decking. The estimator must include it in the takeoff when it is required.

Estimating Reinforcing Bars. The linear footage of rebars can most often be worked up from the concrete calculations. The sections and details must be checked to determine the reinforcing requirements of the various footings. The various footing sizes can generally be taken from the concrete calculations and adapted to the reinforcing takeoff.

EXAMPLE 10-9   FOUNDATION REINFORCING

For this example, determine the quantity of reinforcing steel required for the side A continuous footing of the building found in Appendix A. From Figure 10.3, which was used to quantify the concrete, the dimensions for all of the sides can be found. From that table, side A is 39'10". This footing has reinforcing bars that run both the long and short dimensions. Figure 10.21 is an excerpt of the details found on sheet S8.1 of the project found in Appendix A.

Continuous Footing Short Bars

Number of short bar spaces = 39'10"/11"spacing

Number of short bars = 39.833'/0.9167 = 44 spaces

Add 1 to get the number of bars − Use 45 bars

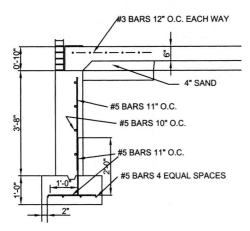

FIGURE 10.21. Footing Detail.

The bar length is 2′10″, which is derived by subtracting the coverage distance (2″ × 2) from the footing width (3′2″).

$$\text{Total bar length (lf)} = \text{Number of bars} \\ \times \text{Length of individual bars}$$

$$\text{Total bar length (lf)} = 45 \text{ bars} \times 2′10″ \text{ per bar}$$

$$\text{Total bar length (lf)} = 45 \text{ bars} \times 2.833′ \text{ per bar} = 127.5 \text{ lf}$$

$$\text{No. 5 bar weighs } 1.043 \text{ pounds per lf}$$

$$\text{Total weight (pounds)} = 127.5 \text{ lf} \times 1.043 \text{ pounds per lf} \\ = 133 \text{ pounds}$$

Continuous Footing Long Bars

$$\text{Length of long bars} = 39′6″$$

$$\text{Total bar length (lf)} = 39′6″ \times 4 \text{ bars}$$

$$\text{Total bar length (lf)} = 39.5′ \times 4 \text{ bars} = 158 \text{ lf}$$

$$\text{Total weight (pounds)} = 158 \text{ lf} \times 1.043 \text{ pounds per lf} \\ = 165 \text{ pounds}$$

$$\text{Total actual weight} = 165 \text{ pounds} + 133 \text{ pounds} = 298 \text{ pounds}$$

$$\text{Add 10 percent for waste and lap} - \text{Use 328 pounds}$$

The reinforcing for the foundation wall is found in virtually the same fashion. From Figure 10.8, side A is 38′10″. Figure 10.21 specifies the vertical bar spacing at 11 inches on center (o.c.). Therefore, the quantity of vertical bars would be found in the following fashion:

Foundation Wall Vertical Bars (Side A)

$$\text{Number of vertical bar spaces} = 38′10″/11″$$

$$\text{Number of vertical bar spaces} = 38.833′/0.9167′ = 42 \text{ spaces}$$

$$\text{Add 1 to get the number of bars} - \text{Use 43 bars}$$

$$\text{Bar length} = \text{Wall height} - \text{Bar coverage}$$

$$\text{Bar length} = 3′8″ - 4″ = 3′4″ \text{ bar length}$$

$$\text{Total bar length (lf)} = \text{Bar count} \times \text{Length of individual bars (ft)}$$

$$\text{Total bar length (lf)} = 43 \text{ bars} \times 3′4″$$

$$\text{Total bar length (lf)} = 43 \text{ bars} \times 3.33′ = 143 \text{ lf}$$

$$\text{Total weight (pounds)} = 143 \text{ lf} \\ \times 1.043 \text{ pounds per lf} = 149 \text{ pounds}$$

$$\text{Add 10 percent for waste and lap} - \text{Use 164 pounds}$$

$$\text{Number of horizontal spaces} = 3′8″/10 \text{ inches}$$

$$\text{Number of horizontal spaces} = 3.667′/0.833′ \\ = 4.4 \text{ spaces} - \text{use 5}$$

$$\text{Number of bars} = \text{Spaces} + 1 - \text{Use 6 bars}$$

$$\text{Bar length} = 38′10″ - 4″ = 38′6″$$

$$\text{Total bar length (lf)} = 38′6″ \times 6$$

$$\text{Total bar length (lf)} = 38.5′ \times 6 = 231 \text{ lf}$$

$$\text{Total weight (pounds)} = 231 \text{ lf} \times 1.043 \text{ pounds per lf} \\ = 241 \text{ pounds}$$

$$\text{Add 10 percent for waste and lap} - \text{Use 265 pounds}$$

Dowels

$$\text{lf of each dowel} = 3′$$

$$\text{Number of dowels} = 43$$

$$\text{Total length of dowels} = 43′ \times 3′ = 129 \text{ feet}$$

$$\text{Total weight (pounds)} = 129 \text{ feet} \times 1.043 \text{ pounds per foot} \\ = 135 \text{ pounds}$$

$$\text{Add 5 percent for lap and waste} - \text{Use 142 pounds} \quad \blacksquare$$

EXAMPLE 10-10 SLAB REINFORCING

If the reinforcing in a slab is done with sized deformed bars, the bars are quantified in the exact manner as the footings and foundation walls. Once again, the quantity of long and short bars needs to be determined. Figure 10.22 is an example of how the slab can be divided into unique areas so that the quantity of reinforcing bars can be determined. Using area A as an example, the long bars will be (39′0″ − 2″) or 38′10″ and the short bars would be (25′4″ − 2″) or 25′2″.

$$\text{Long bar spaces} = 25′4″/12″$$

$$\text{Long bar spaces} = 25.33′/1′ - \text{Use 26 spaces}$$

$$\text{Add 1 to convert to bar count} - \text{Use 27 bars}$$

$$\text{Short bar spaces} = 39′0″/12″$$

$$\text{Short bar spaces} = 39′/1′ = 39 \text{ spaces}$$

$$\text{Add 1 to convert to bar count} - \text{Use 40 bars}$$

$$\text{Total length of long bars (lf)} = 38′10″ \times 27 \text{ bars}$$

$$\text{Total length of long bars (lf)} = 38.833′ \times 27 \text{ bars} = 1,049 \text{ lf}$$

$$\text{Total length of short bars (lf)} = 25′2″ \times 40 \text{ bars}$$

$$\text{Total length of short bars} = 25.167 \times 40 \text{ bars} = 1,007 \text{ lf}$$

$$\text{Total length of bars (lf)} = 1,049 \text{ lf} + 1,007 \text{ lf} = 2,056 \text{ lf}$$

$$\text{Total weight of bars (pounds)} = 2,056 \text{ lf} \times 0.376 \text{ pounds per foot}$$

$$\text{Total weight} = 773 \text{ pounds}$$

$$\text{Add 10 percent for lap and waste} - \text{Use 850 pounds}$$

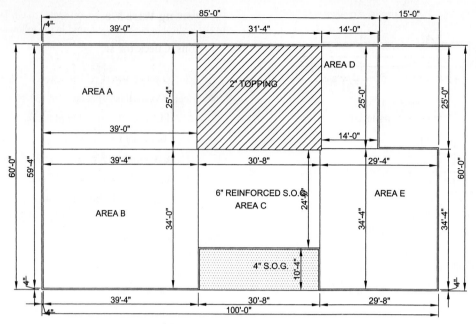

FIGURE 10.22. Reinforcing Steel Layout.

EXAMPLE 10-11   REINFORCING DRILLED PIERS

Estimating the reinforcing in the drilled piers consists of counting the number of vertical bars and determining their length. Since there are three drilled piers, results will be multiplied by three to determine the total.

From the pier detail on sheet S8.1, detail 9 in Appendix A, and from Figure 10.23, there are four No. 5 vertical bars and No. 3 ties that are horizontal at 12 inches on center.

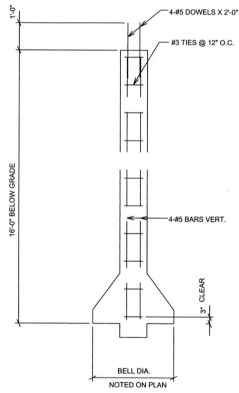

FIGURE 10.23. Drilled Pier Detail.

$$\text{Vertical bars} - \text{No. 5}$$

$$\text{Length of vertical bars} = \text{Pier length} - \text{Coverage}$$

$$\text{Length of vertical bars} = 16' - (2 \times 3'') = 15'6''$$

$$\text{Vertical bars (lf)} = 15'6'' \times 4 \text{ bars}$$

$$\text{Vertical bars} = 15.5' \times 4 \text{ bars} = 62 \text{ lf}$$

$$\text{Vertical dowels (lf)} = 2'0'' \times 4 \text{ dowels} = 8 \text{ lf}$$

$$\text{Total quantity (lf)} = (62 \text{ lf} + 8 \text{ lf}) \times 3 \text{ piers} = 210 \text{ lf}$$

$$\text{Total weight of bars (pounds)} = 210 \text{ lf} \times 1.043 \text{ pounds per foot}$$

$$\text{Total weight} = 219 \text{ pounds}$$

$$\text{Add 10 percent for lap and waste} - \text{Use 241 pounds}$$

$$\text{Horizontal bars} - \text{No. 3}$$

$$\text{Pier diameter} = 18''$$

$$\text{Tie diameter} = \text{Pier diameter} - \text{Coverage}$$
$$= 18'' - (2 \times 2'') = 14''$$

$$\text{Tie length (ft)} = 2\pi r = 2 \times \pi \times 7''/12 \text{ inches per foot}$$

$$\text{Tie length (ft)} = 3.7' \text{ per tie}$$

$$\text{Number of vertical spaces} = 16'0''/1'0'' = 16 \text{ spaces}$$

$$\text{Add 1 to convert to number of ties} - \text{Use 17}$$

$$\text{Total length of ties (lf)} = 17 \text{ ties} \times 3.7 \text{ ft per ties} \times 3 \text{ piers}$$

$$\text{Total length of ties (lf)} = 189 \text{ lf}$$

$$\text{Total weight of bars (pounds)} = 189 \text{ lf} \times 0.376 \text{ pounds per foot}$$

$$\text{Total weight} = 71 \text{ pounds}$$

$$\text{Add 10 percent for lap and waste} - \text{Use 78 pounds}$$

EXAMPLE 10-12   GRADE BEAMS

The grade beam in the front of the building is 30 feet long. The specifics of this grade beam are found in Figure 10.24. There are four No. 5 horizontal bars and No. 3 bars used for stirrups at 12 inches on center (o.c.).

Total quantity of bars No. 5 (lf) = 4 bars × 30′ per bar = 120 lf

Total weight of bars (pounds) = 120 lf × 1.043 pounds per foot

Total weight = 125 pounds

Add 10 percent for lap and waste − Use 138 pounds

Stirrup length = 1′2″ × 4

Stirrup length (lf) = 1.167′ × 4 = 4.7′ per stirrup

Stirrup spaces = 30′/1′ stirrup spacing
= 30 spaces − Use 31 stirrups

Total length of No. 3 stirrups = 31 stirrups × 4.7′ = 146 lf

Total weight of stirrups (pounds) = 146 lf
× 0.376 pounds per foot

Total weight = 55 pounds

Add 10 percent for lap and waste − Use 61 pounds

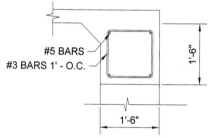

#5 BARS
#3 BARS 1′ - O.C.

1′-6″

1′-6″

FIGURE 10.24. Grade Beam Details. ■

Estimating Wire Mesh.   The square footage of the floor area to be covered may be taken from the slab concrete calculations. Check the sections and details for the size of the mesh required. To determine the number of rolls required, add the lap required to the area to be covered and divide by 750 (the square footage in a roll). Waste averages about 5 percent unless much cutting is required; only full rolls may be purchased in most cases.

Figure 10.25 is a detailed takeoff of all reinforcing bars found in the foundation footings and walls.

EXAMPLE 10-13   WIRE MESH REINFORCING

Using the sample project found in Appendix A as an example, the wire mesh is used for the basement floor and over the precast hollow core planks. Both of these areas are roughly the same square footage. From Figure 10.23, this area is 25′4″ × 30′8″.

sf of concrete requiring mesh = 25′4″ × 30′8″ × 2

sf of concrete requiring mesh = 25.33′ × 30.67′ × 2 = 1,554 sf

Add 15 percent for lap and waste.

Rolls of mesh required = (1,554 sf × 1.15)/750 sf per roll

Order 3 rolls of mesh

Figure 10.25 is the workup sheet for all reinforcing found in the building foundation. ■

Labor.   Figure 10.26 contains sample productivity rates for tying and placing the reinforcing steel. In addition, Figure 10.27 is the priced estimate for the reinforcing steel.

	**ESTIMATE WORK SHEET**

ESTIMATE WORK SHEET

Project: *Little Office Building* — Estimate No. *1234* — Sheet No. *1 of 1*
Location: *Littleville, Tx*
Architect: *U.R. Architects* — **Reinforcing Steel** — Date *11/11/20XX*
Items: *Reinforcing Steel* — By *LHF* Checked *JBC*

Cost Code	Description	Slab Width Ft	In	Bar Spacing In.-O.C.	Pcs	Slab Length Ft	In.	Cover age In.	Bar Length Ea	Total	3	4	5	6	7	8	0.376	0.668	1.043	1.502	2.044	2.67	Quantity	Unit
	Continuous Footing - Long Bars																							
	3′-2″ Wide Footing	3	2	13	4	337	4	2	337.0	1348.0			X						1406					
	3′-0″ Wide Footing	3	2	13	4	78	4	2	78.0	312.0			X						325					
	Continuous Footing - Short Bars																							
	3′-2″ Wide Footing	337	4	11	369	3	2	2	2.8	1045.5			X						1090					
	3′0″ Wide Footing	78	4	11	87	3	2	2	2.8	246.5			X						257					
	Continuous Footings - Dowels	337	4	11	369	3	2	0	3.2	1168.5			X						1219					
	Total Continuous Footings																						4298	Pounds
	Foundation Walls																							
	Short Walls (Short Bars)	303	0	11	332	3	8	2	3.3	1106.7			X						1154					
	Short Walls (Long Bars)	3	8	10	6	303	0	2	302.7	1816.0			X						1894					
	Foundation Walls																							
	Tall Walls (Short Bars)	112	8	11	124	8	4	2	8.0	992.0			X						1035					
	Tall Walls (Long Bars)	8	4	10	11	112	8	2	112.3	1235.7			X						1289					
	Total Foundation Walls																						5372	Pounds
	Grade Beams																							
	Front (Long Bars)				4	30		0	30.0	120.0			X						125					
	Stirrups	30	4	12	32	4	8	0	4.7	149.3	X						56							
	Back (Long Bars)				4	38	10	0	38.8	155.3			X						162					
	Stirrups	38	10	12	40	4	0	0	4.0	160.0	X						60							
	Total Grade Beams																						403	Pounds
	S.O.G.																							
	A Long Bars	39	0	12	40	25	4	2	25.0	1000.0	X						376							
	A Short Bars	25	4	12	27	39	0	2	38.7	1044.0	X						393							
	B Long Bars	39	4	12	41	34	0	2	33.7	1380.3	X						519							
	B Short Bars	34	0	12	35	39	4	2	39.0	1365.0	X						513							
	C Long Bars	30	8	12	32	24	0	2	23.7	757.3	X						285							
	C Short Bars	24	0	12	25	30	8	2	30.3	758.3	X						285							
	D Long Bars	25	0	12	26	14	0	2	13.7	355.3	X						134							
	D Short Bars	14	0	12	15	25	0	2	24.7	370.0	X						139							
	E Long Bars	34	4	12	36	29	4	2	29.0	1044.0	X						393							
	E Short Bars	29	4	12	31	34	4	2	34.0	1054.0	X						396							
	Total Slab On Grade																						3432	Pounds
	Total Weight																3549	0	9957	0	0	0.00		

FIGURE 10.25. Reinforcing Steel Quantity Takeoff.

ESTIMATE WORK SHEET

Project:	Little Office Building												Estimate No.	1234
Location	Littleville, Tx												Sheet No.	1 of 1
Architect	U.R. Architects				**Reinforcing Steel**								Date	11/11/20XX
Items	Reinforcing Steel												By	LHF	Checked	JBC

Cost Code	Description	Slab Width Ft	In	Bar Spacing In. -O.C.	Pcs	Slab Length Ft.	In.	Cover age In.	Bar Length Ea	Total	Bar Size 3	4	5	6	7	8	Bar Weight 0.376	0.668	1.043	1.502	2.044	2.67	Quantity	Unit
	Reinforcing Piers Vertical Bars				12	16	0	3	15.5	186.0			X						194					
	Reinforcing Piers Dowels				12	3	7	0	3.6	43.0			X						45					
	Reinforcing Piers - Stirrups	48	0	12	49	8	3	0	8.3	404.3	X						152							
	Total Drilled Piers																						391	Pounds
	Spread Footing (Long Bars)	3	0	13	4	3	0	2	2.7	10.7		x							11					
	Spread Footing (Short Bars)	3	0	13	4	3	0	2	2.7	10.7		x							11					
	Dowel				4	3	2	0	3.2	12.7		x							13					
	Vertical Bars				4	3	8	2	3.3	13.3		x							14					
	Pier Stirrups	3	8	12	5	2	8	0	2.7	13.3	x						5							
	Total Spread Footings																						54	Pounds
	Front Entrance																							
	Long Bars	30	8	12	32	10	4	2	10.0	320.0	x						120							
	Short Bars	10	4	12	12	30	8	2	30.3	364.0	x						137							
	Exterior Slab On Grade																						257	Pounds
	Total Weight																414	0	288	0	0	0.00		

FIGURE 10.25. (*Continued*)

Type of Placement	Productivity Rate (Labor Hrs./Ton)
Beams #3 to #7	22
Columns #3 to #7	24
Footings #4 to #7	15
Walls #3 to #7	11
Slab on Grade #3 to #7	13
Wire Mesh	1/Roll

FIGURE 10.26. Reinforcing Productivity Rates.

ESTIMATE SUMMARY SHEET

Project	Little Office Building												Estimate No.	1234
Location	Littleville, Tx												Sheet No.	1 of 1
Architect	U.R. Architects, Inc.												Date	11/11/20xx
Items	Reinforcing Steel												By	LHF	Checked	JBC

Cost Code	Description	Q.T.O.	Waste Factor	Purch. Quan.	Unit	Crew	Prod. Rate	Wage Rate	Labor Hours	Unit Cost Labor	Material	Equipment	Labor	Material	Equipment	Total
	Continuous Footings	2.149	10.00%	2.364	Ton		15	14.25	32.2	213.75	550.00		505.28	1,300.15	0.00	1,805.43
	Foundation Walls	2.686	10.00%	2.955	Ton		11	14.25	29.5	156.75	550.00		463.13	1,625.03	0.00	2,088.16
	Grade Beams	0.202	10.00%	0.222	Ton		22	14.25	4.4	313.50	550.00		69.49	121.91	0.00	191.39
	Slab On Grade	1.716	10.00%	1.888	Ton		13	14.25	22.3	185.25	550.00		349.68	1,038.18	0.00	1,387.86
	Drilled Piers	0.196	10.00%	0.215	Ton		24	14.25	4.7	342.00	550.00		73.55	118.28	0.00	191.82
	Spread Footing	0.027	10.00%	0.030	Ton		15	14.25	0.4	213.75	550.00		6.35	16.34	0.00	22.68
	Exterior Slab On Grade	0.129	10.00%	0.141	Ton		13	14.25	1.7	185.25	550.00		26.19	77.74	0.00	103.93
	Wire Mesh	3.000	0.00%	3.000	Roll		1	14.25	3.0	14.25	65.00		42.75	195.00	0.00	237.75
																0.00
									98.29				1,536.41	4,492.62	0.00	6,029.03

FIGURE 10.27. Reinforcing Steel Estimate.

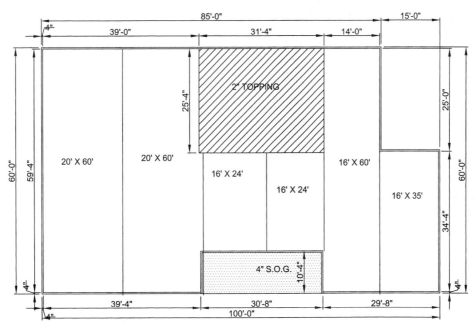

FIGURE 10.28. Vapor Retarder Layout.

EXAMPLE 10-14   LABOR COSTS FOR CONTINUOUS FOOTINGS

Determine the labor cost for placing the reinforcing steel for the continuous footings (refer to Figure 10.26).

Labor hours = (4,298 pounds/2,000 pounds/ton)

$$\times \text{ 15 labor hours/ton}$$

Labor hours = 32.2 labor hours

Labor cost = 32.2 hours × $14.25/hour

Labor cost = $459

Figure 9–27 is a complete estimate for the reinforcing found in the building in Appendix A. ■

10–4 VAPOR RETARDER

The vapor retarder or barrier is placed between the gravel and the slab poured on it and is usually included in the concrete portion of the takeoff. This vapor barrier most commonly consists of polyethylene films or kraft papers. The polyethylene films are designated by the required mil thickness (usually 4 or 6 mil). The material used should be lapped about 6 inches, so an allowance for this must be made depending on the widths available. Polyethylene rolls are available in widths of 3, 4, 6, 8, 10, 12, 16, 18, and 20 feet, and are 100 feet long. Careful planning can significantly cut down on waste, which should average 5 percent plus lapping. Two workers can place 1,000 sf of vapor barrier on the gravel in about one hour, including the time required to get the material from storage and place and secure it. Large areas can be covered in proportionately less time.

Vapor retarders are part of CSI Division 7, Thermal and Moisture Protection (Chapter 13). They are included here since they are part of the concrete slab takeoff.

Estimating Vapor Retarder. The most accurate method of determining the vapor barrier is to sketch a layout of how it might be placed on the job (Figure 10.28). Often several sketches must be made, trying rolls of various sizes before the most economical arrangement is determined.

EXAMPLE 10-15   VAPOR RETARDER (BARRIER)

20′ wide (main level) = 60′ + 60′ = 120′ − Use 2 rolls 4,000 sf

16′ wide (main level) = 24′ + 24′ + 60′ + 35′ = 143′

16′ wide (basement) = 26′ + 26′ = 52′

16′ wide (total) = 143′ + 52′ = 195′ − Use 2 rolls 3,200 sf ■

Labor. The time required for the installation of the vapor barrier has been estimated from the data presented in Figure 10.29. The hourly wages are based on local labor and union conditions; for this example, the wage is $9.25 per hour.

Labor hours = 7.2 thousand sf (actual)

$$\times \text{ 1 labor hour per thousand sf}$$

Labor hours = 7.2 hours

Labor cost = 7.2 hours × $9.25 per hour = $66.60

Sq. Ft. of Vapor Barrier	Labor Hours per 1,000 Sq. Ft.
Up to 1,000	1.5
1,000 to 5,000	1.25
Over 5,000	1

FIGURE 10.29. Vapor Barrier Labor Productivity Rates.

10–5 ACCESSORIES

Any item cast into the concrete should be included in the concrete takeoff. The list of items that might be included is extensive; the materials vary depending on the item and intended usage. The accessory items may include the following:

Expansion Joint Covers. Made of aluminum and bronze, expansion joint covers are available in a wide variety of shapes for various uses. Takeoff by the linear footage is required.

Expansion Joint Fillers. Materials commonly used as fillers are asphalt, fiber, sponge rubber, cork, and asphalt-impregnated fiber. These are available in thicknesses of 1/4, 3/8, 1/2, 3/4, and 1 inch; widths of 2 to 8 inches are most common. Sheets of filler are available and may be cut to the desired width on the job. Whenever possible, filler of the width to be used should be ordered to save labor costs and reduce waste. Lengths of filler strips may be up to 10 feet, and the filler should be taken off by linear feet plus 5 percent waste.

Waterstops. Used to seal construction and expansion joints in poured concrete structures against leakage caused by hydrostatic pressure, waterstops are commonly composed of polyvinyl chloride, rubber, and neoprene in a variety of widths and shapes. The takeoff should be in linear feet, and the estimator must check the roll size in which the specified waterstop is available and add 5 percent for waste. Full rolls only may be purchased (usually 50 feet long).

Manhole Frames and Covers. Manhole frames and covers are available in round or square shapes. The frame is cast in the concrete, and the cover is put in the frame later. The materials used are aluminum (lighter duty) and cast iron (light and heavy duty). The covers may be recessed to receive tile; have a surface that is plain or abrasive; or have holes put in them, depending on the intended usage and desired appearance. The size, material, type, and installation appear in the specifications and on the drawings. They are taken off by the actual number of each type required. The various size frames and types of covers cost varying amounts, so different items are kept separate.

Trench Frames and Covers. The trench frame is cast in the concrete. The linear footage of frames required must be determined, and the number of inside and outside corners must be noted as well as the frame type. Cast-iron frames are available in 3-foot lengths, and aluminum frames are available up to 20 feet long. Covers may also be cast iron or aluminum in a variety of finishes (perforated, abrasive, plain, recessed).

The takeoff is made in linear feet with the widths, material, and finish all noted. Cast-iron covers are most commonly available in 2-foot lengths, while aluminum frames are available in 10-, 12-, and 20-foot lengths, depending on the finish required. Many cast-iron trench manufacturers

will supply fractional sizes of covers and frames to fit whatever size trench length is required. Unless the aluminum frames and covers can be purchased in such a manner, sizable waste may be incurred on small jobs.

Miscellaneous accessories such as anchor bolts, bar supports, screed chairs, screw anchors, screw anchor bolts, plugs, inserts of all types (to receive screws and bolts), anchors, and splices for reinforcing bars must be included in the takeoff. These accessories are taken off by the number required; they may be priced individually or in 10s, 25s, 50s, or 100s, depending on the type and manufacturer. The estimator must carefully note the size required since so many sizes are available. The material (usually steel, cast iron, or plastic) should be listed on the drawings or in the specifications. Some inserts are also available in bronze and stainless steel.

Accessories such as reglets, dovetail anchor slots, and slotted inserts are taken off by the linear footage required. Available in various widths, thicknesses, and lengths, they may be made out of cold rolled steel, galvanized or zinc-coated steel, bronze, stainless steel, copper, or aluminum.

Estimate Expansion Joint Filler. Determine from the plans, details, and sections exactly how many linear feet are required. A 1/2-inch-thick expansion joint filler is required where any concrete slab abuts a vertical surface. This would mean that filler would be required around the outside of the slab, between it and the wall.

10–6 CONCRETE FINISHING

All exposed concrete surfaces require some type of finishing. Basically, finishing consists of the patch-up work after the removal of forms and the dressing up of the surface by troweling, sandblasting, and other methods.

Patch-up work may include patching voids and stone pockets, removing fins, and patching chips. Except for some floor slabs (on grade), there is always a certain amount of this type of work on exposed surfaces. It varies considerably from job to job and can be kept to a minimum with good-quality concrete, with the use of forms that are tight and in good repair, and with careful workmanship, especially in stripping the forms. This may be included with the form-stripping costs, or it may be a separate item. As a separate item, it is much easier to get cost figures and keep a cost control on the particular item rather than "bury" it with stripping costs. Small patches are usually made with a cement-sand grout mix of 1:2; be certain that the type of cement (even the brand name) is the same used in the pour, because different cements are varying shades of gray. The labor hours required will depend on the type of surface, the number of blemishes, and the quality of the patch job required. Scaffolding will be required for work above 6 feet.

The finishes required on the concrete surfaces will vary throughout the project. The finishes are included in the specifications and finish schedules; sections and details should also be checked. Finishes commonly required for

Finish	Method/Location	Labor Hours/ 100 s.f.
Troweling	Machine—Slab on Grade	1.0
	Hand—Slab on Grade	2.0
Broom Finish	Slabs	0.75
	Stairs	4.0
Float Finish	Slab on Grade	3.5
	Slabs Suspended	4.0
	Walls	3.5
	Stairs	3.0
	Curbs	8.5
Bush Hammer	Machine—Green Concrete	3.0
	Machine—Cured Concrete	6.0
	Hand—Green Concrete	9.0
Rubbed	With Burlap and Grout	2.5
	With Float Finish	4.0

FIGURE 10.30. Concrete Finish Labor Productivity.

floors include hand or machine troweled, carborundum rubbing (machine or hand), wood float, broom, floor hardeners, and sealers. Walls and ceilings may also be troweled, but they often receive decorative surfaces such as bush hammered, exposed aggregate, rubbed, sandblasted, ground, and lightly sanded. Finishes such as troweled, ground, sanded, wood float, broom finishes, and bush hammered require no materials to get the desired finish, but require only labor and equipment. Exposed aggregate finishes may be of two types. In the first type, a retarder is used on the form liner, and then the retarder is sprayed off and the surface is cleaned. This finish requires the purchase of a retarding agent, spray equipment to coat the liner, and a hose with water and brushes to clean the surface. (These must be added to materials costs.) The second method is to spray or trowel an exposed aggregate finish on the concrete; it may be a two- or three-coat process, and both materials and equipment are required. For best results, it is recommended that only experienced technicians place this finish. Subcontractors should price this application by the square foot.

Rubbed finishes, either with burlap and grout or with float, require both materials and labor hours plus a few hand tools. The burlap and grout rubbed finish requires less material and more labor than the float finish. A mixer may be required to mix the grout.

Sandblasting requires equipment, labor, and the grit to sandblast the surface. It may be a light, medium, or heavy sandblasting job, with best results usually occurring with green to partially cured concrete.

Bush hammering, a surface finish technique, is done to expose portions of the aggregates and concrete. It may be done by hand, with chisel and hammer, or with pneumatic hammers. The hammers are commonly used, but hand chiseling is not uncommon. Obviously, hand chiseling will raise the cost of finishing considerably.

Other surface finishes may also be encountered. For each finish, analyze thoroughly the operations involved,

material and equipment required, and labor hours needed to do the work.

The finishing of concrete surfaces is estimated by the square foot, except bases, curbs, and sills, which are estimated by the linear foot. Since various finishes will be required throughout, keep the takeoff for each one separately. The labor hours required (approximate) for the various types of finishes are shown in Figure 10.30. Charts of this type should never take the place of experience and common sense, and are included only as a guide.

Materials for most operations (except exposed aggregate or other coating) will cost only 10 to 20 percent of labor. The equipment required will depend on the type of finishing done. Trowels (hand and machine), floats, burlap, sandblasting equipment, sprayers, small mixers, scaffolding, and small hand tools must be included with the costs of their respective items of finishing.

Estimating Concrete Finishing. Areas to be finished may be taken from other concrete calculations, either for the actual concrete required or for the square footage of forms required.

Roof and floor slabs, and slabs on grade, pavements, and sidewalk areas can most easily be taken from the actual concrete required. Be careful to separate each area requiring a different finish. Footing, column, walls, beam, and girder areas are most commonly found in the form calculations.

EXAMPLE 10-16   CONCRETE FINISHING

Using the concrete takeoff from Example 10-7, the needed information for determining the quantity of concrete to be finished can be found. From that takeoff, the following information can be gleaned:

Area of slab (sf) = 4,427 sf of 6″ slab + 735 sf of 2″ topping
= 5,162 sf   ∎

Labor. The time required for finishing the concrete can be estimated from the information provided in Figure 10.30. From this table, the productivity rate for machine troweling concrete is 1 labor hour per 100 sf. Using the locally prevailing wage rate of $13.25 per hour for concrete finishers, the labor costs can be determined.

Labor hours = 51.6 hundred sf × 1 labor hour per 100 sf

Labor hours = 52 hours

Labor cost = 52 hours × $13.25 per hour = $689

10–7 CURING

Proper *curing* is an important factor in obtaining good concrete. The concrete cures with the chemical combinations between the cement and water. This process is referred to as *hydration*. Hydration (chemical combination) requires time, favorable temperatures, and moisture; the period during which the concrete is subjected to favorable temperature and moisture conditions is referred to as the *curing period*. The specified curing period usually ranges from 3 to 14 days. Curing is the final step in the production of good concrete.

Moisture can be retained by the following methods: leaving the forms in place, sprinkling, ponding, spray mists, moisture retention covers, and seal coats. Sometimes combinations of methods may be used, and the forms left on the sides and bottom of the concrete while moisture retention covers or a seal coat is placed over the top. If the forms are left in place in dry weather, wetting of the forms may be required because they have a tendency to dry out; heat may be required in cold weather.

Sprinkling fine sprays of water on the concrete helps keep it moist, but caution is advised. If the surface dries out between sprinklings, it will have a tendency to craze or crack.

Temperature must also be maintained on the concrete. With cold weather construction, it may be necessary to begin with heated materials and provide a heated enclosure. Although the hydration of cement causes a certain amount of heat (referred to as the heat of hydration) and is of some help, additional heat is usually required. The forms may be heated by the use of steam, and the enclosed space itself may be heated by steam pipes or unit heaters of the natural gas, liquefied petroleum gas, or kerosene type. The number required varies for the size of the heater used. Heaters are available from about 50,000 to 300,000 BTUH. Since many unions require that an operator work only with the heaters, local union rules should be checked. With operators, this heating cost is as much as 20 times the cost per cubic yard as without operators, depending on the number of operators required.

During mild weather (40°F to 50°F), simply heating the mixing water may be sufficient for the placement of the concrete, but if temperatures are expected to drop, it may be necessary to apply some external heat. Specifications usually call for a temperature of 60°F to 70°F for the first three days or above 50°F for the first five days.

Hot weather concreting also has its considerations. High temperatures affect the strength and other qualities of the concrete. To keep the temperature down, it may be necessary to wet the aggregate stockpiles with cool water and chill the mixing water. Wood forms should be watered to avoid the absorption of the moisture, needed for hydration, by the forms.

A continuous spray mist is one of the best methods of keeping an exposed surface moist. There is no tendency for the surface to craze or crack with a continuous spray mist. Special equipment can be set up to maintain the spray mist so that it will not require an excessive number of labor hours.

Moisture-retaining covers such as burlap are also used. They should thoroughly cover the exposed surfaces and be kept moist continuously, because they dry out (sometimes rapidly). Straw and canvas are also used as moisture-retaining covers. Burlap and canvas may be reused on many jobs, thus spreading out their initial cost considerably.

Watertight covers may be used on horizontal areas such as floors. Materials employed are usually papers and polyethylene films. Seams should be overlapped and taped, and the papers used should be nonstaining. Often these materials may be reused several times with careful handling and storage.

Sealing compounds are usually applied immediately after the concrete surface has been finished. Both the one- and two-coat applications are colorless, black, or white. If other materials must be bonded to the concrete surface at a later date, they must bond to the type of compound used.

Estimating the cost of curing means that first a determination must be made as to what type of curing will be required and in what type of weather the concrete will be poured. Many large projects have concrete poured throughout the year and thus estimators must consider the problems involved in cold, mild, and hot weather. Smaller projects have a tendency to fall into one or perhaps two of the seasons, but there is a general tendency to avoid the very coldest of weather.

If a temporary enclosure is required (perhaps of wood and polyethylene film), the size and shape of the enclosure must be determined, a material takeoff made, and the number of labor hours determined. For simple enclosures, two workers can erect 100 sf in 30 to 60 minutes, once the materials are assembled in one place. If the wood may be reused, only a portion of the material cost is charged to this portion of the work. The enclosure must be erected and possibly moved during the construction phases and taken down afterward—and each step costs money.

Heating of water and aggregate raises the cost of the concrete by 3 to 10 percent, depending on how much heating is needed and whether ready-mix or job mix concrete is being used. Most ready-mix plants already have the heating facilities and equipment, with the cost spread over their entire production. Job mixing involves the purchase and installation of equipment on the job site as well as its operation in terms of fuel, labor hours, and upkeep.

The cost of portable heaters used to heat the space is usually based on the number of cubic yards of space to be heated. One worker (not an operator) can service the heaters for 100 cy in about one to two labor hours, depending on the type of heater and fuel being used. Fuel and equipment costs are also based on the type used.

If approximately 100,000 BTUH are required per 300 to 400 sf (averaging a height of 8 feet), estimate the number of units needed (depending on the net output of the unit) and determine the amount of fuel that will be consumed per hour, the number of hours the job will require, and the equipment and fuel costs.

The continuous spray requires the purchase of equipment (which is reusable) and the employment of labor to set it up. It will require between one-half and one labor hour to run the hoses and set up the equipment for an area of 100 sf. The equipment should be taken down and stored when it is not in use.

Moisture-retaining and watertight covers are estimated by the square footage of the surface to be covered and separated into slabs or walls and beams. The initial cost of the materials is estimated and is divided by the number of uses expected of them. Covers over slabs may be placed at the rate of 1,000 sf per one to three labor hours, and as many as five uses of the material can be expected (except in the case of canvas, which lasts much longer); wall and beam covers may be placed at the rate of 1,000 sf per two to six labor hours. The sealing of watertight covers takes from three-quarters to one labor hour per 100 lf.

Sealing compounds are estimated by the number of square feet to be covered divided by the coverage (in square feet) per gallon to determine the number of gallons required. Note whether one or two coats are required. If the two-coat application is to be used, be certain to allow for material to do both coats. For two-coat applications, the first coat coverage varies from 200 to 500 sf, while the second coat coverage varies from 350 to 650 sf. However, always check the manufacturer's recommendations. Equipment required will vary according to the job size. Small areas may demand the use of only a paint roller, while medium-size areas may require a pressure-type hand tank or backpack sprayer. Large, expansive areas can best be covered with special mobile equipment. Except for the roller, the equipment has much reuse, usually a life expectancy of five to eight years with reasonable care. The cost of the equipment is spread over the time of its estimated usage. Using the pressure-type sprayer, the estimator figures from one and one-half to three hours will be required per 100 sf. In estimating labor for mobile equipment, the estimator should depend more on the methods in which the equipment is mobilized than on anything else, but labor costs can usually be cut by 50 to 80 percent if mobile equipment is used.

Specifications. The requirements of the architect/engineer regarding curing of the concrete are spelled out in the specifications. Sometimes they detail almost exactly what must be done for each situation, but often they simply state that the concrete must be kept at a certain temperature for a given number of days. The responsibility of protecting the concrete during the curing period is that of the general contractor (who may in turn delegate it to a subcontractor).

10–8 TRANSPORTING CONCRETE

The methods used to transport the concrete from its point of delivery by the ready-mix trucks or the field-batching plant include truck chutes, buggies (with and without power), crane and bucket, crane and hoppers, tower and buggies, forklifts, conveyors, and pumps. Several combinations of methods are available, and no one system is the answer. It is likely that several methods will be used on any one project. Selection of the transporting method depends on the type of pour (floor, wall, curb, etc.), the total volume to be poured in each phase (in cubic yards), the distance above or below grade, equipment already owned, equipment available, the distance from the point of delivery to the point of use, and possible methods of getting the concrete to that point.

Once a decision on the method of transporting the concrete has been made, the cost of equipment must be determined, as well as the anticipated amount of transporting that can be done using that equipment and a given crew of workers. From these, the cost of transporting the concrete may be estimated. Generally speaking, the higher the building and the further the point of use from the point of delivery, the more expensive the cost of transporting.

10–9 FORMS

This portion of the chapter is not a course in form design, but identifies the factors involved in formwork relative to costs. No one design or system will work for all types of formwork. In general, the formwork must be true to grade and alignment, braced against displacement, resistant to all vertical and horizontal loads, resistant to leaking through tight joints, and of a surface finish that produces the desired texture. The pressure on the forms is the biggest consideration in the actual design of the forms.

In the design of wall and column forms, the two most important factors are the rate of placement of the concrete (feet per hour) and the temperature of the concrete in the forms. From these two variables, the lateral pressure (psf) may be determined. Floor slab forms are governed primarily by the actual live and dead loads that will be carried.

Actual construction field experience is a big factor in visualizing exactly what is required in forming and should help in the selection of the form type to be used. The types of forms, form liners, supports, and methods are many; preliminary selections must be made during the bidding period. This is one of the phases in which the proposed job superintendent should be included in the discussions of the methods and types of forms being considered, as well as in the consideration of what extra equipment and work power may be required.

Engineering data relative to forms and the design of forms are available from the American Concrete Institute (ACI), Portland Cement Association (PCA), and most manufacturers. These reference manuals should definitely be included in the estimator's reference library.

The forms for concrete footings, foundations, retaining walls, and floors are estimated by the area (in square feet) of the concrete that comes in contact with the form. The plans should be studied carefully to determine whether it is possible to reuse the form lumber on the building and the number of times it may be reused. It may be possible to use the entire form on a repetitive pour item, or the form may have to be taken apart and reworked into a new form.

On higher buildings (eight stories and more), the cost of forms may be reduced and the speed of construction may be increased if high early strength cements are used instead of ordinary portland cement. The forms could then be stripped in 2 or 3 days instead of the usual 7 to 10 days; however, not all architect/engineers will permit this.

Many types of forms and form liners may be rented. The rental firms often provide engineering services as well as the forms themselves. Often the cost of the forms required for the concrete work can be reduced substantially.

Wood Forms. Wood is one of the most common materials used to build forms. The advantages of wood are that it is readily accessible and easy to work with, and that once used, it may be taken apart and reworked into other shapes. Once it has been decided to use wood, the estimator must determine the quantity of lumber required and the number of uses. This means the construction of the forms must be decided upon with regard to plywood sheathing, wales, studs or joists, bracing, and ties. The estimator can easily determine all of this if the height of the fresh concrete pour (for columns and walls), the temperature of the placed concrete, and the thickness of the slab (for floors) are known. The manufacturer's brochures or ACI formwork engineering data may be used.

Metal Forms. Prebuilt systems of metal forms are used extensively on poured concrete not only on large projects, but even for foundation walls in homes. Advantages are that these systems are reusable several times, easily adaptable to the various required shapes, interchangeable, and require a minimum of hardware and a minimum of wales and ties, which are easily placed. They may be purchased or rented, and several timesaving methods are employed. Curved and battered weights are easily obtained, and while the plastic-coated plywood face liner is most commonly used, other liners are available. Heavy-duty forms are available for heavy construction jobs in which a high rate of placement is desired.

Engineering data and other information pertaining to the uses of steel forms should be obtained from the metal form supplier. The supplier can give information regarding costs (rental and purchase), tie spacing, the number of forms required for the project, and the labor requirements.

Miscellaneous Forms. Column forms are available in steel and laminated plies of fiber for round, square, and rectangular columns. Many manufacturers will design custom forms of steel, fiber, and fiberglass to meet project requirements. These would include tapered, fluted, triangular, and half-rounded shapes. Fiber tubes are available to form voids in cast-in-place (or precast) concrete; various sizes are available. Most of these forms are sold by the linear footage required of a given size. The fiber forms are not reusable, but the steel forms may be used repeatedly.

Estimate. The unit of measurement used for forms is the actual contact area (in square feet) of the concrete against the forms (with the exception of moldings, cornices, sills, and copings, which are taken off by the linear foot). The forms required throughout the project must be listed and described separately. There should be no deductions in the area for openings of less than 30 sf.

Materials in the estimate should include everything required for the construction of the forms except stagings and bridging. Materials that should appear are struts, posts, bracing, bolts, wire, ties, form liners (unless they are special), and equipment for repairing, cleaning, oiling, and removing.

Items affecting the cost of concrete wall forms are the height of the wall (since the higher the wall, the more lumber that will be required per square foot of contact surface) and the shape of the building, including pilasters.

Items affecting the cost of concrete floor forms include the floor-to-floor height, reusability of the forms, length of time the forms must stay in place, type of shoring and supports used, and the number of drop beams required.

The various possibilities of renting or purchasing forms, using gang forms built on ground and lifted into place, slip forming, and so on should be considered during this phase.

Although approximate quantities of materials are given (Figure 10.31) for wood forms, a complete takeoff of materials should be made. The information contained in Figure 10.31 is approximate and should be used only as a check. In addition, Figure 10.32 is a guideline of the labor productivity to build different configurations of formwork.

Estimating Wood Footing Forms. The estimator must first determine whether the entire footing will be poured at one time or whether it will be poured in segments, which would permit the reuse of forms.

Type of Form	Lumber fbm	Labor Hours			
		Assemble	Erect	Strip & Clean	Repair
Footings	200–400	2–6	2–4	2–5	1–4
Walls	200–300	6–12	3–6	1–3	2–4
Floors	170–300	2–12	2–5	1–3	2–5
Columns	170–350	3–7	2–6	2–4	2–4
Beams	250–700	3–8	3–5	2–4	2–4
Stairs	300–800	8–14	3–8	2–4	3–6
Moldings	170–700	4–14	3–8	2–6	3–6
Sill, Coping	150–600	3–12	2–6	2–4	2–6
fbm = Foot Board Measure					

FIGURE 10.31. Wood Forms, Approximate Quantity of Materials, and Labor Hours.

Application	Productivity Rate Labor Hours/SFCA
Foundation Wall (Plywood)	
1 Use	0.44
2 Use	0.28
3 Use	0.25
4 Use	0.22
Footings	
1 Use	0.14
2 Use	0.09
3 Use	0.08
4 Use	0.07
Slab on Grade	
1 Use	1.2
2 Use	1.1
3 Use	1.0
4 Use	0.8
Columns	
1 Use	0.180
2 Use	0.146
3 Use	0.143
4 Use	0.140
Beams	
1 Use	.2
2 Use	0.18
3 Use	0.15
4 Use	0.14

FIGURE 10.32. Approximate Quantity of Labor hours.

EXAMPLE 10-17   FOUNDATION WALL FORMS

Determine the labor costs associated with placing the 3′8″-high foundation wall form from the sample project found in Appendix A. From Example 10-5 and Figure 10.9, the quantity of formwork can easily be found. From that example, there are 303′0″ of 3′8″-high foundation wall. The contact area is found by multiplying these two dimensions and then by doubling that quantity to compensate for both sides being formed.

Contact area (SFCA) = Length (ft) × Width (ft) × Sides formed

Contact area (SFCA) = 303′0″ × 3′8″ × 2 sides

Contact area (SFCA) = 303′ × 3.667′ × 2 sides = 2,222 SFCA

The time required for forms has been estimated from Figure 10.31. The hourly wages are based on local labor and union conditions. Assuming that the forms will be used twice, the following calculation can be performed.

Labor hours = Quantity × Productivity rate

Labor hours = 2,222 SFCA × 0.28 labor hours per SFCA
= 622 labor hours

Labor cost = Labor hours × Wage rate

Labor cost = 622 hours × $13.25 per hour

Labor cost = $8,241.50   ■

10–10 FORM LINERS

The type of liner used with the form will determine the texture or pattern obtained on the surface of the concrete. Depending on the specified finish, formed concrete surfaces requiring little or no additional treatment can be easily obtained. A variety of patterns and textures may be produced by using various materials as liners. Fiberglass liners, plastic-coated plywood, and steel are among the most commonly used. Textures such as wood grain, rough sawn wood, corrugations (of various sizes and shapes), and all types of specialized designs are available. They have liners that will leave a finish resembling sandblast, acid etch, and bush hammered, as well as others. These textures may be used on floors or walls.

Liners are also used to form waffle slabs and tee beam floor systems; they may be fiberglass, steel, or fiber core. When liners of this type are used, the amount of void must be known so that the quantity of concrete may be determined. This information is given in Figure 10.33. Complete information should be obtained from whichever company that supplies the forms for a particular project.

Form liners are often available on a rental or purchase arrangement; specially made form liners of a particular design may have to be purchased. When special designs are required, be certain to get a firm proposal from a manufacturer.

Specifications.   Check the specifications for requirements concerning textures and patterns, liner materials, thicknesses, configurations, and any other liners. It is not uncommon for the specifications to state the type of material from which the liner must be made. The drawings should be checked for types of texture, patterns, or other requirements of the form liner. The form liners required may be already in stock at the manufacturer or may need to be special ordered, often requiring months of work, including shop drawing approvals.

Estimating.   Estimators take off the square footage of the surface requiring a particular type of liner and decide how many liners can be used effectively on the job—this will be the total number of square feet of the liner or the number of

Waffle Slab Form Liners	
Void Size and Depth	Concrete voided c.f.
19" x 19" x 4"	.77
6"	1.09
8"	1.41
10"	1.9
12"	2.14
30" x 30" x 8"	3.85
10"	4.78
12"	5.53
14"	6.54
16"	7.44
20"	9.16

FIGURE 10.33. Void Area in Concrete.

pieces required. Being able to use them several times is what reduces the cost. Dividing the total cost of the liners by the square feet of surface provides a cost per square foot for liners. The same approach is used for rental of liners. Different textures and patterns may be required throughout, so details must be checked carefully.

10–11 CHECKLIST

Forms for:

Footings, walls, and columns
Floors
Piers
Beams
Columns
Girders
Stairs
Platforms
Ramps
Miscellaneous

Forms

Erection
Removal
Repair
Ties
Clamps
Braces
Cleaning
Oiling
Repairs
Liners

Concrete

Footings, walls, and columns
Floors
Toppings
Piers
Beams
Columns
Girders
Stairs
Platforms
Ramps
Curbing
Coping
Walks
Driveways
Architectural
Slabs

Materials, Mixes

Cement
Aggregates
Water
Color
Air-entraining
Other admixtures
Strength requirements
Ready-mix
Heated concrete
Cooled concrete

Finishes

Hand trowel
Machine trowel
Bush hammer
Wood float
Cork float
Broom
Sand
Rubbed
Grouted
Removing fins

Curing

Admixtures
Ponding
Spraying
Straw
Canvas
Vapor barrier
Heat

Reinforcing

Bars
Wire mesh
Steel grade
Bends
Hooks
Ties
Stirrups
Chairs
Cutting

10–12 PRECAST CONCRETE

The term *precast concrete* is applied to individual concrete members that are cast in separate forms and then placed in the structure. They may be cast in a manufacturing plant or on the job site by the general contractor or supplier. The

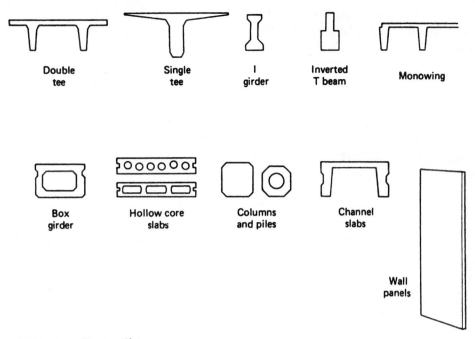

FIGURE 10.34. Precast Shapes.

most common types of structural precast concrete are double and single tees, floor planks, columns, beams, and wall panels (Figure 10.34).

Depending on the requirements of the project, precast concrete is available reinforced or prestressed. Reinforced concrete utilizes reinforcing bars encased in concrete; it is limited in its spans, with 40 feet being about the maximum for a roof. When longer spans are required, prestressed concrete is used, either pretensioned or posttensioned. Prestressing generally utilizes high-strength steel or wire or wire strands as the reinforcing. In pretensioning, the longitudinal reinforcing is put in tension before the concrete is cast. The reinforcing is stretched between anchors and held in this state as the concrete is poured around the steel and cured. With posttensioning, the longitudinal reinforcement is not bonded with the concrete. The reinforcement may be greased or wrapped to avoid bonding with the concrete, or conduits of some type (tubes, hoses) may be cast in the concrete and the reinforcement added later. The reinforcement is then stretched and anchored at the ends; when released, the stretched wires tend to contract and in this manner compress the concrete.

Most precast concrete is priced by the square foot, linear foot, or in a lump sum. It is important to determine exactly what is included in the price. Most suppliers of precast concrete items price them delivered to the job site and installed, especially if they are structural items. When the specifications permit, some contractors will precast and install the pieces required themselves. Unless experienced personnel are available, doing your own precasting may cost more than subcontracting the work to others.

The aggregates used for precast concrete may be heavyweight or lightweight; however, it should be noted that some types of lightweight concrete are not recommended by some

consultant engineers for use on posttensioning work, and care should be taken regarding all materials used in the concrete.

Hole-cutting in tees is subject to the distance between the structural tee portions of the members. The lengths of holes are not as rigidly controlled, but should be approved by the structural engineer.

10–13 SPECIFICATIONS

The specifications must be checked to determine whether a particular manufacturer is specified. All manufacturers who can supply materials and meet the specifications should be encouraged to bid the project. Often there is only limited competition in bidding on precast concrete items. The estimator must also determine the strength, reinforcing, and inserts as well as any special accessories required.

If the project entails any special finish such as sandblasting, filling air holes, colored concrete, special aggregates, or sand finish, this should be noted on the workup sheets. Be certain that there is a clear understanding of exactly what the manufacturer is proposing to furnish. For example, it should be understood who will supply any required anchor bolts, welding, cutting of required holes, and filling of joints. Always check to see whether the manufacturer (or subcontractor) will install the precast. If not, the estimator will have to calculate the cost of the installation.

10–14 ESTIMATING

Floor, wall, and ceiling precast concrete are most commonly taken off by the square foot with the thickness noted. Also be certain to note special requirements such as insulation cast in the concrete, anchorage details, and installation problems.

Beams and columns are taken off by the linear foot with each required size kept separately. Anchorage devices, inserts, and any other special requirements must be noted.

Determine exactly what the suppliers are proposing. If they are not including the cutting of all holes, finishing of concrete, welding, or caulking, then all of these items will have to be figured by the estimator.

10–15   PRECAST TEES

Precast tees are available as simply reinforced or prestressed. The shapes available are double and single tee. For the double tee, the most common widths available are 4 feet and 6 feet with depths of 8 to 26 inches. Spans range up to 75 feet depending on the type of reinforcing and size of the unit. Single tee widths vary from 6 feet to 12 feet with lengths up to about 100 feet. The ends of the tees must be filled with some type of filler. Fillers may be concrete, glass, plastic, and so on. Refer to Figure 10.35.

Most manufacturers who bid this item bid it delivered and installed either on a square-foot or on a lump sum basis. One accessory item often overlooked is the special filler block required to seal the end of the tees.

Concrete fill is often specified as a topping for tees; it may be used for floors or sloped on a roof to direct rainwater. The fill is usually a minimum of 2 inches thick and should have at least a light reinforcing mesh placed in it. Because of the camber in tees, it is sometimes difficult to pour a uniform 2-inch-thick topping and end up with a level floor.

Specifications.   The estimator determines which manufacturers are specified, the strength of concrete, the type and size of reinforcing and aggregate, the finish required, and whether topping is specified. For reinforcing bars, they must determine the type of chairs to be used to hold the bars in place. The bars should be corrosion resistant. The estimator must determine who will cut the holes and caulk the joints, as well as what type of caulking will be used.

Estimating.   The estimator must take off the square footage required. If the supplier made a lump sum bid, check the square-foot price. The square footage required should have all openings deducted. If the project is bid by the square foot, call the supplier to check your square footage against their takeoff. This provides a check for your figures. (However, it is only a check. If they don't agree, recheck the drawings. Never use anyone else's quantities when working up an estimate.)

When an interior finish is required, the shape of the tee is a factor: A tee 4 feet wide and 40 feet long does not have a bottom surface of 160 sf to be finished. The exact amount of square footage involved varies, depending on the depth and design of the unit; it should be carefully checked.

The installation of mechanical and plumbing items sometimes takes some special planning with precast tees in regard to where the conduit, heating, and plumbing pipes will be located and how the fixtures will be attached to the concrete. All of these items must be checked.

The concrete fill (*topping*) is placed by the general contractor in most cases. The type of aggregate must be determined, and the cubic yards required must be taken off. The square footage of the reinforcing mesh and the square footage of the concrete to be finished must be determined. From the specifications, estimators determine the aggregate, strength, type of reinforcing, surface finish, and any special requirements. This operation will probably be done after the rest of the concrete work is completed. The decision must be made as to whether ready-mixed or field-mixed concrete will be used, how it will be moved to the floor, and the particular spot on which it is to be placed.

10–16   PRECAST SLABS

Precast slabs are available in hollow, cored, and solid varieties for use on floors, walls, and roofs. For short spans, various types of panel and channel slabs with reinforcing bars are available in both concrete and gypsum. Longer spans and heavy loads most commonly involve cored units with prestressed wire.

The solid panel and channel slabs are available in heavyweight and lightweight aggregates. The thicknesses and widths available vary considerably, but the maximum span is generally limited to about 10 feet. Some slabs are available tongue-and-grooved and some with metal-edged tongue-and-groove. These types of slabs use reinforcing bars or reinforcing mesh for added tension strengths. These lightweight, easy-to-handle nail, drill, and saw pieces are easily installed on the job over the supporting members. A clip or other special fastener should be used in placing the slabs.

Cored units with prestressed wire are used on roof spans up to about 44 feet. Thicknesses available range from 4 to 16 inches with various widths available, 40 and 48 inches being the most common. Each manufacturer must be contacted to determine the structural limitations of each product. The units generally have high fire resistance ratings and are available with an acoustical finish. Some types are available with exposed aggregate finishes for walls.

Specifications.   The type of material used and the manufacturer specified are the first items to be checked. The materials used to manufacture the plank, type and size of reinforcing, and required fire rating and finish must be checked.

The estimator should also note who cuts the required holes in the planks and who caulks the joints, and the type of caulking to be used. If topping is required, the thickness, reinforcing,

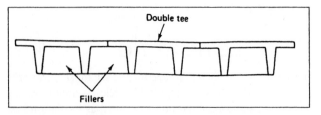

FIGURE 10.35. Double Tees Detail.

aggregates, and strength specified must also be noted. Any inserts, anchors, or special requirements must be noted as well.

Estimating. The precast slabs are generally quoted by the square foot or in a lump sum. Solid panels and channel slabs can often be purchased from the manufacturer and installed by the general contractor. The manufacturers of cored units generally furnish and install the planks themselves. When calculating the square footage, deduct all openings. Also determine who will cut the holes, do any welding and special finishing, and provide anchors and inserts. Concrete topping is commonly used over the cored units, the most commonly specified thickness being 2 inches. Determine the cubic yards of the concrete, square feet of the reinforcing mesh, and square feet of the surface to be finished.

Check the drawings and specifications to determine how the planks will be held in place, special anchorage details, inserts, and any other items that may present a cost or problem on the job. Mechanical and electrical requirements should also be checked.

10–17 PRECAST BEAMS AND COLUMNS

Precast beams and columns are available in square, rectangular, T- and I-shaped sections. They are available simply reinforced, with reinforcing bars, or prestressed with high-strength wire. The sizes and spans depend on the engineering requirements of each project, and the beams and columns are not poured in any particular size or shape. Special forms can easily be made out of wood to form the size and shape required for a particular project.

Specifications. The manufacturers' specified strengths and materials, reinforcing, connection devices, anchors, inserts, and finishes required should all be noted. The different shapes needed throughout and any other special requirements should also be specified.

Estimating. If the contractor intends to precast the concrete, the costs involved are indicated in Section 10–19. Manufacturers who bid this item will bid it per linear foot or in a lump sum. In doing a takeoff, keep the various sizes separate. Take special note of the connection devices required.

Labor. Precast beams (lintels) for door and window openings are generally installed by the mason. If the weight of an individual lintel exceeds about 300 pounds, it may be necessary to have a small lift or crane to put it in place.

10–18 MISCELLANEOUS PRECAST

Precast panels for the exterior walls of homes, warehouses, apartment, and office buildings are available. Their sizes, thicknesses, shapes, designs, and finishes vary considerably. Each individual system must be analyzed carefully to determine the cost in place. Particular attention should be paid to the attachment details at the base, top, and midpoints of each panel, how attachments will be handled at the job site, how much space is required for erection, how much bracing is required for all panels to be securely attached, and how many men are required.

Some of the various methods involve the use of panels 4 feet wide, panels the entire length and width of the house, and precast boxes that are completely furnished before they are installed on the job. A tremendous amount of research goes into precast modules. The higher the cost of labor in the field, the more research there will be to arrive at more economical building methods.

The estimator must carefully analyze each system, consider fabrication costs and time, space requirements, how mechanicals will relate and be installed, and try to determine any hidden costs. New systems require considerable thought and study.

10–19 PRECAST COSTS

If the specifications allow the contractor to precast the concrete shapes required for a project, or if the contractor decides to at least estimate the cost for precasting and compare it with the proposals received, the following considerations will figure into the cost:

1. Precasting takes a lot of space. Is it available on the job site, or will the material be precast off the site and transported to the job site? If it is precast off the site, whatever facilities are used must be charged off against the items being made.

2. Determine whether the types of forms will be steel, wood, fiberglass, or a combination of materials. Who will make the forms? How long will the manufacture take? How much will it cost? The cost of forms must be charged to the precast items being made.

3. Will a specialist be required to supervise the manufacture of the items? Someone will have to coordinate the work and the preparation of shop drawings. This cost must also be included.

4. Materials required for the manufacture must be purchased:
   (a) Reinforcing
   (b) Coarse and fine aggregates
   (c) Cement
   (d) Water
   (e) Anchors and inserts

5. Allowance must be made for the actual cost of labor required to
   (a) Clean the forms.
   (b) Apply oil or retarders to the forms.
   (c) Place the reinforcing (including pretensioning if required).
   (d) Mix and pour the concrete (including troweling the top off).
   (e) Cover the concrete and apply curing method.
   (f) Uncover the concrete after the initial curing.

(g) Strip from form and stockpiling to finish curing.

(h) Erect the concrete on the job. If poured off the site, it will also be necessary to load the precast concrete on trucks, transport it to the job site, unload it, and then erect it.

6. Equipment required may include mixers, lift trucks, and cranes. Special equipment is required to prestress concrete. Equipment to cure the concrete may be required, and miscellaneous equipment such as hoes, shovels, wheelbarrows, hammers, and vibrators will also be required.

7. Shop drawings should be prepared either by a company draftsman or by a consultant.

10-20 PRECAST CHECKLIST

Shapes	Girders
Bearing requirements	Lintels
Accessories	Strength requirements
Walls	Inserts
Floors	Attachment requirements
Ceilings	Finish
Beams	Color
Joists	Special requirements

WEB RESOURCES

www.concrete.org

www.concretenetwork.com

products.construction.com

www.worldofconcrete.com

www.4specs.com

www.tilt-up.org

www.precast.org

www.crsi.org

www.pci.org

REVIEW QUESTIONS

1. What is the difference between plant ready-mixed concrete and job-mixed concrete?

2. Under what circumstances might it be desirable to have a field batching plant for job-mixed concrete?

3. What is the unit of measure for concrete?

4. Why does the estimator have to keep the different places that the concrete will be used separate in the estimate (e.g., concrete sidewalks, floor slabs)?

5. Why should the different strengths of concrete be kept separate?

6. Where would the estimator most likely find the strength of the concrete required?

7. How is rebar taken off? In what unit of measure are large quantities ordered?

8. How does lap affect the rebar quantities?

9. How is wire mesh taken off? How is it ordered?

10. How is vapor barrier taken off? How does the estimator determine the number of rolls required?

11. What unit of measure is used when taking off expansion joint fillers?

12. What unit of measure is used when taking off concrete finishes?

13. Why should each finish be listed separately on the estimate?

14. Where would the estimator look to determine if any curing of concrete is required on the project?

15. Why must the estimator consider how the concrete will be transported to the job site?

16. What unit of measure is used when taking off concrete forms? How can reuse of forms affect the estimate?

17. What unit of measure is used for form liners? When might it be more economical to rent instead of purchasing them?

18. What two methods of pricing might a subcontractor use for precast concrete?

19. Many suppliers take the responsibility of installing the precast units. Why might this be desirable?

20. Under what conditions might it be desirable for a contractor to precast the concrete on the job site?

21. Determine the amount of concrete, reinforcing, forms, and accessories required for the building in Appendix C.

22. Using the drawings of Billy's Convenience Store found in Appendix F, determine the quantity of concrete and reinforcing required for the spread footings.

23. Using the drawings of Billy's Convenience Store found in Appendix F, determine the quantity of concrete, reinforcing, and formwork in the slab on grade.

CHAPTER ELEVEN

MASONRY

11-1 GENERAL

The term *masonry* encompasses all the materials used by masons in a project, such as block, brick, clay, tile, or stone. The mason is also responsible for the installation of lintels, flashing, metal wall reinforcing, weep holes, precast concrete, stone sills and coping, and manhole and catch basin block.

The tremendous amount of varied material available requires that estimators be certain they are bidding exactly what is required. Read the specifications, check the drawings, and call local suppliers to determine the exact availability, costs, and special requirements of the units needed.

11-2 SPECIFICATIONS

Specifications should be carefully checked for the size of the unit, type of bond, color, and shape. Simply because a particular unit is specified does not mean it is available in a particular area. The estimator must check with the local manufacturer and suppliers to ensure the availability of all items, their compliance with the specifications, and the standards established by the local building code. Even though concrete blocks are generally manufactured on the local level, it is not uncommon for the local supplier to make arrangements with a manufacturer 300 or more miles away to provide the units required for a particular project—especially if a small number of special units are involved. If the specifications require a certain fire rating, the supplier must be made aware of the requirements before submitting a proposal for the material.

Many masonry accessories such as lintels and flashing, which are built into the wall, must be installed by the masons. The general contractor needs to verify whether the masonry subcontractor has included these items in his quote. Large precast concrete lintels may require the use of

special equipment, and steel angle lintels often require cutting of the masonry units. In addition, the type of joint "tooling" should be noted, as the different types impact labor costs.

11-3 LABOR

The amount of time required for a mason (with the assistance of helpers) to lay a masonry unit varies with the (1) size, weight, and shape of the unit; (2) bond (pattern); (3) number of openings; (4) whether the walls are straight or have jogs in them; (5) distance the units must be moved (both horizontally and vertically); and (6) the shape and color of the mortar joint. The height of the walls becomes important in estimating labor for masonry units. The masonry work that can be laid up without the use of scaffolding is generally the least expensive; however, that is typically limited to 4 to 5 feet. Labor costs arise from the erection, moving, and dismantling of the scaffolding as the building goes up. The units and mortar have to be placed on the scaffold, which further adds to the labor and equipment costs.

The estimator should check union regulations in the locality in which the project will be constructed, since unions may require that two masons work together where the units weigh more than 35 pounds each.

The weather conditions always affect labor costs, because a mason will lay more brick on a clear, warm, dry day than on a damp, cold day. Winter construction requires the building and maintenance of temporary enclosures and heating.

11-4 BONDS (PATTERNS)

Some types of bonds (Figures 11.1, 11.2, & 11.3) required for masonry units can add tremendously to the labor cost of the project. The least expensive bond (pattern) is the *running*

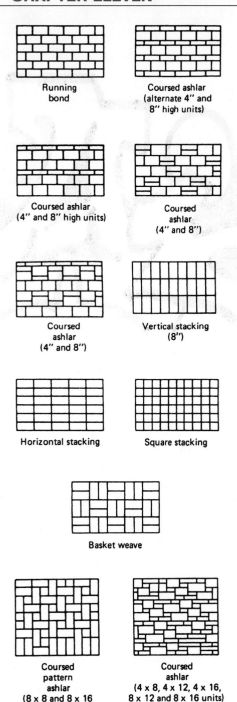

FIGURE 11.1. Typical Concrete Block Patterns.

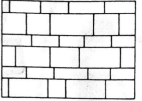

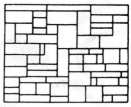

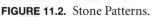

FIGURE 11.2. Stone Patterns.

the quantity takeoff, the estimator must keep the amounts required for the various bonds separate. The types of bonds that may be required are limited only by the limits of the designer's imagination.

The most common shapes of mortar joints are illustrated in Figure 11.4. The joints are generally first struck off flush with the edge of the trowel. Once the mortar has partially set, the tooled joints are molded and compressed with a rounded or V-shaped joint tool. Some joints are formed with the edge of the trowel. The *raked joint* is formed by raking or scratching out the mortar to a given depth, which is generally accomplished with a tool made of an adjustable nail attached to two rollers.

11–5 CONCRETE MASONRY

Concrete masonry comprises all molded concrete units used in the construction of a building and includes concrete brick, hollow and solid block, and decorative types of block. Historically, many of these units are manufactured on the local level, and industry standards are not always followed. There is considerable variation in shapes and sizes available.

Concrete Block

Concrete block has no complete standard of sizes. The standard modular face dimensions of the units are 7⅝ inches high and 15⅝ inches long. Thicknesses available are 3, 4, 6, 8, 10, and 12 inches. (These are nominal dimensions; actual dimensions are 3/8 inch less.) A 3/8-inch mortar joint provides a face dimension of 8 × 16 inches; it requires 112.5 units per 100 sf of wall. Because there are no industry standards, it is important to check with local suppliers to

bond. Another popular bond is the *stacked bond;* this type of bond will increase labor costs by as much as 50 percent if used instead of the running bond. Various *ashlar patterns* may also be required; these may demand several sizes laid up to create a certain effect. The estimator must study the drawings, check the specifications, and keep track of the different bonds that might be required on the project. When doing

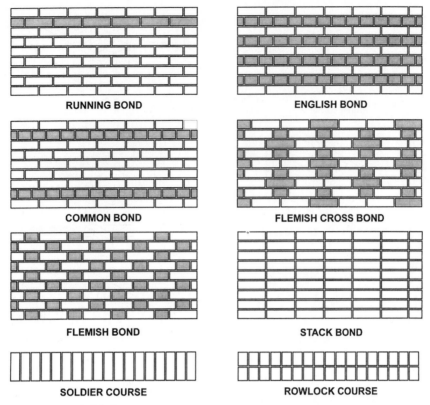

RUNNING BOND

COMMON BOND

FLEMISH BOND

SOLDIER COURSE

ENGLISH BOND

FLEMISH CROSS BOND

STACK BOND

ROWLOCK COURSE

FIGURE 11.3. Brick Patterns.

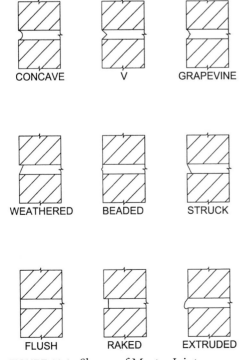

CONCAVE   V   GRAPEVINE

WEATHERED   BEADED   STRUCK

FLUSH   RAKED   EXTRUDED

FIGURE 11.4. Shapes of Mortar Joints.

determine what sizes and shapes are available (Figure 11.5). If the order requires a large number of special units, it is very possible that the local manufacturer will produce the units needed for the project.

Concrete blocks are available either as heavyweight or lightweight units. The heavyweight unit is manufactured out of dense or normal weight aggregates such as sand, gravel and crushed limestone, cement, and water. Lightweight units use aggregates such as vermiculite, expanded slag, or pumice with the cement and water. The lightweight unit may weigh 30 percent less than the heavyweight unit, although it usually costs a few cents more per unit and usually has a slightly lower compressive strength.

The specifications will have to be checked to determine the size, shape, and color of units, as well as the conformance with some set of standards such as that of the American Society for Testing and Materials (ASTM), which sets up strength and absorption requirements. If a specific fire rating is designated for the units, it may require the manufacture of units with thicker face shells than are generally provided by a particular supplier. Check also for the type of bond, mortar, and any other special demands. If the estimator is unfamiliar with what is specified, then he or she can call the local suppliers to discuss the requirements of the specifications with them to be certain that what is specified can be supplied. Some items such as units cured with high-pressure steam curing are unavailable in many areas, and in such a case the estimator (and often the suppliers) may want to call the architect/ engineer to discuss with them the problem of supplying the item.

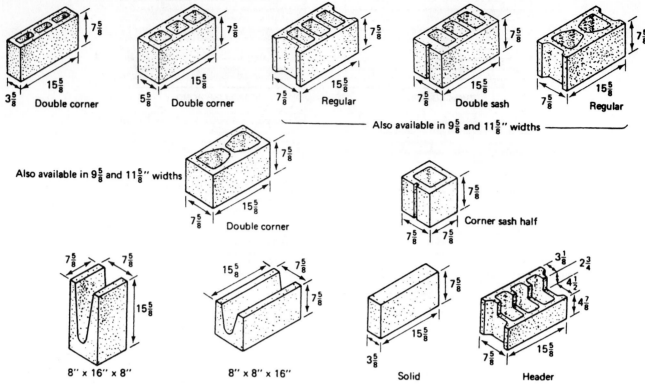

FIGURE 11.5. Typical Block Sizes (All Sizes Are Actual).

11–6 SPECIFICATIONS— CONCRETE MASONRY

The specifications will state exactly which types of units are required in each location. They give the size, shape, color, and any requisite features such as glazed units, strength, and fire ratings. The type, color, thickness, and shape of the mortar joint must be determined, as well as the style of bond required. Also to be checked are the reinforcing, control and expansion joints, wall ties, anchors, flashing, and weep holes needed.

If the specifications are not completely clear as to what is required, the estimator should call the architect/engineer's office to check; they should never guess what the specifications mean.

11–7 ESTIMATING— CONCRETE MASONRY

Concrete masonry should be taken off from the drawings by the square feet of wall required and divided into the different thickness of each wall. The total square footage of each wall, of a given thickness, is then multiplied by the number of units required per square foot (Figure 11.6). For sizes other than those found in Figure 11.5, the following formula can be used:

(There are 144 square inches in 1 square foot.)

$$\text{Units per square foot} = \frac{144}{\text{square inches per unit}}$$

For a nominal size $8'' \times 8'' \times 16''$ block

Materials	Materials (per 100 s.f. face area)			
	Concrete Block			Brick
	8 x 16	8 x 18	5 x 12	2 ²/₃" x 4" x 8"
No. of Units	112.5	100	240	675
Mortar Cu. Ft.				
Face shell bedding	2.3	2.2	3.6	7.2
Face shell and Web bedding	3.2	3.0	2.5	7.2

Values shown are net. Waste for block and brick ranges from 5 to 100%. Waste for mortar may range from 25 percent to 75 percent and actual job experience should be considered on this item. It is suggested that 100% waste be allowed by the inexperienced and actual job figures will allow a downward revision. Figures are for a 3/8" thick joint.

FIGURE 11.6. Materials Required for 100 sf of Face Area.

Mix By Volume	Quantities			
	Masonry Cement (bags)	**Portland Cement (bags)**	**Lime or Putty (c.f.)**	**Sand (c.f.)**
1:3 (masonry)	.33	-	-	.99
1:1:6 (portland)	-	.16	.16	.97

NOTE. Many firms would use the information from past projects to figure the cost per unit of masonry and then use these costs for jobs being bid. This offers a good check for the estimator, but care must be taken to be certain that the working conditions will be about the same and that there has been no wage or benefits increase since the project was completed.

FIGURE 11.7. Mortar Mixes to Mix 1 cf of Mortar.

$$\text{Units per square foot} = \frac{144}{8'' \times 16''}$$

$$\text{Units per square foot} = 1.125 \text{ blocks per sf}$$

When estimating the quantities of concrete masonry, use the exact dimensions shown. Corners should only be taken once, and deductions should be made for all openings in excess of 10 sf. This area is then converted to units, and to this quantity an allowance for waste and breakage must be added.

While performing the takeoff, the estimator should note how much cutting of masonry units will be required. Cutting of the units is expensive and should be anticipated.

In working up the quantity takeoff, the estimator must separate masonry according to

1. Size of the units
2. Shape of the units
3. Colors of the units
4. Type of bond (pattern)
5. Shape of the mortar joints
6. Colors of the mortar joints
7. Any other special requirements (such as fire rating)

In this manner, it is possible to make the estimate as accurate as possible. For mortar requirements, refer to Figure 11.7 and Section 11–16.

EXAMPLE 11-1   CONCRETE BLOCK

Determine the quantity of concrete block required for the west wall of the small commercial building found in Appendix A. Figures 11.8 and 11.9 are excerpts from those drawings and are helpful in determining the quantity of concrete block.

Wall height for concrete block = 16'2" − 10" = 15'4"

West wall length = 60'0" − 9" (both end brick ledge) = 59'3"

Gross wall area (sf) = 15'4" × 59'3"

Gross wall area (sf) = 15.33' × 59.25' = 908 sf

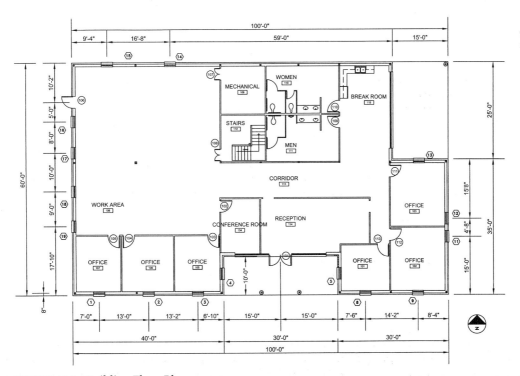

FIGURE 11.8. Building Floor Plan.

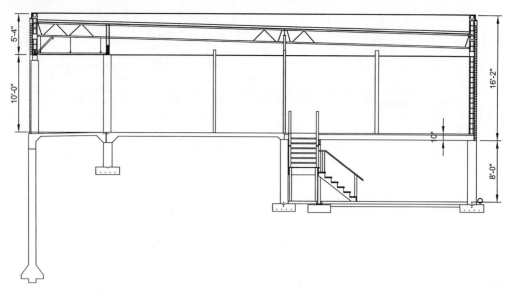

FIGURE 11.9. Wall Section.

Openings

Windows 4 @ 3′ × 7′ = 84 sf

Doors 1 @ 3′ × 7′ = 21 sf

Area of openings = 84 sf + 21 sf = 105 sf

Net west wall area = 908 − 84 − 21 = 803 sf

Using an 8″ × 8″ × 16″ block = 1.125 blocks per sf

Number of blocks = 803 sf × 1.125 blocks per sf

Number of blocks = 903 blocks

Waste @ 6 percent − Use 958 blocks

Header/Lintel Blocks

Top course = 59′3″/16″ = 45 blocks

Windows/doors = 4 per opening

5 openings = 20 blocks

Plain block = 958 − 45 − 20 = 893 blocks

For west wall purchase

893 − 8 × 8 × 16 blocks

65 header blocks

From Figure 11.10, the labor hours can be determined.

Labor Hours per 100 Square Feet										
Wall Thickness	**4"**		**6"**		**8"**		**10"**		**12"**	
Workers	**Mason**	**Laborer**	**Mason**	**Laborer**	**Mason**	**Laborer**	**Mason**	**Laborer**	**Mason**	**Laborer**
Type of Work										
Simple Foundation					4.5-6.0	4.0-7.5	6.0-9.0	7.0-10.5	7.0-10.0	8.0-11.5
Foundation with several corners, openings					5.0-7.5	5.0-7.5	6.5-10.0	7.5-12.0	7.5-10.0	8.5-12.0
Exterior walls 4'-0" high	3.5-5.5	3.5-6.0	4.0-6.0	4.0-6.5	4.5-6.0	5.0-7.5	6.0-9.0	7.0-10.5	7.0-10.0	8.0-11.5
Exterior walls, 4'-0" to 8'-0" above ground or floor	3.5-6.0	4.5-7.5	4.0-6.5	4.5-7.0	4.5-6.5	6.0-9.0	6.5-10.0	7.5-12.0	7.5-10.0	8.5-12.0
Exterior walls, more than 8'-0" above ground or floor	4.5-8.0	6.0-9.5	5.0-9.0	7.0-10.0	5.0-7.0	7.0-10.0	7.0-10.5	7.5-12.0	7.5-10.0	8.5-12.0
Interior partitions	3.0-6.0	3.5-7.0	3.5-6.5	4.5-7.5	4.5-6.0	5.0-7.5				

Note:
1.   The more corners and openings in the masonry wall, the more work hours it requires.
2.   When lightweight units are used the work hours should be decreased by 10 percent.
3.   Work hours include simple pointing and cleaning required.
4.   Special bonds and patterns may increase the work hours by 20 to 50 percent.

FIGURE 11.10. Labor Hours Required for Concrete Masonry.

Using 6 mason labor hours per 100 sf and 8 laborer labor hours per 100 sf

$$8.03 \text{ hundred sf} \times 6 = 48 \text{ mason labor hours}$$

$$8.03 \text{ hundred sf} \times 8 = 64 \text{ laborer labor hours}$$

Assuming a labor rate of $14.25 for masons and $10.00 for laborers per hour, the labor costs can be determined.

$$
\begin{aligned}
\text{Mason labor cost} &= 48 \text{ hours} \times \$14.25 \text{ per hour} \\
&= \$684.00
\end{aligned}
$$

$$
\begin{aligned}
\text{Laborer labor cost} &= 64 \text{ hours} \times \$10.00 \text{ per hour} \\
&= \$640.00
\end{aligned}
$$

$$
\begin{aligned}
\text{Total labor cost} &= \$684 + \$640 \\
&= \$1,324 \quad\blacksquare
\end{aligned}
$$

11-8 CLAY MASONRY

Clay masonry includes brick and hollow tile. Brick (Figure 11.11) is considered a *solid masonry unit,* whereas tile (Figure 11.12) is a *hollow masonry unit.* The faces of each type of unit may be given a ceramic glazed finish or a variety of textured faces.

Brick is considered a solid masonry unit even if it contains cores, as long as the core unit is less than 25 percent of the cross-sectional area of the unit. Bricks are available in a large variety of sizes. They are usually classified by material, manufacturer, kind (common or face), size, texture, or finish on the face.

Hollow tile is classified as either *structural clay tile* (Figure 11.12) or *structural facing tile.* When the units are designed to be laid up with the cells in a horizontal position, they are referred to as *side construction;* with *end construction* the units are laid with the cells in a vertical position. Structural clay tiles are available in load-bearing and nonload-bearing varieties, the sizes and shapes of which vary considerably.

Structural facing tile is available either glazed or unglazed. Unglazed units may have rough or smooth finishes, and the ceramic glazed units are available in a wide variety of colors. These types of units are often used on the exteriors of buildings and on the interiors of walls of dairy and food producers, laboratories, and other such locations that require hard, nonporous, easy-to-clean surfaces.

11-9 SPECIFICATIONS— BRICK

The specifications must be checked to determine exactly the type of material and the type of mortar as well as the shape, thickness, and color of the joint itself. The style of bond must also be determined. From the specifications, the estimator also

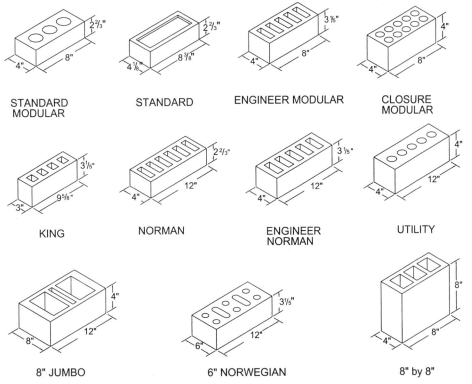

FIGURE 11.11. Nominal Brick Sizes.

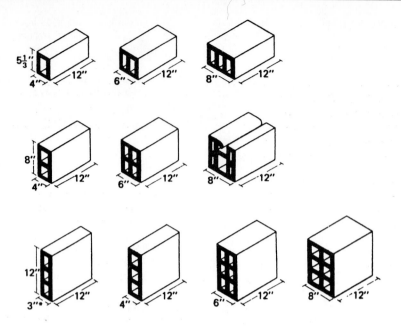

Note: All units shown are horizontal cell tile, also available in vertical cell tile.
*Actual dimension, all other dimensions are nominal ($\frac{1}{2}$ in. less than shown).

FIGURE 11.12. Structural Clay Tile.

(Courtesy of Brick Institute of America)

determines the types of lintels, flashing, reinforcing, and weep holes required, and who supplies and who installs each item.

Glazed-face brick with special shapes for stretchers, jambs, corners, sills, wall ends, and for use in other particular areas is available. Because these special pieces are relatively expensive, the estimator must allow for them. Also, the materials must meet all ASTM requirements of the specification.

Many specifications designate an amount, *cash allowance,* for the purchase of face brick. The estimator then allows the special amounts in the estimate. This practice allows the owner and architect/engineer to determine the exact type of brick desired at a later date.

11–10 ESTIMATING BRICK

The first thing to be determined in estimating the quantity of brick is the size of the brick and the width of the mortar joint. They are both necessary to determine the number of bricks per square foot of wall area and the quantity of mortar. Brick is sold by the thousand units, so the final estimate of materials required must be in the number of units required.

To determine the number of bricks required for a given project, the first step is to obtain the length and height of all walls to be faced with brick and then calculate the area of wall. Make deductions for all openings so that the estimate will be as accurate as possible. Check the jamb detail of the opening to determine whether extra brick will be required for the reveal; generally if the reveal is over 4 inches deep, extra brick will be required.

Once the number of square feet has been determined, the number of bricks can be calculated. This calculation varies depending on the size of the brick, width of the mortar joint, and style of bond required. The figures must be extremely accurate, as actual quantities and costs must be determined. It is only in this manner that estimators will increase their chances of getting work at a profit.

Figure 11.13 shows the number of bricks required per square foot of wall surface for various patterns and bonds. Special bond patterns require that the estimator analyze the style of bond required and determine the number of bricks. One method of analyzing the amount of brick required is to make a drawing of several square feet of wall surface, determine the brick to be used, and divide that into the total area drawn. Sketches are often made right on the workup sheets by the estimator.

Labor costs will be affected by lengths of straight walls, number of jogs in the wall, windows, piers, pilasters, and anything else that might slow the mason's work, such as the weather conditions. Also to be calculated are the amount of mortar required and any lintels, flashing, reinforcing, and weep holes that may be specified. Any special requirements such as colored mortar, shape of joint, and type of flashing are noted on the workup sheet.

7th Course bonding

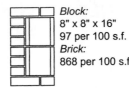

12" wall

Block:
8" x 8" x 16"
97 per 100 s.f.
Brick:
868 per 100 s.f.

Block:
4" x 8" x 16"
97 per 100 s.f.
Brick:
773 per 100 s.f.

8" wall

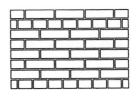

6th Course bonding

12" wall

Block regular:
8" x 8" x 16", 57 per 100 s.f.
Block header:
57 per 100 s.f.
Brick:
778 per 100 s.f.

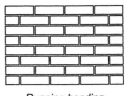

Running bonding

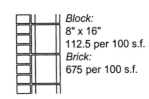

8" or 12" wall

Block:
8" x 16"
112.5 per 100 s.f.
Brick:
675 per 100 s.f.

FIGURE 11.13. Bricks Required per Square Foot of Wall Surface.

EXAMPLE 11-2   FACE BRICK

Using the west wall that was quantified in the previous example, most of the dimensions can be reused. The gross area will be slightly different, as the exterior face dimensions will be used rather than the face dimensions of the concrete block.

$$\text{Gross wall area} = 16'2'' \times 60' = 970 \text{ sf}$$

$$\text{Area of openings} = 105 \text{ sf (from Example 11-1)}$$

$$\text{Net wall area} = 970 \text{ sf} - 105 \text{ sf} = 865 \text{ sf}$$

Assuming modular brick, there are 675 bricks per 100 sf of wall (refer to Figure 11.6).

$$\text{Number of bricks} = 8.65 \text{ hundred sf of wall} \times 675$$

$$\text{Number of bricks} = 5,839 \text{ bricks}$$

$$\text{Waste @ 5 percent} - \text{Use } 6,131 \text{ bricks}$$

The labor hours and labor costs can be determined using the productivity rates found in Figure 11.14.

Type of Bond	Mason	Laborer
Common	10.0 to 15.0	11.0 to 15.5
Running	8.0 to 12.0	9.0 to 13.0
Stack	12.0 to 18.0	10.0 to 15.0
Flemish	11.0 to 16.0	12.0 to 16.0
English	11.0 to 16.0	12.0 to 16.0

Note:
1. The more corners and openings in the wall, the more labor hours it will require.
2. Add for the following brick types:

Norman Brick	+30%
Roman Brick	+25%
Jumbo Brick	+25%
Modular Brick	Same as Standard
Jumbo Utility Brick	+30%
Spartan Brick	+25%

FIGURE 11.14. Labor Hours Required to Lay 1,000 Standard Size Bricks.

11–11 SPECIFICATIONS—TILE

Check the specifications for the type, size, and special shapes of the tile. The type, thickness, and color of mortar must be noted as well as the type of lintels, flashing, and reinforcing.

A wide variety of glazed finishes and special shapes are available. The specifications and drawings must be studied to determine the shapes required for the project. Because special shapes tend to be expensive, they should be estimated carefully. Be certain that the materials specified meet all ASTM specifications.

11–12 ESTIMATING TILE

Hollow masonry units of clay tile are estimated first by the square feet of wall area required, each thickness of wall being kept separate. Then the number of units required is determined.

To determine the amount of tile required for a given project, the first step is to take off the length, height, and thickness of units required throughout. All openings should be deducted. Check the details for any special shapes or cuts that might be specified.

When the total square footage of a given thickness has been determined, calculate the number of units. Figure the size of the unit plus the width of mortar joint; this total face dimension is divided into square feet of masonry to determine the number of units.

The labor costs will be affected by jogs in the wall, openings, piers, pilasters, weather conditions, and the height of the building. Items that need to be included are mortar, lintels, flashing, reinforcing, weep holes, and any special shapes.

11–13 STONE MASONRY

Stone masonry is primarily used as a veneer for interior and exterior walls; it is also used for walkways, riprap, and trim on buildings. Stone masonry is usually divided into that which is laid up dry with no mortar being used—such as on some low walls, sloping walls, walkways, and rip-rap—and wet masonry, in which mortar is used.

Stone is used in so many sizes, bonds, and shapes that a detailed estimate is required. The types of stone most commonly used are granite, sandstone, marble, slate, limestone, and trap. The finishes available include various split finishes and tooled, rubbed, machine, cross-broached, and brushed finishes. Stone is generally available in random, irregular sizes, sawed-bed stone, and cut stone, which consists of larger pieces of cut and finished stone pieces. Random, irregular-sized stone is used for rubble masonry and rustic and cobblestone work; it is often used for chimneys, rustic walls, and fences. Sawed-bed stones are used for veneer work on the interior and exterior of buildings. Patterns used include random and coursed ashlar. Coursed ashlar has regular courses, whereas random ashlar has irregular-sized pieces and generally will require fitting on the job.

11–14 SPECIFICATIONS—STONE

Check exactly the type of stone required, the coursing, thickness, type, and color of mortar as well as wall ties, flashing, weep holes, and other special requirements. Not all types and shapes of stone are readily available. The supplier of the stone must be involved early in the bidding period. For large projects, the stone required for the job will actually be quarried and finished in accordance with the specifications. Otherwise, the required stone may be shipped in from other parts of the country or from other countries. In these special cases, estimators place a purchase order as soon as they are awarded the bid.

11–15 ESTIMATING STONE

Stone is usually estimated by the area in square feet, with the thickness given. In this manner, the total may be converted to cubic feet and cubic yards easily, while still providing the estimator with the basic square foot measurement. Stone trim is usually estimated by the linear foot.

Stone is sold in various ways—sometimes by the cubic yard, often by the ton. Cut stone is often sold by the square foot; of course, the square foot price goes up as the thickness increases. Large blocks of stone are generally sold by the cubic foot. It is not unusual for suppliers to submit lump-sum proposals whereby they will supply all of a certain type of stone required for a given amount of money. This is especially true for cut stone panels.

In calculating the quantities required, note that the length times height equals the square footage required; if the number must be in cubic feet, multiply the square footage by the thickness. Deduct all openings but usually not the corners. This calculation gives the volume of material required. The stone does not consume all the space; the volume of mortar must also be deducted. The pattern in which the stone is laid and the type of stone used will greatly affect the amount of mortar required. Cut stone may have 2 to 4 percent of the total volume as mortar, ashlar masonry 6 to 20 percent, and random rubble 15 to 25 percent. Waste is equally hard to anticipate. Cut and dressed stone has virtually no waste, whereas ashlar patterns may have 10 to 15 percent.

Dressing a stone involves the labor required to provide a certain surface finish to the stone. Dressing and cutting stone require skill on the part of the mason, which varies considerably from person to person. There is an increased tendency to have all stone dressed at the quarry or supplier's plant rather than on the job site.

The mortar used should be nonstaining mortar cement mixed in accordance with the specifications. When cut stone is used, some specifications require that the mortar joints be raked out and that a specified thickness of caulking be used. The type and quantity of caulking must then be taken off. The type, thickness, and color of the mortar joint must also be taken off.

Wall ties are often used for securing random, rubble, and ashlar masonry to the backup material. The type of wall tie specified must be noted, as must the number of wall ties per square foot. Divide this into the total square footage to determine the total number of ties required.

Cleaning must be allowed for as outlined in Section 11–19. Flashing should be taken off by the square foot and lintels by the linear foot. In each case note the type required, the supplier, and the installer.

Stone trim is used for door and windowsills, steps, copings, and moldings. The supplier prices trim by the linear foot or as a lump sum. Some type of anchor or dowel arrangement is often required for setting the pieces. The supplier will know who is supplying the anchors and dowels, and who will provide the anchor and dowel holes. The holes must be larger than the dowels being used.

Large pieces of cut stone may require cranes or other lifting devices to move it and set it in place. When large pieces of stone are being used as facing on a building, special inserts and attachments to hold the stone securely in place will be required.

All the aforementioned factors or items will also affect the labor hours required to put the stone in place.

11–16 MORTAR

The *mortar* used for masonry units may consist of portland cement, mortar, sand, and hydrated lime; or of masonry cement and sand. The amounts of each material required vary depending on the proportions of the mix selected, the thickness of the mortar joint (3/8 inch is a common joint thickness, but 1/4- and 1/2-inch joints are also used), and the color of the mortar. Dry mortar mix (a premixed combination of cement, lime, and sand) may be used on a project. There has been an increase in the use of colored mortar on the work; colors commonly used are red, brown, white, and black, but almost any color may be specified. Pure white mortar may require the use of white cement and white mortar sand; the use of regular mortar sand will generally result in a creamy color. The other colors are obtained by adding color pigments to the standard mix. Considerable trial and error may be required before a color acceptable to the owner and architect/engineer is found.

When both colored and gray mortars are required on the same project, the mixer used to mix the mortar must be thoroughly cleaned between mixings. If both types will be required, it may be most economical to use two mixers: one for colored mortar and the other for gray. White mortar should never be mixed in the mixer that is used for colored or gray mortar unless the mixer has been thoroughly cleaned.

EXAMPLE 11-3 MORTAR TAKEOFF

Example 11-1 contains 803 sf of block. Figure 11.6 details mortar at 3.2 cf per 100 sf of face area. Therefore, the required mortar

would be

Mortar for block = 8.03 squares of face area

$$\times\ 3.2\ \text{cf per square} = 26\ \text{cf}$$

$$\text{Waste @ 40 percent} = 36\ \text{cf}$$

From Example 11-2, there were 865 sf of brick. Using Figure 11.6, the required mortar is 7.2 cf per 100 sf of the face wall area. Therefore, the required mortar for the brick would be

Mortar for brick = 8.65 squares of face area

$$\times\ 7.2\ \text{cf per square} = 62\ \text{cf}$$

$$\text{Waste @ 40 percent} = 87\ \text{cf} \quad\blacksquare$$

Labor. The amount of labor required for mortar is usually considered a part of the labor to lay the masonry units.

11–17 ACCESSORIES

Masonry wall reinforcing. Steel reinforcing, which comes in a wide variety of styles and wire gauges, is placed continuously in the mortar joints (Figures 11.15 and 11.16). It is used primarily to minimize shrinkage, temperature, and settlement cracks in masonry, as well as to provide shear transfer to the steel. The reinforcing is generally available in lengths up to 20 feet. The estimator must determine the linear footage required. The drawings and specifications must be checked to determine the spacing required (sometimes every course, often every second or third course). The reinforcement is also used to tie the outer and inner wythes together in cavity wall construction. The reinforcing is available in plain or corrosion-resistant wire. The estimator should check the specifications to determine what is required and check with local suppliers to determine prices and availability of the specified material.

In calculating this item for 8″ high block, the easiest method for larger quantities (over 1,500 sf) is to multiply the square footage by the factor given:

Reinforcement every course 1.50

Second course 0.75

Third course 0.50

To this amount, add 5 percent waste and 5 percent lap.

For small quantities, the courses involved and the length of each course must be figured. Deduct only openings in excess of 50 sf.

Control Joints. A control joint is a straight vertical joint that cuts the masonry wall from top to bottom. The horizontal distance varies from 1/2 to about 2 inches. The joint must also be filled with some type of material; materials usually specified are caulking, neoprene and molded rubber, and copper and aluminum. These materials are sold by the linear foot and in a variety of shapes. Check the specifications to find the types required and check both the drawings and the specifications for the locations of control joints. Extra labor is involved in laying the masonry, since alternate courses utilizing half-size units will be required to make a straight vertical joint.

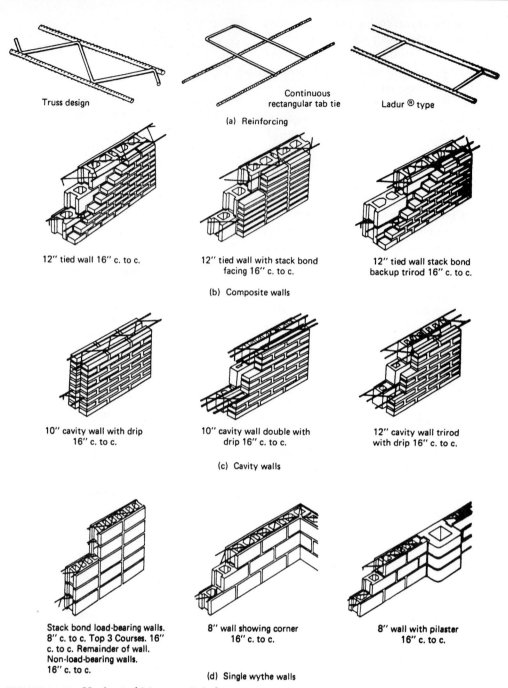

Truss design

Continuous
rectangular tab tie

Ladur ® type

(a) Reinforcing

12" tied wall 16" c. to c.

12" tied wall with stack bond
facing 16" c. to c.

12" tied wall stack bond
backup trirod 16" c. to c.

(b) Composite walls

10" cavity wall with drip
16" c. to c.

10" cavity wall double with
drip 16" c. to c.

12" cavity wall trirod
with drip 16" c. to c.

(c) Cavity walls

Stack bond load-bearing walls.
8" c. to c. Top 3 Courses. 16"
c. to c. Remainder of wall.
Non-load-bearing walls.
16" c. to c.

8" wall showing corner
16" c. to c.

8" wall with pilaster
16" c. to c.

(d) Single wythe walls

FIGURE 11.15. Horizontal Masonry Reinforcement.

(Courtesy of Dur-O-Wall)

Wall Ties.   Wall ties (Figure 11.17) are used to tie the outer wythe with the inner wythe. They allow the mason to construct one wythe of wall to a given height before working on the other wythe, resulting in increased productivity. Wall ties are available in a variety of sizes and shapes, including corrugated strips of metal 1¼ inches wide and 6 inches long (about 22 gauge) and wire bent to a variety of shapes. Adjustable wall ties are among the most popular, as they may be used where the coursing of the inner and outer wythes is not lined up. Noncorrosive metals or galvanized steel may be used. Check the specifications for the type of ties required and their spacing. To determine the amount required, take the total square footage of masonry and divide it by the spacing. A spacing of 16 inches vertically and 24 inches horizontally requires one tie for every 2.66 sf; a spacing of 16 inches vertically and 36 inches horizontally requires one tie for every 4.0 sf. Often, closer spacings are required. Also, allow for extra ties at control joints, wall intersections, and vertical supports as specified.

Flashing.   (Flashing is also discussed in Chapter 13, Thermal and Moisture Protection.) The flashing (Figure 11.18) built into the walls is generally installed by the mason. It is installed to keep moisture out and to divert any moisture

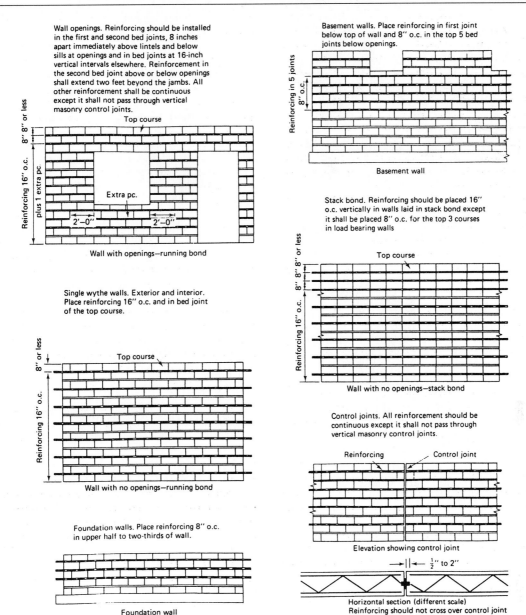

FIGURE 11.16. Reinforcement Layout.

(Courtesy of Dur-O-Wall)

that does get in back to the outside of the building. Flashing may be required under sills and copings, over openings for doors and windows, at intersections of roof and masonry wall, at floor lines, and at the bases of the buildings (a little above grade) to divert moisture. Materials used include copper, aluminum, copper-backed paper, copper and polyethylene, plastic sheeting (elastomeric compounds), wire and paper, and copper and fabric. Check the specifications to determine the type required. The drawings and specifications must also be checked to determine the locations in which the flashing must be used. Flashing is generally sold by the square foot or by the roll. A great deal of labor may be required to bend metal flashing into shape. Check carefully as to whether the flashing is to be purchased and installed under this section of the estimate, or whether it is to be pur-

chased under the roofing section and installed under the masonry section.

Weep Holes. In conjunction with the flashing at the base of the building (above grade level), weep holes are often provided to drain any moisture that might have gotten through the outer wythe. Weep holes may also be required at other locations in the construction. The maximum horizontal spacing for weep holes is about 3 feet, but specifications often require closer spacing. The holes may be formed by using short lengths of cord inserted by the mason or by well-oiled rubber tubing. The material used should extend upward into the cavity for several inches to provide a drainage channel through the mortar droppings that accumulate in the cavity.

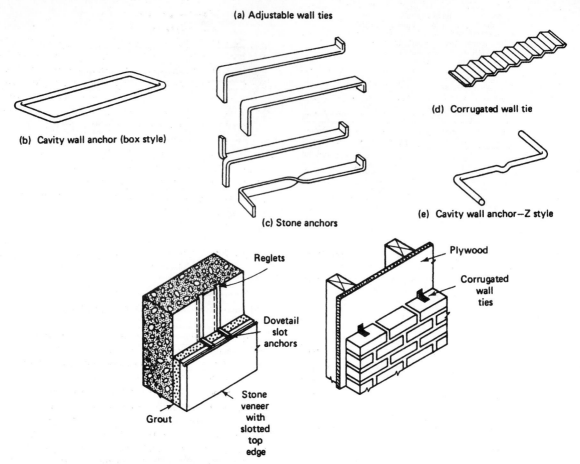

FIGURE 11.17. Typical Wall Ties and Wall Tie Installation.
(Courtesy of Dur-O-Wall)

Lintels. A *lintel* is the horizontal member that supports the masonry above a wall opening. In other words, it spans the opening. Materials used for lintels include steel angle iron, composite steel sections, lintel block (shaped like a U) with reinforcing bars and filled with concrete, and precast concrete. The lintels are usually set in place by masons as they lay up the wall. Some specifications require that the lintel materials be supplied under this section, whereas other specifications require the steel angles and composite steel section to be supplied under "structural steel" or "miscellaneous accessories." Precast lintels may be supplied under "concrete"; the lintel block will probably be included under "masonry," as will the reinforcing bars and concrete used in conjunction with it.

It is not unusual for several types of lintels, in a variety of sizes, to be required on any one project. They must be separated into the types, sizes, and lengths for each material used. Steel lintels may require extra cutting on the job so that the masonry will be able to fit around them. If a lintel is heavy, it may be necessary to use equipment (such as a lift truck or a crane) to put it in place. In determining length, be certain to take the full masonry opening and add the required lintel bearing on each end. Lintel bearing for steel is generally a minimum of 4 inches, whereas lintel block and precast lintels are often required to bear 8 inches on each end. Steel is purchased by the pound, precast concrete by the linear foot, and lintel block by the unit (note the width,

height, and length). Precast concrete lintels are covered in Section 10–18.

Sills. *Sills* are the members at the bottoms of window or door openings. Materials used are brick, stone, tile, and precast concrete. These types of sills are installed by the mason, although the precast concrete may be supplied under a different portion of the specifications. The brick and tile sills are priced by the number of units required, and the stone and precast concrete sills are sold by the linear foot. The estimator should check the maximum length of stone and precast concrete sills required and note it on the takeoff.

Also to be checked is the type of sill required: A *slip sill* is slightly smaller than the width of the opening and can be installed after the masonry is complete. A *lug sill*, which extends into the masonry at each end of the wall, must be built into the masonry as the job progresses. Some specifications require special finishes on the sill and will have to be checked. Also, if dowels or other inserts are required, that fact should be noted.

Coping. The *coping* covers the top course of a wall to protect it from the weather. It is most often used on parapet walls. Masonry materials used include coping block, stone, tile, and precast concrete. Check the specifications for the exact type required and who supplies it. The drawings will show the locations in which it is used, its shape, and how it is

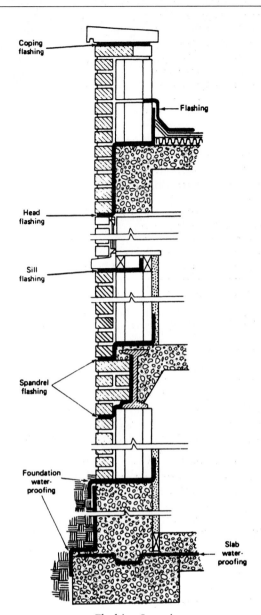

Coping
flashing

Flashing

Head
flashing

Sill
flashing

Spandrel
flashing

Foundation
water-
proofing

Slab
water-
proofing

FIGURE 11.18. Flashing Location.

to be attached. The coping block and tile are sold by the unit, and the stone and precast coping are sold by the linear foot. Figure 11.19 gives the labor productivity rates for installing masonry accessories.

Special colors, finishes, dowels, dowel holes, and inserts may be required. Check the drawing and specifications for these items and note all requirements on the workup sheet.

Accessories	Unit	Labor Hours
Wall Reinforcing	1,000 l.f.	1.0 to 1.5
Lintels – Precast Concrete (up to 300 pounds)	100 l.f.	4.5 to 7.0
Coping, Sills and Precast Concrete (up to 150 pounds)	100 l.f.	5.0 to 8.0
Wall Ties	100 Pieces	1.0 to 1.5

FIGURE 11.19. Labor Hours Required for Installing Accessories.

11–18   REINFORCING BARS

Reinforcing bars are often used in masonry walls to create multistory bearing wall construction; they are also used in conjunction with bond beam block and grout in bond beams used to tie the building together. Reinforcing bars are sold by weight: The various lengths required of each size are taken off, and the total weight of each size required is multiplied by the price.

The specifications should be checked to determine the type of steel required. If galvanized reinforcing bars are required, the cost for materials will easily double. Also, galvanized reinforcing bars must be special ordered, thus they are ordered quickly after the contract has been awarded (Section 10–3 also discusses reinforcing).

11–19   CLEANING

The specifications must be checked to determine the amount of cleaning required and the materials that must be used to clean. The materials exposed inside and outside of the building will probably require cleaning, while the concealed masonry, such as block used as a backup, generally receives no cleaning.

Clay Masonry.   For brickwork, there should be no attempt to clean for a minimum of 48 hours after the completion of the wall. After the minimum time, soap powder (or other mild solutions) with water and stiff brushes may be tried. When cleaning unglazed brick and tile, first use plain water and stiff brush. If these solutions do not work, the surface should be thoroughly wetted with clear water, scrubbed with a solution of acid and water, and thoroughly rinsed. Always try the acid solution on an inconspicuous area prior to using it on the entire wall. Acids should not be used on glazed facing tile.

Concrete Masonry.   Acid is not used on concrete masonry. If mortar droppings fall on the units, the droppings should be allowed to dry before removal to avoid smearing the face of the unit. When the droppings have dried, they can be removed with a trowel, and a final brushing will remove most of the mortar.

The estimator must determine the type of materials required for cleaning, the area of the surfaces to be cleaned, the equipment required, the amount of cleaning that will actually have to be done, and the number of labor hours that will be required. The better the workmanship on the job, the less money that has to be allowed for cleaning. When the color of the mortar is different from that of the masonry unit, the cost for cleaning will be higher because all mortar droppings must be cleaned off to get an unblemished facing.

Stone Masonry.   Clean stone masonry with a stiff fiber brush and clear water (soapy water may be used, if necessary). Then rinse with clear water to remove construction and mortar stains. Machine cleaning processes should be approved by the stone supplier before they are used. Wire brushes, acids, and sandblasting are not permitted for cleaning stonework.

11-20   EQUIPMENT

The equipment required for laying masonry units includes the mason's hand tools, mortar boxes, mortar mixer, hoes, hoses, shovels, wheelbarrows, mortar boards (tubs), pails, scaffolding, power hoist, hand hoist, elevator tower, hoisting equipment, telehandlers (long-reach forklifts), and lift trucks. The estimator must decide what equipment is required on the project, how much of each type is required, and the cost that must be allowed. The estimator must also remember to include the costs of ownership (or rental), operating the equipment, and mobilization to and from the project, erection, and dismantling. Items to be considered in determining the amount of equipment required are the height of the building, the number of times the scaffolding will be moved, the number of masons and helpers needed, and the type of units being handled.

11-21   COLD WEATHER

Cold weather construction is more expensive than warm weather construction. Increased costs stem from the construction of temporary enclosures so the masons can work, higher frequency of equipment repair, thawing materials, and the need for temporary heat.

Masonry should not be laid if the temperature is 40°F and falling or less than 32°F and rising at the place where the work is in progress, unless adequate precautions against freezing are taken. The masonry must be protected from freezing for at least 48 hours after it is laid. Any ice on the masonry materials must be thawed before use.

Mortar also has special requirements. Its temperature should be between 70°F and 120°F. During cold weather construction, it is common practice to heat the water used to raise the temperature of the mortar. Moisture present in the sand will freeze unless heated; upon freezing, it must be thawed before it can be used.

11-22   SUBCONTRACTORS

In most localities, masonry subcontractors are available. The estimator will have to decide whether it is advantageous to use a subcontractor on each project. The decision to use a subcontractor does not mean that the estimator does not have to prepare an estimate for that particular item; the subcontractor's bid must be checked to be certain that it is neither too high nor too low. Even though a particular contractor does not ordinarily subcontract masonry work, it is possible that the subcontractor can do the work for less money. There may be a shortage of masons, or the contractor's masonry crews may be tied up on other projects.

If the decision is made to consider the use of subcontractors, the first thing the estimator should decide upon is which subcontractors she wants to submit a proposal for the project. The subcontractors should be notified as early in the bidding period as possible to allow them time to make a thorough and complete estimate. Often the estimator will meet with the subcontractors to discuss the project in general and go over exactly which items are to be included in the proposal. Sometimes the proposal is for materials and labor, other times for labor only. Both parties must clearly understand the items that are to be included.

EXAMPLE 11-4   MASONRY TAKEOFF

Figures 11.20 and 11.21 are the completed masonry takeoffs for the commercial building found in Appendix A.

ESTIMATE WORK SHEET

Project: Little Office Building
Location: Littleville, Tx
Architect: U.R. Architects
Items:

Gross Wall Area

Estimate No. 1234
Sheet No. 1 of 1
Date 11/11/20xx
By LHF   Checked JBC

Cost Code	Description	Gross Length Ft.	Gross Length In.	Brick Ledge Width Ft.	Brick Ledge Width In.	Brick Ledge Count	Corner Thickness Ft.	Corner Thickness In	Corner Count	Gross Length (L.F.)	Gross Brick Ledge (L.F.)	Gross Corners (L.F.)	Net Length (L.F.)	Wall Height (F.T.)	Gross Side Total (S.F.)	Quantity	Unit
	Concrete Block																
	West Side	60	0	4.5	2				0	60	0.75	0	59.25	15.33	908.30	908	S.F.
	Front Side (North Wall)	40	0	4.5	2			8	2	40	0.75	1.3333	37.92	15.33	581.26		
		10	0	4.5	1			8	1	10	0.75	0.6667	8.58	15.33	131.58		
		2	0							2	0	0	2.00	15.33	30.66		
		2	0							2	0	0	2.00	15.33	30.66		
		10	0	4.5	1			8	1	10	0.75	0.6667	8.58	15.33	131.58		
		30	0	4.5	2			8	2	30	0.75	1.3333	27.92	15.33	427.96	1,334	S.F.
	East Side	35	0	4.5	2					35	0.75	0	34.25	15.33	525.05		
		15	0	4.5	1					15	0.75	0	14.25	15.33	218.45		
		2								2	0	0	2.00	15.33	30.66		
		1								1	0	0	1.00	15.33	15.33	789	S.F.
	Back Side (South)	100		4.5	2			8	1	100	0.75	0.6667	98.58	15.33	1,511.28	1,511	S.F.
													296.33				
																4,543	S.F.
	Brick Veneer																
	West Side	60								60	0	0	60.00	16.17	970.00		
	Front Side (North Wall)	40								40	0	0	40.00	16.17	646.67		
		10								10	0	0	10.00	16.17	161.67		
		2								2	0	0	2.00	16.17	32.33		
		2								2	0	0	2.00	16.17	32.33		
		10								10	0	0	10.00	16.17	161.67		
		30								30	0	0	30.00	16.17	485.00		
	East Side	35								35	0	0	35.00	16.17	565.83		
		15								15	0	0	15.00	16.17	242.50		
		2								2	0	0	2.00	16.17	32.33		
		1								1	0	0	1.00	16.17	16.17		
	Back Wall	85								85	0	0	85.00	16.17	1,374.17		
																4,721	S.F.

FIGURE 11.20. Gross Wall Area.

Project:	Little Office Building	ESTIMATE WORK SHEET	Estimate No.	1234
Location	Littleville, Tx		Sheet No.	1 of 1
Architect	U.R. Architects	**Masonry Quantities**	Date	11/11/20xx
Items	Foundation Concrete		By LHF	Checked JBC

Cost Code	Description	Comments / Calculations	Sub Totals	Quantity	Unit
	Lintel Blocks				
	Lintel Blocks Top Course	296 Feet / 16" per Block	200.00		
	Lintel Block For Bar Joists	40' + 30' + 15' + 85' = 170' Divide by 16	127.82		
	Lintel Blocks For Openings	4 per Opening - 17 Openings	68.00		
	Total Lentel Blocks		395.82		Lintel
	Required Lentel Blocks	Waste @ 5%		416	Blocks
	Concrete Blocks				
	Concrete Block Gross Area		4,543.00		
	Windows	16 @ 3' X 7'	-126.00		
	Doors	1 @ 3'-7'	-21.00		
	Net Wall Area		4,396.00		
	Gross Blocks	112.5 per 100 S.F.	4,945.50		
	Less Lentel Blocks		396.00		
			4,549.50		
	Required Blocks	Waste @ 6%		4,822	Blocks
	Mortar	3.2 C.F. per 100 S.F. (4543 S.F.)	145.38		
		Waste at 40 %		204	C.F. Mortar
	Horizontal Reinforcing	Every 3rd Course (.5 Factor)	2,271.50		
		Waste @ 5%		2,385	L.F.
	Vertical Reinforcing	# 3 Bars Every 6' (296/6) Spaces	49.33		
		Nominal 15 Ft Ht (L.F.)	740.00		
		Waste @ 20% (High because of Splices)		888	L.F.
	Face Brick				
	Gross Area		4,721.00		
	Windows	16 @ 3' X 7'	-126.00		
	Doors	1 @ 3'-7'	-21.00		
			4,700.00		
		675 per Square	28,999.00		
		Waste @ 5%		30,449	Face Bricks
	Mortar	7.2 C.F. per Square	338.40		
		Waste @ 40%		474	C.F. Mortar

FIGURE 11.21. Masonry Quantities.

11–23 CHECKLIST

Masonry:

type (concrete, brick, stone, gypsum)

kind

size (face size and thickness)

load-bearing

nonload-bearing

bonds (patterns)

colors

special facings

fire ratings

amount of cutting

copings

sills

steps

walks

Reinforcing:

bars

wall reinforcing

galvanizing (if required)

Mortar:

cement

lime (if required)

fine aggregate

water

admixtures

coloring

shape of joint

Miscellaneous:

inserts

anchors

bolts

dowels

reglets

wall ties

flashing

lintels

expansion joints

control joints

weep holes

WEB RESOURCES

www.masonryresearch.org

www.gobrick.com

www.masoncontractors.org

www.cement.org

www.masonrysociety.org

www.ncma.org

www.aecinfo.com

REVIEW QUESTIONS

1. What factors affect the costs of labor when estimating masonry?

2. How may the type of bond (pattern) affect the amount of materials required?

3. Why is high accuracy required with an item such as masonry?

4. Why should local suppliers be contacted early in the bidding process when special shapes or colors are required?

5. Why must the estimator separate the various sizes of masonry units in the estimate?

6. What is a cash allowance and how does it work?

7. How are stone veneer quantities estimated?

8. How may cold weather affect the cost of a building?

9. Determine the masonry, mortar, accessories, and any other related items required for the building in Appendix B.

10. Assume that the building in Appendix C has all exterior walls of 4-inch brick with 8-inch concrete block backup. Estimate the masonry, mortar, and accessories required. Note: Brick begins at elevation 104'4".

11. Using Billy's Convenience Store found in Appendix F, estimate the quantity of brick veneer and split-faced block.

CHAPTER TWELVE

METALS

12–1 GENERAL

This chapter covers structural metal framing, metal joists, metal decking, cold-formed framing and metal fabrications, ornamental metal, and expansion control assemblies. The structural products are most commonly made of steel and aluminum. The rest of the products are available in a wide range of metals, including steel, aluminum, brass, and bronze.

12–2 STRUCTURAL METAL

General contractors typically handle structural metal in one of two ways: either they purchase it fabricated and erect it with their own construction crew, or they have the company fabricate and erect or arrange erection by another company. Many contractors do not have the equipment and skilled personnel required to erect the structural metals. The structural metal includes all columns, beams, lintels, trusses, joists, bearing plates, girts, purlins, decking, bracing, tension rods, and any other items required.

When estimating structural metals, the estimator should quantify each item such as column bases, columns, trusses or lintels, separately. Structural metals are purchased by the ton. The cost per ton varies depending on the type and shape of metal required. Labor operations are different for each type.

The estimate of the field cost of erecting structural metals will vary depending on weather conditions, delivery of materials, equipment available, size of the building, and the amount of riveting, bolting, and welding required.

12–3 STRUCTURAL METAL FRAMING

The metals used for the framing of a structure primarily include wide flange beams (W series), light beams, I-beams, plates, channels, and angles. The wide flange shapes are the most commonly used today and are designated as shown in Figure 12.1.

Special shapes and composite members require customized fabrication and should be listed separately from the standard mill pieces. The estimator needs to be meticulous when quantifying each size and length. It is also a good practice to check not only the structural drawings, but also the sections and details. The specifications should include the fastening method required and the drawings of the fastening details.

Estimating. Both aluminum and steel are sold by weight, so the takeoff is made in pounds and converted into tons. The takeoff should first include a listing of all metals required for the structure. A definite sequence for the takeoff should be maintained. A commonly used sequence is columns and details, beams and details, and bracing and flooring (if required). Floor by floor, a complete takeoff is required.

Structural drawings, details, and specifications do not always show required items. Among the items that may not be shown specifically, but are required for a complete job, are various field connections, field bolts, ties, beam separators, anchors, turnbuckles, rivets, bearing plates, welds, setting plates, and templates. The specification may require conformance with American Institute of Steel Construction (AISC) standards, with the exact methods to be determined by the fabricator and erector. When this situation occurs, a complete understanding is required of the AISC and code requirements to do a complete estimate. Without this thorough

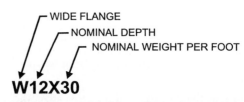

FIGURE 12.1. Wide Flange Beam Designation.

understanding, the estimator can make an approximate estimate for a check on subcontractor prices, but should not attempt to use those figures for compiling a bid price.

Once a complete takeoff of the structural metals has been made, they should be grouped according to the grade required and the shape of the structural piece. Special built-up shapes must be listed separately; in each area, the shapes should be broken down into small, medium, and large weights per foot. For standard mill pieces, the higher the weight per foot, the lower the cost per pound. The weights of the various standard shapes may be obtained from any structural steel handbook or from the manufacturing company. To determine the weight of built-up members, the estimator adds each of the component shapes used to make the special shape.

EXAMPLE 12-1   STRUCTURAL FRAMING TAKEOFF

Determine the structural metal framing requirements for the building in Appendix A (columns and beams). Figure 12.2 (Sheet S2.2 in Appendix A) is the roof framing plan and Figure 12.3 is the column details (Sheet S8.1 detail 6 in Appendix A).

From Figures 12.2 and 12.3 and details, calculations for structural metals can be found in Figure 12.4.

Weight of W12 × 30s that are 16′2″ long:

Weight (pounds) = 16′2″ × 30 pounds per foot × 3

Weight (pounds) = 16.167′ × 30 × 3 = 1,455 pounds

Figure 12.5 is a spreadsheet that contains the completed structural metals quantification for the roof framing (does not include basement metals).

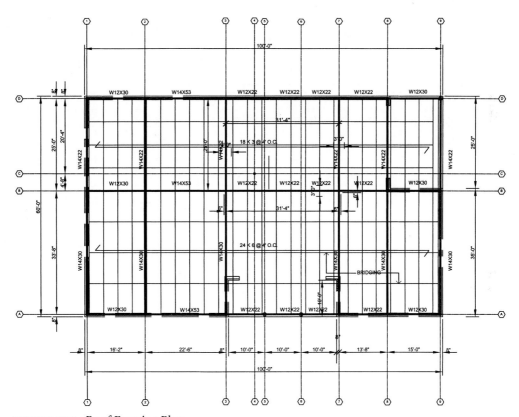

FIGURE 12.2. Roof Framing Plan.

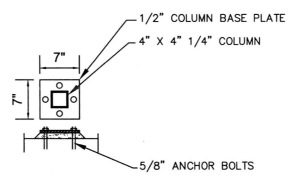

FIGURE 12.3. Column Details.

Designation	Length	Count
W12X30	16'-2"	3
W12X30	15'-0"	3
W12X22	10'-0"	9
W12X22	13'-8"	3
W14X22	25'-0"	6
W14X30	35'-0"	6
W14X53	22'-6"	3
4"X4"X1/4"		22

FIGURE 12.4. Structural Metals Takeoff.

ESTIMATE WORK SHEET
STRUCTURAL STEEL

Project:	Little Office Building									Estimate No.		1234
Location	Littleville, Tx									Sheet No.		1 of 1
Architect	U.R. Architects									Date		11/11/20XX
Items										By	LHF	Checked	JBC

Cost Code	Description	Designation	Pounds / Foot	Length Ft.	Length In.	Length Ft.	Count			Quantity	Unit
	Roof Framing (Beams)	W12X30	30.00	16.00	2.00	16.16667	3			1,455	Pounds
	Roof Framing (Beams)	W12X30	30	15	0	15	3			1,350	Pounds
	Roof Framing (Beams)	W12X22	22	10	0	10	9			1,980	Pounds
	Roof Framing (Beams)	W12X22	22	13	8	13.66667	3			902	Pounds
	Roof Framing (Beams)	W14X22	22	25	0	25	6			3,300	Pounds
	Roof Framing (Beams)	W14X30	30	35	0	35	6			6,300	Pounds
	Roof Framing (Beams)	W14X53	53	22	6	22.5	3			3,578	Pounds
	Columns	4" X 4" X 1/4"	12.02	12	0	12	22			3,173	Pounds
										22,038	Pounds

FIGURE 12.5. Structural Metals Quantification.

12–4 METAL JOISTS

Metal joists, also referred to as *open web steel joists,* are prefabricated lightweight trusses. Figure 12.6 shows various types of metal joists. The weight per foot is typically found in the manufacturer's catalog. Figures 12.7 and 12.8 are excerpts from a bar joist catalog that show weights and anchoring.

To determine the quantity of bar joists, calculate the linear footage of joists required and multiply this number by its weight per foot; the product represents the total weight. The cost per pound is considerably greater than that for some other types of steel (such as reinforcing bars or wide flange shapes) due to the sophisticated shaping and fabrication required.

Metal joists by themselves do not make an enclosed structure—they are one part of an assembly. The other materials included in this assembly are varied, and selection may be made based on economy, appearance, sound control, fire-rating requirements, or any other criterion. When doing the takeoff of joists, the estimator should be aware of the other parts in the assembly; they must all be included somewhere in the estimate. The mechanical and electrical requirements are generally quite compatible with steel joists; only large ducts have to be carefully planned for job installation.

Specifications. Specifications will list the type of joists required, the type of attachment they will have to the rest of

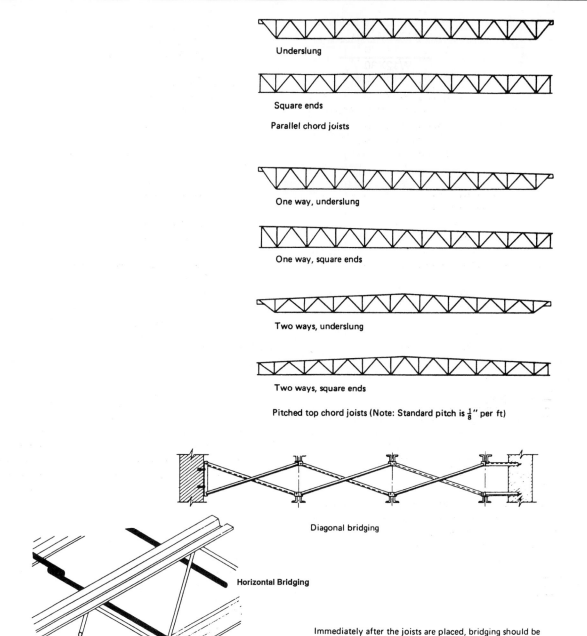

Undersling

Square ends

Parallel chord joists

One way, underslung

One way, square ends

Two ways, underslung

Two ways, square ends

Pitched top chord joists (Note: Standard pitch is $\frac{1}{8}$" per ft)

Diagonal bridging

Horizontal Bridging

Immediately after the joists are placed, bridging should be installed and welded at the intersections with the joists. This holds the joists in alignment and provides necessary lateral bracing during construction.

FIGURE 12.6. Common Steel Joist Shapes and Joist Bridging.

the structure, and the accessories and finish. Many specifications will enumerate industry standards that must be met for strength and for the type of steel used in the joists, erection and attachment techniques, and finishes. Read the standards carefully. Accessories that may be specified are bridging, ceiling extensions, masonry wall anchors, bridging anchors, and header angles.

Estimating.   Because structural metal is sold by the pound, the total number of pounds required must be determined. First, take off the linear feet of each different type of

joist required. Then multiply the lineal footage of each type times the weight per foot to determine the total weight of each separate type. Also, estimate the accessories required—both their type and the number needed.

If the contractor is to erect the joists, an equipment list, costs, and the labor hours are required. On small jobs, it is not unusual for the contractor to use his own forces to set the joists. Remember that all accessories must also be installed. If the joist erection is to be subcontracted, the general contractor's responsibilities must be defined, as well as who will install the accessories.

STANDARD LOAD TABLE / OPEN WEB STEEL JOISTS, K-SERIES

Based on a Maximum Allowable Tensile Stress of 30,000 psi

JOIST DESIGNATION	18K3	18K4	18K5	18K6	18K7	18K9	18K10	20K3	20K4	20K5	20K6	20K7	20K9	20K10	22K4	22K5	22K6	22K7	22K9	22K10	22K11
DEPTH (IN.)	18	18	18	18	18	18	18	20	20	20	20	20	20	20	22	22	22	22	22	22	22
APPROX. WT. (lbs./ft.)	6.6	7.2	7.7	8.5	9.0	10.2	11.7	6.7	7.6	8.2	8.9	9.3	10.8	12.2	8.0	8.8	9.2	9.7	11.3	12.6	13.8
SPAN (ft.) ↓																					
18	550/550	550/550	550/550	550/550	550/550	550/550	550/550														
19	514/494	550/523	550/523	550/523	550/523	550/523	550/523														
20	463/423	550/490	550/490	550/490	550/490	550/490	550/490	517/517	550/550	550/550	550/550	550/550	550/550	550/550							
21	420/364	506/426	550/460	550/460	550/460	550/460	550/460	468/453	550/520	550/520	550/520	550/520	550/520	550/520							
22	382/316	460/370	518/414	550/438	550/438	550/438	550/438	426/393	514/461	550/490	550/490	550/490	550/490	550/490	550/548	550/548	550/548	550/548	550/548	550/548	550/548
23	349/276	420/323	473/362	516/393	550/418	550/418	550/418	389/344	469/402	529/451	550/468	550/468	550/468	550/468	518/491	550/518	550/518	550/518	550/518	550/518	550/518
24	320/242	385/284	434/345	473/382	526/396	550/396	550/396	357/302	430/353	485/396	528/430	550/448	550/448	550/448	475/431	536/483	550/495	550/495	550/495	550/495	550/495
25	294/214	355/250	400/281	435/305	485/337	550/377	550/377	329/266	396/312	446/350	486/380	541/421	550/426	550/426	438/381	493/427	537/464	550/474	550/474	550/474	550/474
26	272/190	328/222	369/249	402/271	448/299	538/354	550/361	304/236	366/277	412/310	449/373	500/405	550/405	550/405	404/338	455/379	496/411	550/454	550/454	550/454	550/454
27	252/169	303/198	342/222	372/241	415/267	498/315	550/347	281/211	339/247	382/277	416/301	463/333	550/389	550/389	374/301	422/337	459/367	512/406	550/432	550/432	550/432
28	234/151	282/177	318/199	346/216	385/239	463/282	548/331	261/189	315/221	355/248	386/269	430/298	517/375	550/375	348/270	392/302	427/328	475/364	550/413	550/413	550/413
29	218/136	263/159	296/179	322/194	359/215	431/254	511/298	243/170	293/199	330/223	360/242	401/268	482/317	550/359	324/242	365/272	398/295	443/327	532/387	550/399	550/399
30	203/123	245/144	276/161	301/175	335/194	402/229	477/269	227/153	274/179	308/201	336/218	374/242	450/286	533/336	302/219	341/245	371/266	413/295	497/349	550/385	550/385
31	190/111	229/130	258/146	281/158	313/175	376/207	446/243	212/138	256/162	289/182	314/198	350/219	421/259	499/304	283/198	319/222	347/241	387/267	465/316	550/369	550/369
32	178/101	215/118	242/132	264/144	294/159	353/188	418/221	199/126	240/147	271/165	295/179	328/199	395/235	468/276	265/180	299/201	326/219	363/242	436/287	517/337	549/355
33	168/92	202/108	228/121	248/131	276/145	332/171	393/201	187/114	226/134	254/150	277/163	309/181	371/214	440/251	249/164	281/183	306/199	341/221	410/261	486/307	532/334
34	158/84	190/98	214/110	233/120	260/132	312/156	370/184	176/105	212/122	239/137	261/149	290/165	349/195	414/229	235/149	265/167	288/182	321/202	386/239	458/280	516/314
35	149/77	179/90	202/101	220/110	245/121	294/143	349/168	166/96	200/112	226/126	246/137	274/151	329/179	390/210	221/137	249/153	272/167	303/185	364/219	432/257	494/292
36	141/70	169/82	191/92	208/101	232/111	278/132	330/154	157/88	189/103	213/115	232/125	259/139	311/164	369/193	209/126	236/141	257/153	286/169	344/201	408/236	467/269
37								148/81	179/95	202/106	220/115	245/128	294/151	349/178	198/116	223/130	243/141	271/156	325/185	386/217	442/247
38								141/74	170/87	191/98	208/106	232/118	279/139	331/164	187/107	211/119	230/130	256/144	308/170	366/200	419/228
39								133/69	161/81	181/90	198/98	220/109	265/129	314/151	178/98	200/110	218/120	243/133	292/157	347/185	397/211
40								127/64	153/75	172/84	188/91	209/101	251/119	298/140	169/91	190/102	207/111	231/123	278/146	330/171	377/195
41															161/85	181/95	197/103	220/114	264/135	314/159	359/181
42															153/79	173/83	188/96	209/106	252/126	299/148	342/168
43															146/73	165/82	179/89	200/99	240/117	285/138	326/157
44															139/68	157/76	171/83	191/92	229/109	272/128	311/146

FIGURE 12.7. Joist Loading Table.

(Courtesy of Steel Joist Institute)

STANDARD LOAD TABLE / OPEN WEB STEEL JOISTS, K-SERIES

Based on a Maximum Allowable Tensile Stress of 30,000 psi

JOIST DESIGNATION	24K4	24K5	24K6	24K7	24K8	24K9	24K10	24K12	26K5	26K6	26K7	26K8	26K9	26K10	26K12	28K6	28K7	28K8	28K9	28K10	28K12
DEPTH (in.)	24	24	24	24	24	24	24	24	26	26	26	26	26	26	26	28	28	28	28	28	28
APPROX. WT. (lbs./ft.)	8.4	9.3	9.7	10.1	11.5	12.0	13.1	16.0	9.8	10.6	10.9	12.1	12.2	13.8	16.6	11.4	11.8	12.7	13.0	14.3	17.1
SPAN (ft.) ↓																					
24	520	550	550	550	550	550	550	550													
	516	544	544	544	544	544	544	544													
25	479	540	550	550	550	550	550	550													
	456	511	520	520	520	520	520	520													
26	442	499	543	550	550	550	550	550	542	550	550	550	550	550	550						
	405	453	493	499	499	499	499	499	535	541	541	541	541	541	541						
27	410	462	503	550	550	550	550	550	502	547	550	550	550	550	550						
	361	404	439	479	479	479	479	479	477	519	522	522	522	522	522						
28	381	429	467	521	550	550	550	550	466	508	550	550	550	550	550	548	550	550	550	550	550
	323	362	393	436	456	456	456	456	427	464	501	501	501	501	501	541	543	543	543	543	543
29	354	400	435	485	536	550	550	550	434	473	527	550	550	550	550	511	550	550	550	550	550
	290	325	354	392	436	436	436	436	384	417	463	479	479	479	479	486	522	522	522	522	522
30	331	373	406	453	500	544	550	550	405	441	492	544	550	550	550	477	531	550	550	550	550
	262	293	319	353	387	419	422	422	346	377	417	457	459	459	459	439	486	500	500	500	500
31	310	349	380	424	468	510	550	550	379	413	460	509	550	550	550	446	497	550	550	550	550
	237	266	289	320	350	379	410	410	314	341	378	413	444	444	444	397	440	480	480	480	480
32	290	327	357	397	439	478	549	549	356	387	432	477	519	549	549	418	466	515	549	549	549
	215	241	262	290	318	344	393	393	285	309	343	375	407	431	431	361	400	438	463	463	463
33	273	308	335	373	413	449	532	532	334	364	406	448	488	532	532	393	438	484	527	532	532
	196	220	239	265	289	313	368	368	259	282	312	342	370	404	404	329	364	399	432	435	435
34	257	290	315	351	388	423	502	516	315	343	382	422	459	516	516	370	412	456	496	516	516
	179	201	218	242	264	286	337	344	237	257	285	312	338	378	378	300	333	364	395	410	410
35	242	273	297	331	366	399	473	501	297	323	360	398	433	501	501	349	389	430	468	501	501
	164	184	200	221	242	262	308	324	217	236	261	286	310	356	356	275	305	333	361	389	389
36	229	258	281	313	346	377	447	487	280	305	340	376	409	486	487	330	367	406	442	487	487
	150	169	183	203	222	241	283	306	199	216	240	263	284	334	334	252	280	306	332	366	366
37	216	244	266	296	327	356	423	474	265	289	322	356	387	460	474	312	348	384	418	474	474
	138	155	169	187	205	222	260	290	183	199	221	242	260	308	315	232	257	282	305	344	344
38	205	231	252	281	310	338	401	461	251	274	305	337	367	436	461	296	329	364	396	461	461
	128	143	156	172	189	204	240	275	169	184	204	223	241	284	299	214	237	260	282	325	325
39	195	219	239	266	294	320	380	449	238	260	289	320	348	413	449	280	313	346	376	447	449
	118	132	144	159	174	189	222	261	156	170	188	206	223	262	299	198	219	240	260	306	308
40	185	208	227	253	280	304	361	438	227	247	275	304	331	393	438	266	297	328	357	424	438
	109	122	133	148	161	175	206	247	145	157	174	191	207	243	269	183	203	222	241	284	291
41	176	198	216	241	266	290	344	427	215	235	262	289	315	374	427	253	283	312	340	404	427
	101	114	124	137	150	162	191	235	134	146	162	177	192	225	256	170	189	206	224	263	277
42	168	189	206	229	253	276	327	417	205	224	249	275	300	356	417	241	269	297	324	384	417
	94	106	115	127	139	151	177	224	125	136	150	164	178	210	244	158	175	192	208	245	264
43	160	180	196	219	242	263	312	406	196	213	238	263	286	339	407	230	257	284	309	367	407
	88	98	107	118	130	140	165	213	116	126	140	153	166	195	232	147	163	179	194	228	252
44	153	172	187	209	231	251	298	387	187	204	227	251	273	324	398	220	245	271	295	350	398
	82	92	100	110	121	131	154	199	108	118	131	143	155	182	222	137	152	167	181	212	240
45	146	164	179	199	220	240	285	370	179	194	217	240	261	310	389	210	234	259	282	334	389
	76	86	93	103	113	122	144	185	101	110	122	133	145	170	212	128	142	156	169	198	229
46	139	157	171	191	211	230	272	354	171	186	207	229	250	296	380	201	224	248	270	320	380
	71	80	87	97	106	114	135	174	95	103	114	125	135	159	203	120	133	146	158	186	219
47	133	150	164	183	202	220	261	339	164	178	199	219	239	284	369	192	214	237	258	306	372
	67	75	82	90	99	107	126	163	89	96	107	117	127	149	192	112	125	136	148	174	210
48	128	144	157	175	194	211	250	325	157	171	190	210	229	272	353	184	206	227	247	294	365
	63	70	77	85	93	101	118	153	83	90	100	110	119	140	180	105	117	128	139	163	201
49									150	164	183	202	220	261	339	177	197	218	237	282	357
									78	85	94	103	112	131	169	99	110	120	130	153	193
50									144	157	175	194	211	250	325	170	189	209	228	270	350
									73	80	89	97	105	124	159	93	103	113	123	144	185
51									139	151	168	186	203	241	313	163	182	201	219	260	338
									69	75	83	91	99	116	150	88	97	106	115	136	175
52									133	145	162	179	195	231	301	157	175	193	210	250	325
									65	71	79	86	93	110	142	83	92	100	109	128	165
53																151	168	186	203	240	313
																78	87	95	103	121	156
54																145	162	179	195	232	301
																74	82	89	97	114	147
55																140	156	173	188	223	290
																70	77	85	92	108	139
56																135	151	166	181	215	280
																66	73	80	87	102	132

FIGURE 12.7. (*Continued*)

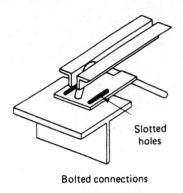

Bolted connections

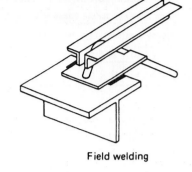

Field welding

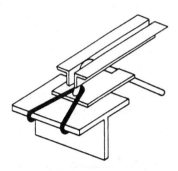

Beam anchors

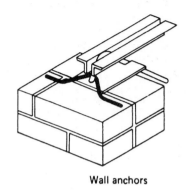

Wall anchors

FIGURE 12.8. End Anchorage.

EXAMPLE 12-2   BAR JOISTS

Determine the quantity of bar joists required to frame the roof in drawing S2.2 in Appendix A (Figure 12.2). From these drawings, the bar joist quantity table, as shown in Figure 12.9, can be developed.

Designation	Length	Count
18K3	25'-0"	23
24K6	35'-8"	23

FIGURE 12.9. Bar Joist Takeoff.

From Figure 12.7, the 18K3 weighs 6.6 pounds per foot and the 24K6 weighs 9.7 pounds per foot.

$$\text{Linear feet of 18K3} = 25' \times 23 = 575 \text{ ft}$$

$$\text{Linear feet of 24K6} = 35'8'' \times 23 = 820 \text{ ft}$$

$$\text{Pounds of 18K3} = 575 \text{ ft} \times 6.6 \text{ pounds per foot} = 3,795 \text{ pounds}$$

$$\text{Pounds of 24K6} = 820 \text{ ft} \times 9.7 \text{ pounds per foot} = 7,954 \text{ pounds}$$

$$\text{Total weight of bar joist} = 3,795 \text{ pounds} + 7,954 \text{ pounds}$$

$$\text{Total weight of bar joist} = 11,749 \text{ pounds}/2,000 \text{ pounds per ton}$$

$$\text{Total weight of bar joist} = 5.9 \text{ tons} \quad \blacksquare$$

Labor.   The time required for the installation of steel joists may be taken from Figure 12.10. The hourly wages are based on local labor and union conditions. A crane is usually required for joists over 300 pounds. The crane would reduce installation time on large projects and justify the cost of the crane, even if the joist weights were less than 300 pounds.

12-5   METAL DECKING

Metal decking is used for floor and roof applications. Depending on the particular requirements of the job, a wide selection of shapes, sizes, thicknesses, and accessories are available.

Roof applications range from simple decks over which insulation board and built-up roofing are applied, to forms and reinforcing over which concrete may be poured, and to decking that can receive recessed lighting and has acoustical properties. Depending on the type used and the design of the deck, allowable spans range from 3 feet to about 33 feet.

Decking for floor applications is equally varied from the simplest type used as a form and reinforcing for concrete to elaborate systems that combine electrical and telephone outlets, electrical raceways, air ducts, acoustical finishes, and recessed lighting.

Joist Type	Labor Hours per Ton
J & K, up to 30 Ft.	5.5 to 9.0
J & K, over 30 Ft.	4.5 to 8.0
LJ & LH	4.0 to 6.0
DLJ & DLH	4.0 to 6.0

FIGURE 12.10. Bar Joist Installation Productivity Rates.

Decking is generally available either unpainted, primed, painted, or galvanized. Accessories available include flexible rubber closures to seal the flutes, clips that fasten the decking to the purlins, lighting, and acoustical finishes.

Specifications. Determine the type of decking required, thickness, gauge of metal, finish required on the decking as received from the supplier, method of attachment, accessories required, and the specified manufacturers.

Items that are necessary for the completion of the decking and which must be included in the specifications include painting of the underside of the deck, acoustical treatment, openings, and insulation.

Estimating. Because metal decking is priced by the square (100 square feet), the first thing the estimator must determine is how many squares are required. Again, a systematic plan should be used: Start on the floor on which the decking is first used and work up through the building, keeping the estimates for all floors separate.

Decking is usually installed by welding directly through the bottom of the rib, usually a maximum of 12 inches on center, with side joints mechanically fastened not more than 3 feet on center. The estimator will have to determine approximately how many weld washers will be required and how long it will take to install them. Otherwise, fastening is sometimes effected by clips, screws, and bolts.

The estimator should consult local dealers and suppliers for material prices to be used in preparing the estimate. Materials priced f.o.b. the dealer will require that the estimator add the cost of transporting the materials to the job site. Once at the job site, they must be unloaded, perhaps stored, and then placed on the appropriate floor for use. In most cases, one or two workers can quickly and easily position the decking and make preparations for the welder to make the connections.

The estimator must be especially careful on multifloor buildings to count the number of floors requiring the steel deck and keep the roof deck and any possible poured concrete for the first floor separate. Most estimators make a small sketch of the number of floors to help avoid errors. In checking the number of floors, be aware that many times there may also be a lobby level, lower lobby level, and basement.

EXAMPLE 12-3 DECKING

Determine decking for the sample project in Appendix A (Figure 12.2).

Roof area (sf) = 60′ × 100′ = 6,000 sf

Roof area (sq) = 6,000 sf/100 sf per sq = 60 sq

Add 5 percent for waste — Use 63 sq

From the labor productivity rates in Figure 12.11 and assuming a prevailing wage rate of $20.65 per hour, the following labor cost calculations can be performed.

Using 22 gauge decking and 2 labor hours per square:

Labor hours = 6.3 sq × 2 labor hours per sq = 12.6 labor hours

Labor cost ($) = 12.6 hours × $20.65 per hour = $260.19

Steel Decking (gauge)	Labor Hours per Square
22	1.0 to 2.5
18	2.0 to 4.0
14	2.5 to 4.5

FIGURE 12.11. Steel Decking Productivity Rates. ■

12-6 MISCELLANEOUS STRUCTURAL METAL

Other types of structural framing are sometimes used. They include structural aluminum or steel studs, joists, purlins, and various shapes of structural pipe and tubes. The procedure for estimating each of these items is the same as that outlined previously for the rest of the structural steel.

1. Take off the various types and shapes.
2. Determine the pounds of each type required.
3. The cost per ton times the tonnage required equals the material cost.
4. Determine the labor hours and equipment required and their respective costs.

12-7 METAL ERECTION SUBCONTRACTORS

Most of the time, these subcontractors can erect structural steel at considerable savings compared with the cost to the average general contractor. They have specialized, well-organized workers to complement the required equipment that includes cranes, air tools, rivet busters, welders, and impact wrenches. These factors, when combined with the experienced organization that specializes in one phase of construction, are hard to beat.

Using subcontractors never lets the estimator off the hook; an estimate of steel is still required because cost control is difficult to maintain and improve upon.

12-8 MISCELLANEOUS METALS

Metal Fabrications. The metal stairs, ladders, handrails, railings, gratings, floor plates, and any castings are all considered part of metal fabrications. These items are typically manufactured to conventional (standard) details.

The estimator should carefully review the drawings and specifications of each item. Next, possible suppliers should be called in to discuss pricing and installation. It may be possible to install with the general contractor's crews, or special installation or equipment may be needed, requiring the job to be subcontracted.

Ornamental Metal. Ornamental metals include ornamental stairs, prefabricated spiral stairs, ornamental handrails and railings, and ornamental sheet metal. These types of products are most commonly made out of steel, aluminum, brass, and bronze and are often special ordered.

The estimator should make notes on the types of materials required, catalog numbers (if available), and quantities required and call a supplier that can provide them. Installation may be with the general contractor's work crews, or a subcontractor may be used.

Expansion Control Covers. Expansion control covers include the manufactured, cast or extruded metal expansion joint frames and covers, slide bearings, anchors, and related accessories. These materials are taken off by the linear foot, with the materials and any special requirements noted. Many times they are installed by a subcontractor doing related work. For example, an expansion joint cover being used on an exterior block wall might be installed by the masonry subcontractor (if one is used). In other cases, the general contractor may assume the responsibility. The key is to be certain that both material and installation are included in the estimate somewhere.

12-9 METAL CHECKLIST

Shapes:
sections
weights
locations

fasteners

Engineering:
fabrication

shop drawings
shop painting
testing
inspection
unloading, loading
erection
plumbing up

Installation:
riveting
welding
bracing (cross and wind)
erection
bolts

Miscellaneous:
clips
ties
rods
lengths
quantities
painting

hangers
plates
anchor bolts

Fabrications:
metal stairs
ladders
handrails and railings
(pipe and tube)
gratings and floor plates
castings

Ornamental:
stairs
prefab spiral stairs
handrails and railings
metal castings
sheet metal

Expansion control:
expansion joint cover
assemblies
slide bearings

WEB RESOURCES

www.sdi.org

www.daleincor.com

www.asce.org

www.mca1.org

www.steeltubeinstitute.org

www.steeljoist.com

www.vulcraft-in.com

REVIEW QUESTIONS

1. What two materials are most commonly used for structural framing metals, and how are they priced?

2. Under what conditions might it be desirable for the contractor to use a structural subcontractor to erect the structural metal frame of the project?

3. Why should the estimator list each of the different shapes (such as columns and steel joists) separately?

4. What is the unit of measure for metal decks? What type of information needs to be noted?

5. How are fabricated metal and ornamental metal usually priced?

6. Estimate the metal decking requirements for the building in Appendix C.

7. Using the drawings for Billy's Convenience Store (Appendix F), prepare an estimate for the structural steel found in the roof structure.

CHAPTER THIRTEEN

WOOD

13-1 FRAME CONSTRUCTION

The wood frame construction discussed in this chapter relates primarily to light construction of ordinary wood buildings. It covers the rough carpentry work, which includes framing, sheathing, and subfloors. Flooring, roofing, drywall and wet wall construction, and insulation are all included in their respective chapters, and discussions of them are not repeated here.

The lumber most commonly used for framing is yard or common lumber; the size classification of lumber most commonly used is dimensional (2 to 4 inches thick and in a variety of widths). The allowable spans of wood depend on the loading conditions and the allowable working stresses for the wood. As higher working stresses are required, it may become necessary to go to a stress-grade lumber that has been assigned working stresses. This type of lumber is graded by machine testing and is stamped at the mill. This information is available in the pamphlet from your local lumber supplier.

Rough lumber is wood that has been sawed, edged, and trimmed, but has not been dressed. When lumber is surfaced by a planing member, it is referred to as *dressed*. This process gives the piece a uniform size and a smooth surface. Before the lumber is dressed, it has the full given dimension (i.e., 2 × 4 inches), but once it has been surfaced on all four sides (S4S), it will actually measure less. For example, a 2 by 4 is actually 1½ inches by 3½ inches. The full size of lumber is referred to as its nominal size and is the size by which it is quantified and purchased.

The only safe way to estimate the quantity of lumber required for any particular job is to do a takeoff of each piece of lumber needed for the work. Since the time needed to do such an estimate is excessive, tables are included to provide as accurate material quantities as are necessary in as short a time as possible. These are the methods generally used.

13-2 BOARD MEASURE

When any quantity of lumber is purchased, it is priced and sold by the thousand feet board measure, abbreviated *mbm* or more commonly *mbf* (thousand board feet). The estimator must calculate the number of board feet required on the job.

One board foot is equal to the volume of a piece of wood 1 inch thick and 1 foot square (Figure 13.1A). By using Formula 13-1, the number of board feet can be quickly determined. The nominal dimensions of the lumber are used even when calculating board feet.

$$N = P \times \frac{T \times W}{12} \times L \qquad \textbf{Formula 13-1}$$

Board Foot Measurement

N = Number of feet (board measure)
P = Number of pieces of lumber
T = Thickness of the lumber (in inches)
W = Width of the lumber (in inches)
L = Length of the pieces (in feet)
12 = inches (one foot), a constant—it does not change

EXAMPLE 13-1 CALCULATING BOARD FEET

Estimate the board feet of 10 pieces of lumber 2 × 6 inches and 16 feet long (it would be written 10–2 × 6s @ 16′0″ (see Figure 13.1B).

$$\text{Board Feet} = 10 \times \frac{2'' \times 6''}{12''} \times 16' = 160 \text{ bf}$$

This formula can be used for any size order and any number of pieces that are required. Figure 13.2 shows the board feet, per piece, for some typical lumber sizes and lengths. Multiply the board feet by the number of pieces required for the total fbm. ∎

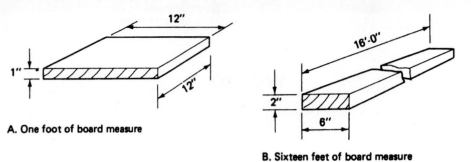

FIGURE 13.1. Board Measures, Example.

Nominal Size Of Lumber	Length of pieces in feet						
	8	10	12	14	16	18	20
2 x 4	5 1/3	6 2/3	8	9 1/3	10 2/3	12	13 1/3
2 x 6	8	10	12	14	16	18	20
2 x 8	10 2/3	13 1/3	16	18 2/3	21 1/3	24	26 2/3
2 x 10	13 1/3	16 2/3	20	23 1/3	26 2/3	30	33 1/3
2 x 12	16	20	24	28	32	36	40
2 x 14	18 2/3	23 1/3	28	32 2/3	37 1/3	42	46 2/3
2 X 16	21 1/3	26 2/3	32	37 1/3	42 2/3	48	53 1/3

FIGURE 13.2. Typical Board Feet.

13–3 FLOOR FRAMING

In beginning a wood framing quantity takeoff, the first portion estimated is the floor framing. As shown in Figure 13.3, the floor framing generally consists of a girder, sill, floor joist, joist headers, sheathing, and subflooring. The first step in the estimate is to determine the grade (quality) of lumber required and to check the specifications for any special requirements. This information is then noted on the workup sheets.

Girders. When the building is of a width greater than that which the floor joists can span, a beam of some type is required. A built-up wood member, referred to as a *girder,* is often used. The sizes of the pieces used to build up the girder must be listed, and the length of the girder must be noted. If the span becomes very long, a flitch beam may be required. This type of beam consists of a steel plate that is "sandwiched" between two pieces of framing lumber. In addition, many mills have developed engineered wood beams that have shapes with webs and flanges that can accommodate large spans and loads.

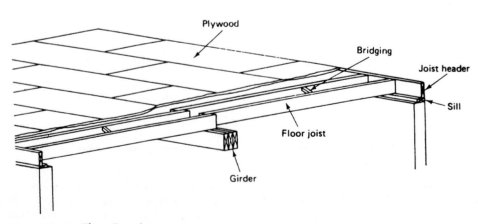

FIGURE 13.3. Floor Framing.

EXAMPLE 13-2   GIRDER TAKEOFF

From Figures 13.4 and Figures 13.5, the inside foundation wall dimension and girder length can be determined.

Inside dimension = Outside dimension

− Foundation wall thickness

Inside dimensions = $50'0'' - 0'8'' - 0'8''$

Inside dimensions = $50' - 0.67' - 0.67' = 48.66'$

Inside dimensions = $48'8''$

Girder length = Inside dimension + Bearing distance

Girder length = $48'8'' + 6'' + 6''$

Girder length = $48.67' + 0.5' + 0.5' = 49.67'$

Girder length = $49'8''$

Because the girders consist of three 2

$\times$ 10s, there are 149 lf of 2 $\times$ 10s.

Use 10′ lengths to minimize waste.

Purchase Quantity

Size	Length (lf)	Pieces
2 × 10s	10	15

$$\text{Board feet} = 15 \times \frac{2'' \times 10''}{12} \times 10' = 250 \text{ bf}$$

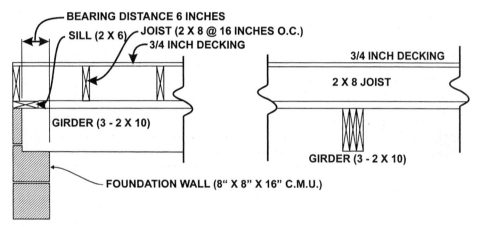

FIGURE 13.4. Bearing Distance.

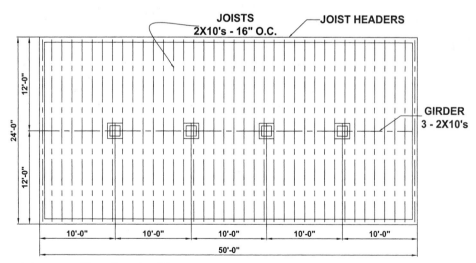

FIGURE 13.5. Floor Framing Plan.

Sills.   Sill plates are most commonly 2 × 4s, 2 × 6s, 2 × 8s, and 2 × 10s and are placed on the foundation so that the length of sill plate required is the distance around the perimeter of the building. Lengths ordered will depend on the particular building. Not all frame buildings require a sill plate, so the details should be checked; but generally, where there are floor joists, there are sill plates.

The length of sill is often taken off as the distance around the building. An exact takeoff would require that the estimator allow for the overlapping of the sill pieces at the corners. This type of accuracy is required only when the planning of each piece of wood is involved on a series of projects, such as mass-produced housing, and the amount becomes significant.

EXAMPLE 13-3   SILL QUANTITY

The exact sill length is shown in the following calculation. However, the perimeter of 148′ is sufficient for this example.

Sill length = 50′0″ + 50′0″ + 23′1″ − 23′1″

Sill length = 50′ + 50′ + 23.083′ + 23.083′ = 146.166′

Using the building perimeter of 148′, the following quantity of sill material would be purchased. Since most stock lumber comes in even increments, the purchase quantities have been rounded up to the nearest stock length. Furthermore, the sill lumber is typically treated and should be kept separate on the quantity takeoff since treated lumber is more expensive.

Purchase Quantity

Size	Length (lf)	Pieces
2 × 6	10	10
2 × 6	12	4

Total Length = 10 × 10′ + 4 × 12′ = 148′

$$\text{Board feet} = 1 \times \frac{2'' \times 6''}{12} \times 148' = 148 \text{ bf} \quad \blacksquare$$

Wood Floor Joists.   The wood joists should be taken off and separated into the various sizes and lengths required. The spacing most commonly used for joists is 16 inches on center, but spacings of 12, 20, 24, 30, and 36 inches on center are also found. The most commonly used sizes for floor joists are 2 × 6s, 2 × 8s, 2 × 10s, and 2 × 12s, although wider and deeper lumber is sometimes used. Wood I-joists (such as TJIs), constructed of engineered lumber, are commonly used for joists. Wood I-joists are taken off in the same manner as 2 × joists, except additional time and materials are needed to provide web stiffeners (Figure 13.6) at the joist ends and other bearing points.

To determine the number of joists required for any given area, the length of the floor is divided by the joist spac-

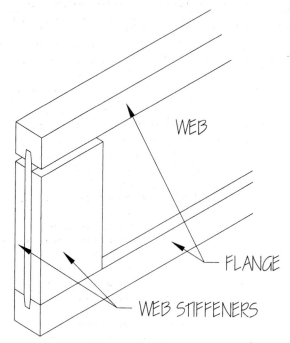

FIGURE 13.6. Wood I-Joist.

ing, and then one joist is added for the extra joist that is required at the end of the span. If the joists are to be doubled under partitions or if headers frame into them, one extra joist should be added for each occurrence. Factors for various joist spacings are given in Figure 13.7.

The length of the joist is taken as the inside dimension of its span plus 4 to 6 inches at each end for bearing on the wall, girder, or sill.

Joist Estimating Steps

1. From the foundation plan and wall section, determine the size of the floor joists required.

2. Determine the number of floor joists required by first finding the number of spaces, then by adding one extra joist to enclose the last space.

3. Multiply by the number of bays.

4. Add one extra for partitions that run parallel to the joists.

5. Determine the required length of the floor joists.

O.C. spacing (inches)	Divide length to be framed by	OR	Multiply length to be framed by
12	1		1
16	1.33		0.75
20	1.67		0.60
24	2		0.50
30	2.5		0.40
36	3		0.33

Answer gives number of spaces involved. Add one (1) to obtain the number of framing members required.
This table makes no allowance for waste, doubling of members or intersecting walls (for stud takeoff).

FIGURE 13.7. Joist Spacing Table.

EXAMPLE 13-4 JOIST QUANTITY

Joist length 12′ (refer to Figures 13.8, 13.9, 13.10, and 13.11)

Spaces = 50′0″ × 0.75 spaces/ft = 38 spaces

(add 1 to convert from spaces to joists)

Use 39 joists

Joists (2 bays) = 39 × 2 = 78 joists

Extra joists under walls (refer to Figure 13.8)

7 additional joists

Purchase Quantity

Size	Length (lf)	Pieces
2 × 10	12	85

$$\text{Board feet} \;=\; 85 \times \frac{2'' \times 10''}{12} \times 12' \;=\; 1{,}700 \text{ bf}$$

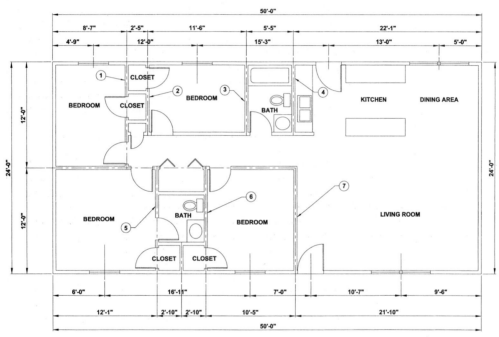

FIGURE 13.8. Floor Plan.

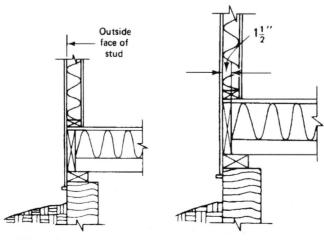

FIGURE 13.9. Joist Headers.

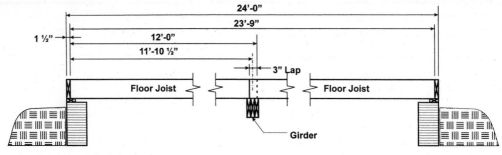

FIGURE 13.10. Joist Dimensions.

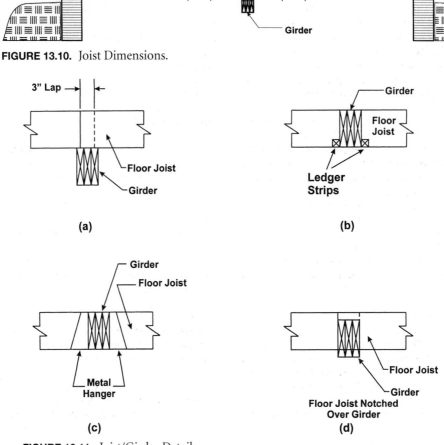

(a)

(b)

(c)

(d)

FIGURE 13.11. Joist/Girder Details.   ■

Trimmers and Headers.   Openings in the floor, such as for stairs or fireplaces, are framed with trimmers running in the direction of the joists and headers which support the cut-off "tail beams" of the joist (Figure 13.12).

Unless the specifications say otherwise, when the header length (Figure 13.13) is 4 feet or less, most codes allow single trimmers and headers to be used. For header lengths greater than 4 feet, codes usually require double trimmers and headers (Figure 13.14), and under special conditions some codes even require them to be tripled.

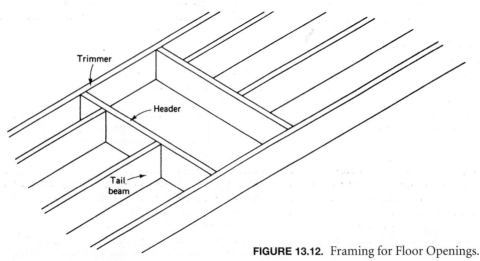

FIGURE 13.12. Framing for Floor Openings.

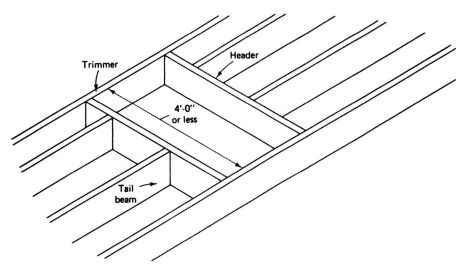

FIGURE 13.13. Single Trimmers and Headers.

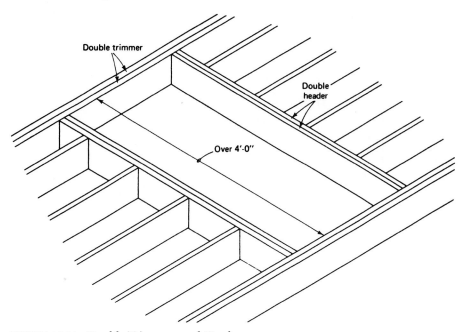

FIGURE 13.14. Double Trimmers and Headers.

To determine the extra material required for openings, two situations are investigated:

EXAMPLE 13-5   HEADER AND TRIMMER JOISTS

1. 3'0" × 4'0" opening as shown in Figure 13.15 (access to attic)
2. 3'0" × 8'0'' opening as shown in Figure 13.15 (stairway opening)

The material is determined by the following:

1. Sketching the floor joists without an opening (Figure 13.16).
2. Locating the proposed opening on the floor joist sketch (Figure 13.17).
3. Sketching what header and trimmer pieces are required and how the cut pieces of the joist may be used as headers and trimmers (Figure 13.18).

Purchase Quantity

4-Foot Opening

Size	Length (lf)	Pieces	Comments
2 × 10	12	3	Extra trimmer joist
2 × 10	8	2	Extra header joist

8-Foot Opening

Size	Length (lf)	Pieces	Comments
2 × 10	12	3	Extra trimmer joist

Double header cut from waste

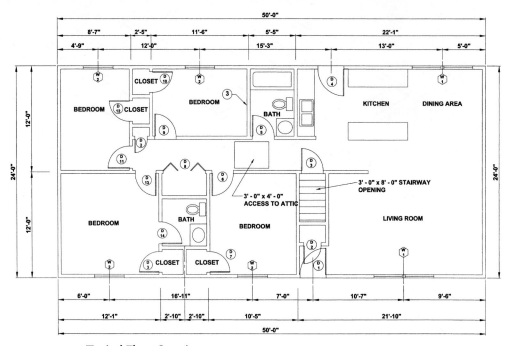

FIGURE 13.15. Typical Floor Openings.

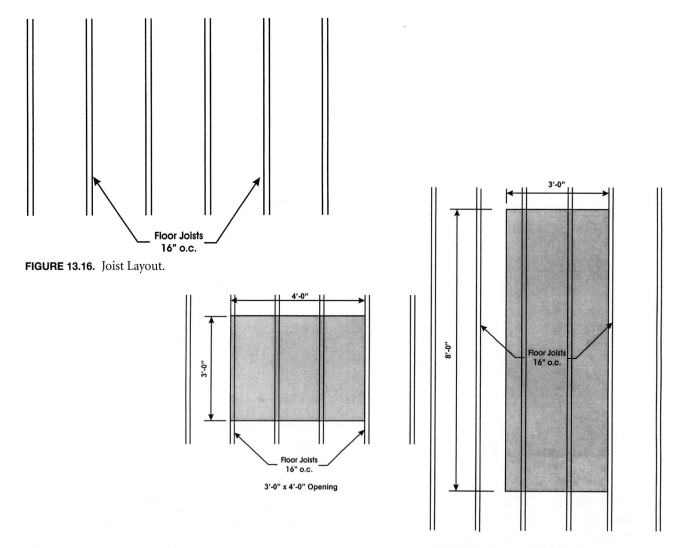

FIGURE 13.16. Joist Layout.

3'-0" x 4'-0" Opening

FIGURE 13.17. Joist Openings.

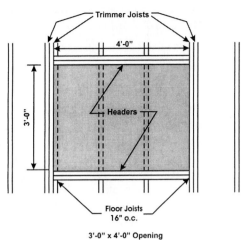

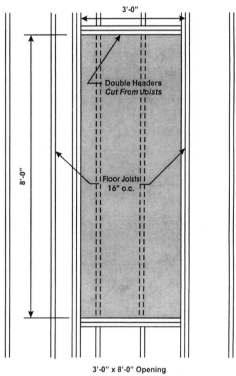

FIGURE 13.18. Headers/Trimmers.

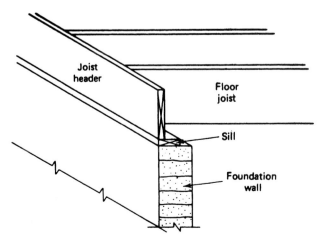

FIGURE 13.19. Joist Header Detail.

Joist headers are taken off next. The header runs along the ends of the joist to seal the exposed edges (Figure 13.19). The headers are of the same size as the joists, and the length required will be two times the length of the building (Figure 13.8).

EXAMPLE 13-6   JOIST HEADER QUANTITY

From Figures 13.5 and 13.8, the joist header runs along the 50-foot dimension of the building. Therefore, there are 100 lf of joist headers.

Purchase Quantity

Size	Length (lf)	Pieces
2 × 10	10	10

$$bf = 10 \times \frac{2'' \times 10''}{12} \times 10' = 167 \text{ bf} \quad \blacksquare$$

Bridging is customarily used with joists (except for glued nailed systems) and must be included in the costs. Codes and specifications vary on the amount of bridging required, but at least one row of cross-bridging is required between the joists. The bridging may be wood, 1 × 3s or 1 × 4s, metal bridging, or solid bridging (Figure 13.20). The wood bridging must be cut, while the metal bridging is obtained ready for installation and requiring only one nail at each end, half as many nails as would be needed with the wood bridging. I-joist required solid bridging with I-joist material.

In this estimate, the specifications require metal bridging with a maximum spacing of 8 feet between bridging or bridging and bearing. A check of the joists' lengths shows that they are approximately 12 feet long and that one row of bridging will be required for each row of joists.

Because each space requires two pieces of metal bridging, determine the amount required by multiplying the joist *spaces* (not the number of joists) times two (two pieces each space) times the number of rows of bridging required.

EXAMPLE 13-7 METAL BRIDGING

38 spaces between joists (refer to Example 13-4)

Metal Bridging = 38 spaces × 2 pieces per space × 2 bays
= 152 pieces ∎

Sheathing/decking is taken off next. This sheathing is most commonly plywood or waferboard. First, a careful check of the specifications should provide the sheathing information required. The thickness of sheathing required may be given in the specifications or drawings (Figure 13.4). In addition, the specifications will spell out any special installation requirements, such as the glue-nailed system. The sheathing is most accurately estimated by doing a sketch of the area to be covered and by planning the sheathing layout. Using 4 × 8 sheets, a layout for the residence is shown in Figure 13.21.

Another commonly used method of estimating sheathing is to determine the square footage to be covered and divide the total square footage to be covered by the area of one sheet of sheathing (Example 13-8). Although both methods give almost the same answer, the use of square feet alone does not allow planning of sheet layout on irregular plans and does not allow properly for waste. This is particularly true when using tongue and groove sheathing where waste without the proper tongue or groove cannot be used even if it is of a usable size. List the quantity of sheathing required on the workup sheet, being certain to list all related information.

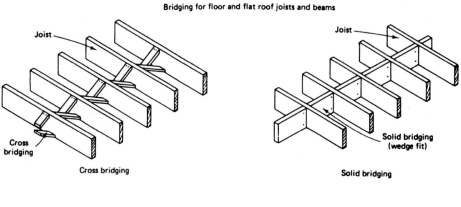

Bridging for floor and flat roof joists and beams

Cross bridging

Solid bridging

Minimum sizes—wood cross bridging, 1 inch by 3 inches. Solid wood bridging, 2 inches thick, and same depth as members bridged. Metal bridging, minimum 18 gage.

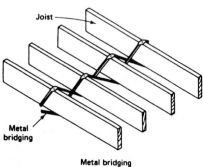

Metal bridging

FIGURE 13.20. Bridging.

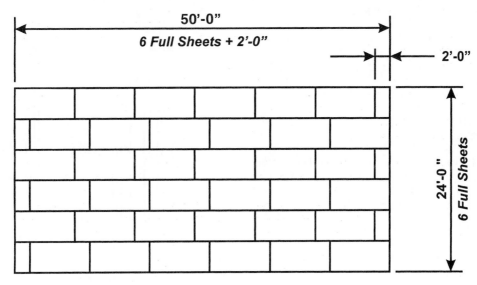

50'-0"
6 Full Sheets + 2'-0"

2'-0"

24'-0"   **6 Full Sheets**

36 Full Sheets Plus 6 - 2'-0" Pieces = 37 ½ Sheets
Order 38 Sheets

FIGURE 13.21. Plywood Layout.

EXAMPLE 13-8   PLYWOOD QUANTITY

$$\text{Sheets of plywood} = \frac{\text{square feet of coverage}}{32 \text{ sf/sheet}}$$

$$\text{Sheets of plywood} = \frac{50' \times 24'}{32 \text{ sf/sheet}} = 37.5 \text{ sheets — Use 38} \quad \blacksquare$$

Subflooring is sometimes used over the sheathing. Subflooring may be a pressed particleboard (where the finished floor is carpet), or it may be plywood or waferboard (where the finished floor is resilient tile, ceramic tile, slate, etc.). A careful review of the specifications and drawings will show whether subflooring is required, as well as the type and thickness and location where it is required. It is taken off in sheets the same as sheathing.

EXAMPLE 13-9   FLOORING SYSTEM WORKUP SHEET

Figure 13.22 is the workup sheet for the flooring system.

		ESTIMATE WORK SHEET							

ESTIMATE WORK SHEET
FLOORING SYSTEM

Project:	Residence				Estimate No.	1234	
Location	Littleville, Tx				Sheet No.	1 of 1	
Architect	U.R. Architects				Date	11/11/20XX	
Items	Floor System				By LHF	Checked JBC	

Cost Code	Description		pcs	l.f.	width inches	Thickness inches	Quantity	Unit
	Girder	Inside Dimensions = 48'-8" Bearing Length = 6"						
		Actual L.F. of Girder is 49'-8"	15	10	2	10	250	B.F.
		Girder is 3 - 2 x 10's nailed and glued						
		Order 15 - 2 x 10's @ 10'						
	Sill							
	2 x 6 #2 y.p. Wolmanized	Foundation perimeter = 147 L.F.						
		Order 4 - 2 x 6's @ 12' -- Treated	4	12	2	6	48	B.F.
		10 - 2 x 6's @ 10' -- Treated	10	10	2	6	100	B.F.
	Floor Joists							
	2 x 10 #2 y.p.	Use 12' Joists						
		50' X .75 = 38 spaces + 1 for end = 39 X 2 bays = 78 joists	78	12	2	10	1,560	B.F.
		Extra joists under partitions	7	12	2	10	140	B.F.
		Order 85 - 2 x 10's @ 12'						
	Joist Header	Both long dimensions -- Use 50' per side	10	10	2	10	167	B.F.
		Order 10 - 2 x 10's @10'						
	Bridging	38 spaces on each side = 76 spaces total						
	Use metal bridging	2 pcs. oer space = 152 pcs						152	pcs
	Plywood Decking	1200 S.F. of floor area / 32 s.f. per sheet = 38 sheets					38	sheets
		Order 38 sheets of 4' X 8' 3/4" B-C Plywood, T & G						

FIGURE 13.22. Flooring Workup Sheet.

■

13–4 WALL FRAMING

In this section, a quantity takeoff of the framing required for exterior and interior walls is done. Because the exterior and interior walls have different finish materials, they will be estimated separately. The exterior walls are taken off first, then the interior.

Exterior Walls

Basically, most of the wall framing consists of sole plates, studs, double plates, headers, and finish materials (Figure 13.23).

Plates.   The most commonly used assembly incorporates a double-top plate and a single-bottom plate, although other combinations may be used. The estimator first begins by reviewing the specifications and drawings for the thickness of materials (commonly 2 × 4 or 2 × 6), the grade of lumber to be used, and information on the number of plates required. The assembly in Figure 13.23 provides an 8-foot ceiling height, which is most commonly used and works economically with 4′ × 8′ sheets of plywood and gypsum board. When an 8′2″ ceiling height is required, a double-bottom sole plate is used (Figure 13.24).

Estimating Steps

1. Plates are required around the perimeter of the building. Since this is the same perimeter used to determine the sill material, the perimeter is already known. However, the sill is typically constructed of treated lumber, and caution needs to be exercised to ensure that the treated lumber is not combined with the untreated lumber.

2. The total linear feet of exterior plates are determined by multiplying the linear feet of wall times the number of plates. If the wall is to be placed on a concrete slab, the bottom plate will need to be treated.

3. List this information on the workup sheet, and calculate the board feet required.

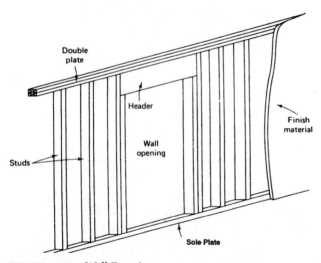

FIGURE 13.23. Wall Framing.

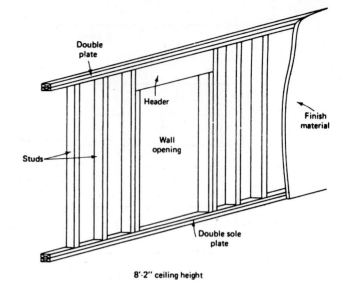

8′-2″ ceiling height

FIGURE 13.24. Double Top and Bottom Plates.

EXAMPLE 13-10   TOP AND BOTTOM PLATES

Perimeter = 148 lf
Assume sole plates are treated 2 × 4s and top plates (double) are untreated 2 × 4.

Purchase Quantity

Size	Length (lf)	Pieces
2 × 4 treated	10	10 (sole plate)
2 × 4 treated	12	4 (sole plate)
2 × 4	10	30

Sole Plate

$$\text{Board feet} = 10 \times \frac{2'' \times 4''}{12} \times 10' = 67 \text{ bf}$$

$$\text{Board feet} = 4 \times \frac{2'' \times 4''}{12} \times 12' = 32 \text{ bf}$$

$$\text{Board feet} = 67 \text{ bf} + 32 \text{ bf} = 99 \text{ bf}$$

Top Double Plate

$$\text{Board feet} = 30 \times \frac{2'' \times 4''}{12} \times 10' = 200 \text{ bf}$$

Note: Interior plates will be done with the interior walls later in this portion of the takeoff, but the same basic procedure shown here will be used.   ■

Studs.   The stud takeoff should be separated into the various sizes and lengths required. Studs are most commonly 2 × 4s at 16-inches or 24-inches on center, or 2 × 6s at 24- inches on center. The primary advantage of using 2 × 6s is that it allows for 5½ inches of insulation as compared with 3½ inches of insulation with 2 × 4s.

Estimating Steps

1. Review the specifications and drawings for the thickness and spacing of studs and lumber grade.

2. The exterior studs will be required around the perimeter of the building; the perimeter used for the sills and plates may be used.

3. Divide the perimeter (length of the wall) by the spacing of the studs to determine the number of spaces. Then add one to close off the last space.

4. Add extra studs for corners, wall intersections (where two walls join), and wall openings.

EXAMPLE 13-11   EXTERIOR WALL STUDS

Exterior walls 2 × 4s at 16″ on center

Using the 148 lf perimeter

Studs = 148 lf × 0.75 spaces per foot = 111 spaces — Use 112 studs

Purchase Quantity

Size	Length (lf)	Pieces
2 × 4	8	112

∎

(a) Corners using 2 × 4 studs: a corner is usually made up of three studs (Figure 13.25). This requires two extra studs at each corner. Estimate the extra material required by counting the number of corners and by multiplying the number of corners by two.

(b) Corners using 2 × 6 studs: a corner is usually made up of two studs (Figure 13.26), requiring one extra stud. Drywall clips are used to secure the drywall panel to the studs. Some specifications and details require that three studs be used for 2 × 6 corners (similar to the 2 × 4 corner).

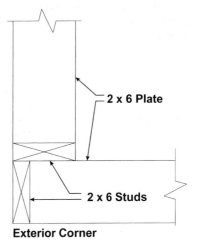

FIGURE 13.26. 2 × 6 Corner Stud Detail.

EXAMPLE 13-12   CORNER STUDS

Add 2 studs per corner

Studs (corners) = 4 corners × 2 studs per corner = 8 studs

Purchase Quantity

Size	Length (lf)	Pieces
2 × 4	8	8

∎

(c) Wall intersections: using 2 × 4 studs, the wall intersection is made up of three studs (Figure 13.27). This requires two extra studs at each intersection. The extra material is estimated by counting the number of wall intersections (on the exterior wall) and by multiplying the number of intersections by two.

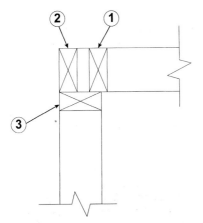

FIGURE 13.25. Corner Stud.

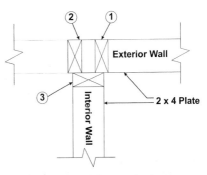

FIGURE 13.27. Wall Intersection Detail (2 × 4s).

EXAMPLE 13-13   INTERIOR WALLS INTERSECTING EXTERIOR WALLS

Refer to Figures 13.27 and 13.28

9 interior walls intersect with the exterior

Studs (intersections) = 9 intersections × 2 studs per intersection

Studs (intersections) = 18 studs

Purchase Quantity

Size	Length (lf)	Pieces
2 × 4	8	18

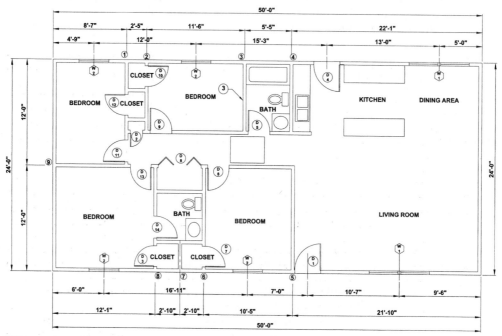

FIGURE 13.28. Floor Plan.

■

(d) Wall intersections: using 2 × 6 exterior studs, the wall intersection may be made up of one or two studs (Figure 13.29) with drywall clips used to secure the gypsum board (drywall). Some specifications and details require that three studs be used for 2 × 6 intersections (similar to the 2 × 4 intersections).

(e) Wall openings: typically, extra studs are required at all openings in the wall (Figure 13.30). However, when the openings are planned to fit into the framing of the building, they require less extra material. Figure 13.31a shows the material required for a window, which has not been worked into the module of the framing. Figure 13.31b shows the material for a window, which has been worked into the module of the framing. The layout in Figure 13.31a actually requires four extra studs; in 13.31b, three extra studs; and in 13.32a, two extra studs.

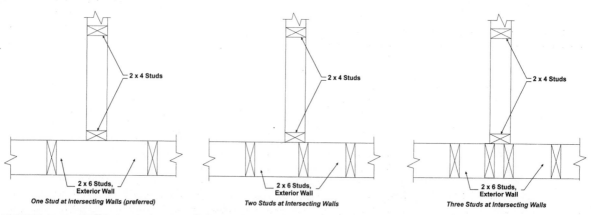

FIGURE 13.29. Wall Intersection Detail (2 × 6s).

In this example, there is no indication that openings have been worked into the framing module. For this reason, an average of three extra studs was included for all windows and door openings.

EXAMPLE 13-14   CRIPPLE AND EXTRA STUDS FOR WALL OPENINGS

Refer to Figure 13.28

Studs (openings) = 8 openings × 3 studs per opening = 24 studs

Purchase Quantity

Size	Length (lf)	Pieces
2 × 4	8	24

■

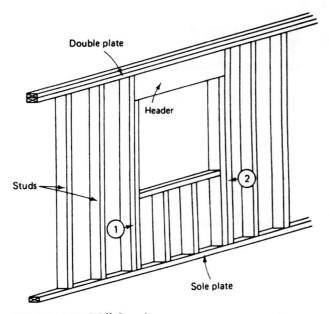

FIGURE 13.30. Wall Opening.

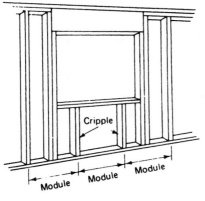

(a)   Extra stud material
4 extra studs

Window not on module

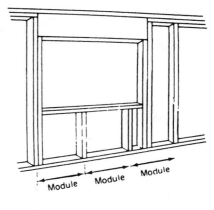

(b)   Extra stud material
3 extra studs

Window partially on module

FIGURE 13.31. Wall Openings.

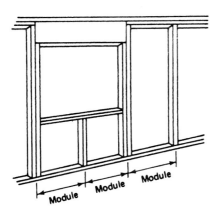

(a) Extra stud material
2 extra studs

FIGURE 13.32. Wall Openings.

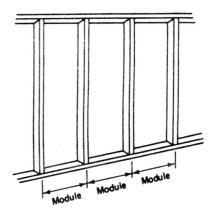

(b) Framing without window

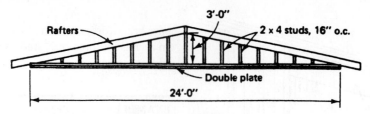

FIGURE 13.33. Gable End Frame.

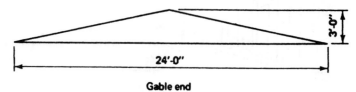

FIGURE 13.34. Gable End Sketch.

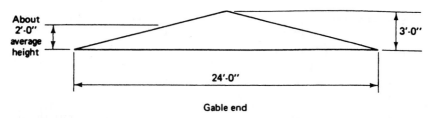

FIGURE 13.35. Gable Average Height.

5. Additional studs are also required on the gable ends of the building (Figure 13.33), which will have to be framed with studs between the double top plate and the rafter unless trusses are used (Section 13–10). The specifications and elevations should be checked to determine whether the gable end is plain or has a louver.

Estimating the studs for each gable end is accomplished by first drawing a sketch of the gable end and by noting its size (Figure 13.34). Find the number of studs required by dividing the length by the stud spacing. Then, from the sketch find the approximate average height of the stud required and record the information on the workup sheet. Multiply the quantity by two since there are two gable ends on this building (Figure 13.35).

EXAMPLE 13-15   GABLE END STUDS

Spaces = 24′ gable end length × 0.75 spaces per foot
= 18 spaces — Use 19 studs

Average height = 2/3 × h

Average height = 2/3 × 3′ = 2′

Quantity must be doubled to compensate for the gable ends at both ends of the building. Each 10′ stud will produce an average of five studs for the gable ends.

Studs = 19 × 2/5 = 8 studs

Purchase Quantity

Size	Length (lf)	Pieces
2 × 4	10	8

Headers.   Headers are required to support the weight of the building over the openings. A check of the specifications and drawings must be made to determine whether the headers required are solid wood, headers and cripples, or plywood headers (Figure 13.36). For ease of construction, many carpenters and home builders feel that a solid header provides best results and they use 2 × 12s as headers throughout the project, even in nonload-bearing walls. Shortages and higher costs of materials have increased the usage of plywood and smaller size headers. When using 2 × materials for the headers, 2 × 4 walls

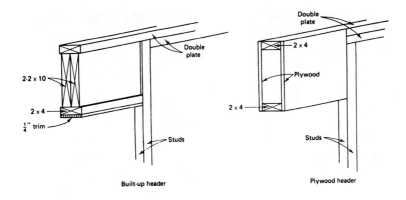

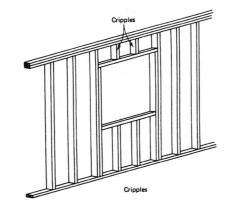

FIGURE 13.36. Headers.

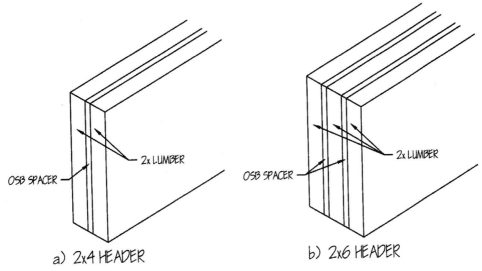

a) 2x4 HEADER

b) 2x6 HEADER

FIGURE 13.37. 2 × Header Construction.

require 2 each 2 × headers with a 1/2″ plywood or 5/16″ OSB spacer (Figure 13.37a), and 2 × 6 walls require 3 each 2 × headers with two 1/2″ plywood or 5/16″ OSB spacers (Figure 13.37b).

The header length must also be considered. As shown in Figure 13.38, the header extends over the top of the studs and it is wider than the opening. Most specifications and building codes require that headers for openings 6 feet wide or less must extend over one stud at each end (Figure 13.38), and headers for openings over 6 feet must extend over two studs at each end.

List the number of openings, their width, the number of headers, their length, and the linear feet and board feet required.

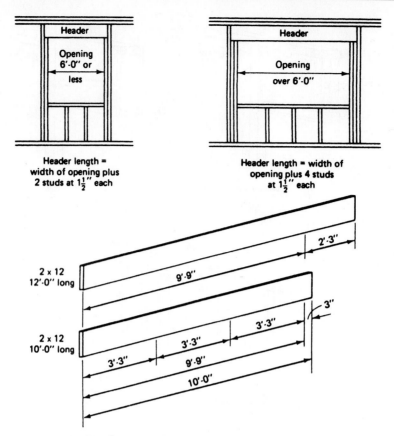

FIGURE 13.38. Header Example.

EXAMPLE 13-16

Header Quantity

No. of Openings	Size	Header Length	Pieces
9	3′0″	3′3″	18
1	2′8″	2′11″	2

Purchase Quantity

Size	Length (lf)	Pieces
2 × 12	10	6
2 × 12	8	1

∎

Wall Sheathing.   Exterior wall sheathing may be a fiberboard material soaked with a bituminous material, insulation board (often urethane insulation covered with an aluminum reflective coating), waferboard, or plywood (Figure 13.39). Carefully check the specifications and working drawings to determine what is required (insulation requirements, thickness). Fiberboard and insulation

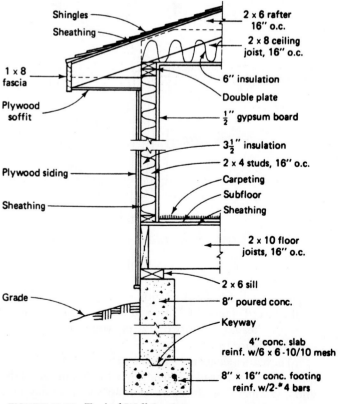

FIGURE 13.39. Typical Wall Section.

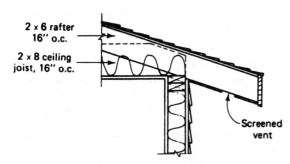

FIGURE 13.40. Sloped Soffit.

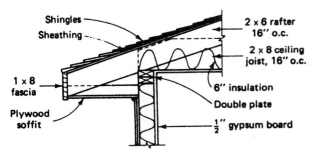

FIGURE 13.41. Boxed Soffit.

board sheathing must be covered by another material (such as brick, wood, or aluminum siding), while the plywood may be covered or left exposed. All of these sheathing materials are taken off first by determining the square feet required and then by determining the number of sheets required. The most accurate takeoff is made by sketching a layout of the material required (as with the sheathing in floor framing). The estimator must check the height of sheathing carefully, as a building with a sloped soffit (Figure 13.40) may require a 9-foot length, while 8 feet may be sufficient when a boxed-in soffit is used (Figure 13.41). Remember that the sheathing must cover the floor framing (joist and joist headers) as well as the exposed parts of the wall. When the wall is a shear wall, the sheathing must go to the top of the wall.

Openings in the exterior wall are neglected unless they are large, and the sheathing that would be cut out can be used elsewhere. Otherwise, it is considered waste.

EXAMPLE 13-17   EXTERIOR SHEATHING

Wall = 148 lf of exterior wall × 8′ average height = 1,184 sf

Gable = 0.5 × 24′ × 3′ × 2 gables = 72 sf

Area (sf) = 1,184 sf + 72 sf = 1,256 sf

Sheets = 1,256 sf/32 sf per sheet = 39.25 sheets — Use 40 sheets

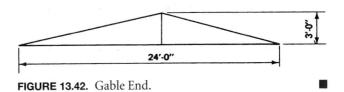

FIGURE 13.42. Gable End. ■

Interior Walls

Interior walls are framed with studs, top and bottom plates, and a finish on both sides of the wall. When estimating the material for interior walls, the first step is to determine from the specifications and drawings what thickness(es) of walls is (are) required. Most commonly, 2 × 4 studs are used, but 2 × 3 studs and metal studs (Section 15-2) can be used. Also, the stud spacing may be different from the exterior walls.

Next, the linear feet of interior walls must be determined. This length is taken from the plan by:

1. Using dimensions from the floor plan.
2. Scaling the lengths with a scale.
3. Using a distance measurer over the interior walls.
4. Using a takeoff software package, such as On-Screen Takeoffs.

On large projects, extreme care must be taken when using a scale, distance measurer, or software package since the drawing may not be drawn to the exact scale shown.

Any walls of different thicknesses (such as a 6-inch-thick wall, sometimes used where plumbing must be installed) and of special construction (such as a double or staggered wall) (Figure 13.43) may require larger stud or plate sizes.

Plates.   Refer to the discussion under "Exterior Walls" earlier in this section.

Studs.   Refer to the discussion under "Exterior Walls" earlier in this section. As in exterior walls, deduct only where there are large openings and take into account all corners, wall openings, and wall intersections.

Headers.   Refer to the discussion under "Exterior Walls" earlier in this section.

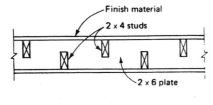

Staggered wall

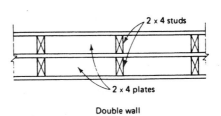

Double wall

FIGURE 13.43. Special Wall Construction.

EXAMPLE 13-18   INTERIOR AND EXTERIOR WALL WORKUP SHEETS

Following are the workup sheets (Figure 13.44) for the exterior and interior wall framing for the building found in Figure 13.8.

ESTIMATE WORK SHEET
GENERAL

Project:	Residence		Estimate No.	1234
Location	Littleville, Tx		Sheet No.	1 of 1
Architect	U.R. Architects		Date	11/11/20XX
Items	Walls		By	LHF Checked JBC

Cost Code	Description		Pcs	l.f.	W inches	Thickness Inches	Quantity	Unit
	Sole Plates	Exterior - Perimeter #2 Treated — 148 L.F.						
		Order 10 - 2 x 4's @ 10'	10	10	2	4	67	b.f.
		4 - 2 x 4's @ 12'	4	12	2	4	32	b.f.
		Interior = 149 l.f.						
		Order 15 - 2 x 4's @ 10'	15	10	2	4	100	b.f.
	Top Double Plates	Exterior = 148 l.f.	30	10	2	4	200	b.f.
		Interior = 149 l.f.	30	10	2	4	200	b.f.
		Order 60 - 2 x 4's @ 10'						
	Studs	Exterior = 148 l.f.						
		148L.F. X .75 spaces / ft. = 111 spaces USE 112 studs	112	8	2	4	597	b.f.
		Exterior Corners = 4 USE 8 studs	8	8	2	4	43	b.f.
		Exterior / Interior Intersections = 9 USE 18 studs	18	8	2	4	96	b.f.
		Exterior Openings = 8 USE 24 studs (Cripples)	24	8	2	4	128	b.f.
		Interior = 149 l.f.						
		149 l.f. X .75 spaces / ft. = 112 spaces USE 113 studs	113	8	2	4	603	b.f.
		Interior Corners = 1 USE 2 studs	2	8	2	4	11	b.f.
		Interior / Interior Intersections = 16 USE 32 studs	32	8	2	4	171	b.f.
		Interior Openings = 12 USE 36 studs	36	8	2	4	192	b.f.
		Order 321 - 2 x 4's @ 8'						
	Misc Studs	Gabel Ends						
		38 pcs @ 2' ea = 76 l.f.						
		Order 8 - 2 x 4's @ 10'	8	10	2	4	53	b.f.

ESTIMATE WORK SHEET
GENERAL

Project:	Residence		Estimate No.	1234
Location	Littleville, Tx		Sheet No.	1 of 1
Architect	U.R. Architects		Date	11/11/20XX
Items	Walls		By	LHF Checked JBC

Cost Code	Description		Pcs	l.f.	Thick inches	Width Inches	Quantity	Unit
	Headers	Exterior count size l.f.						
		9 openings @ 3'-3" Use 2 x 12 9 3.25 29.25						
		1 Opening @ 2'-11" Use 2 x 12 1 2.92 2.92						
		Order 6 - 2 x 12's @ 10'	6	10	2	12	120	b.f.
		1 - 2 x 12 @ 8'	1	8	2	12	16	b.f.
		Interior						
		1'-9" Use 2 x 12 1 1.75 1.75						
		2'-7" Use 2 x 12 2 2.583 5.166						
		2'-9" Use 2 x 12 8 2.75 22						
		5'-3" Use 2 x 12 1 5.25 5.25						
		Total 34.166						
		Order 6 - 2 x 12's @ 12'	6	12	2	12	144	b.f.
		Total l.f.						
	Exterior Sheathing	Walls 148 l.f. X 8' high = 1184 s.f.						
		Gables = 72 s.f.						
		1256 s.f. / 32 s.f. per sheet = USE 40 Sheets					40	Sheets

FIGURE 13.44. Stud Workup Sheet.

13–5   CEILING ASSEMBLY

In this section, a quantity takeoff of the framing required for the ceiling assembly of a wood frame building is done. The ceiling assembly will require a takeoff of ceiling joists, headers, and trimmers that is quite similar to the takeoff done for the floor assembly.

First, a careful check of the specifications and drawings must be made to determine whether the ceiling and roof are made up of joists and rafters (Figure 13.45), often called "stick construction," or prefabricated wood trusses which are discussed in Section 13–10.

Ceiling joists, from the contract documents, determine the following:

1. The size, spacing, and grade of framing required.

2. The number of ceiling joists required by dividing the spacing into the length and by adding one (the same as done for floor joists).

3. The length of each ceiling joist. (Don't forget to add one-half of any required lap.)

EXAMPLE 13-19   CEILING JOISTS

Spaces = $50' \times 0.75$ spaces per foot = 38 spaces — Use 39 joists

Purchase Quantity

Size	Length (lf)	Pieces
2 × 8	12	78

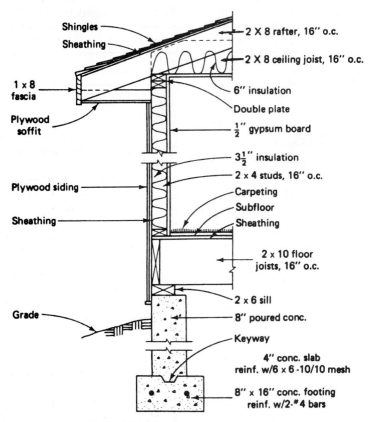

FIGURE 13.45. Typical Wall Section.

The headers and trimmers required for any openings (such as stairways, fireplace, and attic access openings) are considered next. This is taken off the same as for floor framing (Section 13–3).

EXAMPLE 13-20   ATTIC OPENING

A 3'0" by 3'9" attic opening is required. Refer to Figure 13.46.

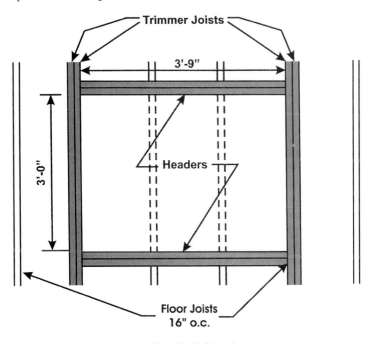

3'-0" x 3'-9" Opening

FIGURE 13.46. Headers and Trimmers.

Purchase Quantity Trimmer Joists

Size	Length (lf)	Pieces
2 × 8	12	2

13–6 ROOF ASSEMBLY

In this section, a quantity takeoff of the framing required for the roof assembly of a wood frame building is done. The roof assembly will require a takeoff of rafters, headers and trimmers, collar ties (or supports), ridge, lookouts, sheathing, and the felt that covers the sheathing. If trusses are specified, a separate takeoff for roof and ceiling assembly will not be made, and the trusses will be estimated as discussed in Section 13–10.

Rafters.   Roof rafters should be taken off and separated into the sizes and lengths required. The spacings most commonly used for rafters are 16 and 24 inches on center, but 12, 20, 32, and 36 inches on centers are also used. Rafter sizes of 2 × 6, 2 × 8, 2 × 10, and 2 × 12 are most common. The lengths of rafters should be carefully taken from the drawings or worked out by the estimator if the drawings are not to scale. Be certain to add any required overhangs to the

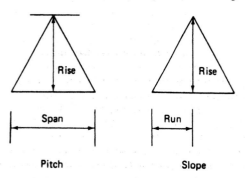

Pitch = $\dfrac{\text{Rise}}{\text{Span}}$

Slope = $\dfrac{\text{Rise}}{\text{Run}}$

Example: Span 40'-0", Rise 10'-0"

Pitch = $\dfrac{10}{40}$ = $\dfrac{1}{4}$ Pitch

Slope = $\dfrac{10}{20}$ = $\dfrac{6}{12}$ = 6 inches per foot

FIGURE 13.47. Roof Pitch and Slope Calculations.

Pitch of Roof	Slope of Roof	For Length of Rafter Multiply Length of Run by
1/12	2 in 12	1.015
1/8	3 in 12	1.03
1/6	4 in 12	1.055
5/24	5 in 12	1.083
1/4	6 in 12	1.12
Note: Run measurement should include any required overhang.		

FIGURE 13.48. Pitch/Slope Rafter Length Factors.

Purchase Quantity Header Joists

Size	Length (lf)	Pieces
2 × 8	8	2

The ceiling finish material (drywall or wetwall) is estimated in Chapter 16.   ■

lengths of the rafters. The number of rafters for a pitched roof can be determined in the same manner as the number of joists. The principle of pitch versus slope should be understood to reduce mistakes. Figure 13.47 shows the difference between pitch and slope, and Figure 13.48 shows the length of rafters required for varying pitches and slopes.

Rafters, from the contract documents, determine the following:

1. The size, spacing, and grade of framing required.
2. The number of rafters required (divide spacing into length and add one).
    If the spacing is the same as the ceiling and floor joists, the same amount will be required.
3. The length of each rafter (Figure 13.49). (Don't forget to allow for slope or pitch and overhang.)

EXAMPLE 13-21   RAFTER QUANTITY

3 on 12 slope length factor is 1.03 (Figure 13.48)

Run = 12' run to center line + 1'6" projection = 13'6"

Rafter length = 13.5' total run length × 1.03 = 13.9'

Use 14' rafters

Ridge length = 50' building length + 2' overhang = 52'

Spaces = 52' × 0.75 spaces per foot = 39 spaces

Use 40 rafters per side

Purchase Quantity

Size	Length (lf)	Pieces
2 × 8	14	80

■

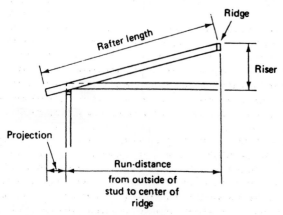

FIGURE 13.49. Rafter Length.

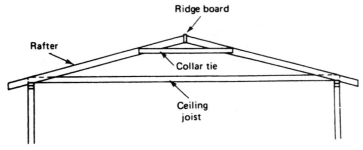

FIGURE 13.50. Collar Tie and Ridge Board.

Headers and Trimmers. Headers and trimmers are required for any openings (such as chimneys) just as in floor and ceiling joists. For a complete discussion of headers and trimmers, refer to Section 13–3, "Floor Framing."

Collar Ties. Collar ties are used to keep the rafters from spreading (Figure 13.50). Most codes and specifications require them to be a maximum of 5 feet apart or every third rafter, whichever is less. A check of the contract documents will determine the following:

a. Required size (usually 1 × 6 or 2 × 4)

b. Spacing

c. Location (how high up)

Determine the number of collar ties required by dividing the total length of the building (used to determine the number of rafters) by the spacing, and add one to close up the last space.

The length required can be a little harder to determine. Most specifications do not spell out the exact location of the collar ties (up high near ridge, low closer to joists, halfway in between), but the typical installation has the collar ties about one-third down from the ridge with a length of 5 to 8 feet, depending on the slope and span of the particular installation.

EXAMPLE 13-22 COLLAR TIES

78 rafter pairs without gable end overhang

39 rafter pairs

39/3 = 13 spaces

Use 14 collars

From Figure 13.51, collars are 8′ long

Purchase Quantity

Size	Length (lf)	Pieces
2 × 4	8	14

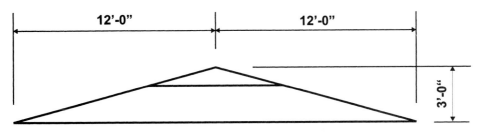

SIMILAR TRIANGLES

$$\frac{3}{12} = \frac{1}{X}$$

$$X = 4'$$

Collar Length = 8′

FIGURE 13.51. Collar Length.

Ridge. The *ridge board* (Figure 13.50) is taken off next. The contract documents must be checked for the size of the ridge board required. Quite often, no mention of ridge board size is made anywhere on the contract documents. This will have to be checked with the architect/engineer (or whoever has authority for the work). Generally, the ridge is

1 inch thick and one size wider than the rafter. In this case, a 2 × 8 rafter would be used with a 1 ×10 ridge board. At other times, a 2-inch thickness may be required.

The length of the ridge board required will be the length of the building plus any side overhang.

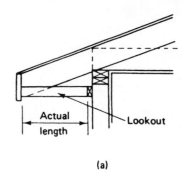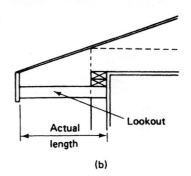

FIGURE 13.52. Lookouts.

EXAMPLE 13-23   RIDGE

There is 52 lf of ridge from the edge of the overhang to the far overhang.

Purchase Quantity

Size	Length (lf)	Pieces
1 × 10	8′	7

$$bf = 7 \times \frac{1'' \times 10''}{12} \times 8' = 47 \text{ bf} \qquad ■$$

Lookouts. Lookouts are often required when a soffit is boxed in as shown in Figure 13.52. The number of lookouts is found by dividing the total linear feet of boxed-in soffit by the spacing and by adding one extra lookout for each side of the house on which it is needed. (If required, front and back add two extra lookouts.) The length of the lookout is found by reviewing the sections in detail. The detail in Figure 13.52a indicates that the lookout would be the width of the overhang minus the actual thickness of the fascia board and the supporting pieces against the wall. In Figure 13.52b, the lookout is supported by being nailed to the stud wall; its length is equal to the overhang plus stud wall minus the actual thickness of the fascia board.

Sheathing Materials.   Sheathing for the roof is taken off next. A review of the contract documents should provide the sheathing information needed:

1. Thickness or identification index (maximum spacing of supports)
2. Veneer grades
3. Species grade
4. Any special installation requirements

The sheathing is most accurately estimated by making a sketch of the area to be covered and by planning the sheathing sheet layout. For roof sheathing, be certain that the rafter length that is required is used and that it takes into consideration any roof overhang and the slope (or pitch) of the roof. The roof sheathing for the residence is then compiled on the workup sheets.

Because the carpenters usually install the roofing felt (which protects the plywood or OSB until the roofers come), the roofing felt is taken off here or under roofing, depending on the estimator's preference. Because felt, shingles, and built-up roofing materials are estimated in Chapter 14, that is where it will be estimated for this residence.

EXAMPLE 13-24   ROOF DECKING

Refer to Figure 13.53.

$$\text{Actual rafter length} = 13.9'$$
$$\text{Ridge length} = 52'$$

$$\text{Square feet of roof} = 2 \text{ sides} \times 13.9' \text{ rafter length}$$
$$\times 52' \text{ ridge length} = 1{,}446 \text{ sf}$$

$$\text{Sheets} = 1{,}446 \text{ sf of deck}/32 \text{ sf per sheet}$$
$$= 45.2 \text{ sheets} - \text{Use } 46 \text{ sheets}$$

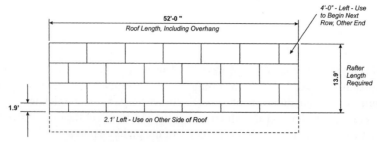

FIGURE 13.53. Plywood Sheet Layout.

13–7 TRIM

The trim may be exterior or interior. Exterior items include moldings, fascias, cornices, and corner boards. Interior trim may include base, moldings, and chair rails. Other trim items may be shown on the sections and details or may be included in the specifications. Trim is taken off by the linear foot required and usually requires a finish (paint, varnish, etc.), although some types are available prefinished (particularly for use with prefinished plywood panels).

EXAMPLE 13-25 EXTERIOR TRIM

The exterior trim for the residence is limited to the fascia board and soffit.

Fascia Board. 1×10, fir

$$Length = 52 + 52 + 13.9 + 13.9 + 13.9 + 13.9$$
$$= 159.6 \text{ or } 160 \text{ lf}$$

Order 170 lf

Soffit, Plywood. 1/2″ thick, A-C, exterior

$$Area = 1'6''\text{ wide} \times 104 \text{ lf} = 156 \text{ sf}/32 \text{ sf per sheet} = 4.87$$

Order five $4' \times 8'$ sheets

Baseboard. Exterior wall = 148 lf; interior wall = 149 lf $\times$ 2 (both sides) − doors (25 lf). 4″ ranch mold, 421 lf required—Order 440 lf. ■

EXAMPLE 13-26 ROOF ESTIMATE WORKUP SHEET

The workup sheet in Figure 13.54 is for the roofing system for the residence in Figure 13.8.

Project:	Residence						
Location	Littleville, Tx						
Architect	U.R. Architects						
Items	Roof Structure						

ESTIMATE WORK SHEET — GENERAL

Estimate No. 1234; Sheet No. 1 of 1; Date 11/11/20XX; By LHF; Checked JBC

Cost Code	Description		Pcs	l.f.	W inches	Thick Inches	Quantity	Unit
	Ceiling Joists	50 ft X .75 spaces / ft. = 38 spaces -- USE 39 Joists / side						
		Order 78 - 2 x 8's @ 12'	78	12	8	2	1,248	b.f.
		Attic Opening 3'-0" X 3'-9"						
		Order 2 - 2 x 8's @ 12'	2	12	8	2	32	b.f.
		2 - 2 x 8's @ 8'	2	8	8	2	21	b.f.
	Rafters	Actual Rafter Length = 13.9' USE 14' rafters						
		Building Length + Overhang = 52'						
		52 l.f. X .75 spaces per ft. = 39 spaces USE 80 rafters						
		Order 80 - 2 x 8's @ 14'	80	14	8	2	1,493	b.f.
	Collar Ties	39 Spaces w/ 1 collar tie per 3rd rafter						
		39 Spaces / 3 = 13 spaces USE 14 collar ties						
		Collar Ties at 1/3 from Ridge = 8'						
		Order 14 - 2 x 4's @ 8'	14	8	4	2	75	b.f.
	Ridge	Ridge 52 l.f.						
		Order 7 - 1 x 10's @ 8'	7	8	10	1	47	b.f.
	Lookouts	Same Number of 2 x 4's for Lookouts as Rafters -- USE 80						
		Each Lookout is 1' - 4" - Use 8' - 2 x 4's						
		Order 14 - 2 x 4's @ 8' -	14	8	4	2	75	b.f.
	Roof Decking	13.9' X 52' = s.f. of decking per side = 723 s.f. per side						
		1446 s.f. of deck on roof / 32 s.f. per sheet = 45.1 USE 46						46	Sheets
	Exterior Trim	Facia Board						
		52' + 52' + 14' + 14' + 14' + 14' = 160 ft						
		Order 12 - 1 x 10's @ 14'	12	14	10	1	140	b.f.
		Soffit Plywood						
		1' - 6" X 104 l.f. = 156 s.f.						
		Order 5 Sheets						5	Sheets
		Baseboard						
		148 l.f. + 149 l.f. + 149 l.f. - 25 l.f. = 421 l.f.						
		Order 31 pcs @ 14'						31	Pcs.

FIGURE 13.54. Roof Framing Workup Sheet. ■

13–8 LABOR

Labor may be calculated at the end of each portion of the rough framing, or it may be done for all of the rough framing at once. Many estimators will use a square foot figure for the rough framing based on the cost of the past work and taking into consideration the difficulty of work involved. A job such as the small residence being estimated would be considered very simple to frame and would receive the lowest square foot cost. The cost would increase as the building became more complicated. Many builders use framing subcontractors for this type of work. The subcontractors may price the job by the square foot or as a lump sum. All of these methods provide the easiest approach to the estimator.

When estimators use their own workforce and want to estimate the time involved, they usually use records from past jobs, depending on how organized they are. The labor would be estimated for the framing by using the appropriate portion of the table for each portion of the work to be done from Figure 13.55.

Light Framing	Unit	Labor Hours
Sills	100 l.f.	2.0 to 4.5
Joists, floor and ceiling	MBM	16.0 to 24.0
Walls, interior, exterior (including plates)	MBM	18.0 to 30.0
Rafters, gable roof	MBM	18.5 to 30.0
hip roof	MBM	22.0 to 35.0
Cross bridging, wood	100 sets	4.0 to 6.0
metal	100 sets	3.0 to 5.0
Plywood, floor	100 s.f.	1.0 to 2.0
wall	100 s.f.	1.2 to 2.5
roof	100 s.f.	1.4 to 2.8
Trim, fascia	100 l.f.	3.5 to 5.0
soffit	100 l.f.	2.0 to 3.5
baseboard	100 l.f.	1.5 to 2.5
molding	100 l.f.	2.0 to 4.0

FIGURE 13.55. Labor Hours Required for Framing.

EXAMPLE 13-27   PRICED FRAMING ESTIMATE

The summary sheet in Figure 13.56 is for all of the framing in the residence in Figure 13.8.

ESTIMATE SUMMARY SHEET

Project: Small House Project
Location: Littleville, Texas
Architect: U.R. Architects
Items: Framing

Estimate No. 1234
Sheet No. 1 of 1
Date 1/1/20XX
By LHF   Checked ABC

Cost Code	Description	Q.T.O.	Waste Factor	Purch. Quan.	Unit	Crew	Prod. Rate	Wage Rate	Labor Hours	Unit Cost Labor	Unit Cost Material	Unit Cost Equipment	Labor	Material	Equipment	Total
	FOUNDATION															
	Girders 2 x 10	250	0.00%	250	FBM		0	$14.25	5.0	$0.29	$0.65		71.25	162.50	0.00	233.75
	Foundation Wall Sill 2 x 6 Treated	148		148	FBM			$14.25	3.0	$0.29	$0.99		42.18	146.52	0.00	188.70
	Floor Joists 2 x 10	1,700		1700	FBM			$14.25	34.0	$0.29	$0.65		484.50	1,105.00	0.00	1,589.50
	Joist Headers 2 x 10	167		167	FBM			$14.25	3.3	$0.29	$0.65		47.60	108.55	0.00	156.15
	Metal Bridging	152		152	Pcs			$14.25	4.6	$0.43	$2.00		64.98	304.00	0.00	368.98
	Plywood Decking 3/4" B-C T&G	38		38	Sheets			$14.25	18.2	$6.84	$21.00		259.92	798.00	0.00	1,057.92
	WALL FRAMING															
	2x4 - 10'	67		67	FBM			$14.25	1.3	$0.29	$0.95		19.10	63.65	0.00	82.75
	2x4 - 12'	32		32	FBM			$14.25	0.6	$0.29	$0.95		9.12	30.40	0.00	39.52
	Untreated Sole Plates 2x4 - 10'	100		100	FBM			$14.25	2.0	$0.29	$0.58		28.50	58.00	0.00	86.50
	Top Double Plates @ 10'	400		400	FBM			$14.25	8.8	$0.31	$0.58		125.40	232.00	0.00	357.40
	Studs 2x4 @ 8'	1,841		1841	FBM			14.25	40.5	$0.31	$0.58		577.15	1,067.78	0.00	1,644.93
	Misc. Studs Gabel Ends 2x4 @ 10'	53		53	FBM			14.25	1.2	$0.31	$0.58		16.62	30.74	0.00	47.36
	Headers															
	2x12 @ 10'	120		120	FBM			14.25	2.6	$0.31	$0.80		37.62	96.00	0.00	133.62
	2x12 @ 8'	16		16	FBM			14.25	0.4	$0.31	$0.80		5.02	12.80	0.00	17.82
	2x12 @ 12'	144		144	FBM			14.25	3.2	$0.31	$0.80		45.14	115.20	0.00	160.34
	Exterior Sheating	40		40	Sheets			14.25	0.0	$9.11	$32.00		364.50	1,280.00	0.00	1,644.50
	ROOF CEILING															
	Ceiling Joists															
	2x8 @ 12'	1,280		1280	FBM			14.25	0.0	$0.28	$0.62		364.50	793.60	0.00	1,158.10
	2x8 @ 8'	21		21	FBM			14.25	0.0	$0.29	$0.62		5.99	13.02	0.00	19.01
	Rafters 2 x 8 @ 14'	1,493		1493	FBM			14.25	0.0	$0.36	$0.62		531.88	925.66	0.00	1,457.54
	Collar Ties 2x4 @ 8'	75		75	FBM			14.25	0.0	$0.29	$0.58		21.38	43.50	0.00	64.88
	Roof Decking	46		46	Sheets			14.25	0.0	$10.26	$12.50		471.96	575.00	0.00	1,046.96
	RIDGE & TRIM								0.0							
	Soffit 1/4" Plywood	5		5	Sheets			14.25	0.0	$9.98	$22.00		49.88	110.00	0.00	159.88
	1 x 10 @ 14'	140		140	FBM			14.25	0.0	$9.98	$1.30		1,396.50	182.00	0.00	1,578.50
	1 x 12 @ 8'	47		47	FBM			14.25	0.0	$9.96	$1.30		468.33	61.10	0.00	529.43
				0					0.0							0.00
									128.71				5,509.02	8,315.02	0.00	13,824.04

FIGURE 13.56. Framing Estimate.

13-9 WOOD SYSTEMS

Wood is used as a component in several structural systems, among them wood trusses, laminated beams, wood decking, and box beams. Wood is used for a variety of reasons: The wood trusses are economical, whereas the laminated beams and box beams are both economical and used for appearance.

13-10 WOOD TRUSSES

Wood joists with spans in excess of 150 feet are readily available. The cost savings of these types of trusses compared with steel range as high as 25 percent: Less weight is involved, and the trusses are not as susceptible to transportation and erection damage. Decking is quickly and easily nailed directly to the chords. Trusses of almost any

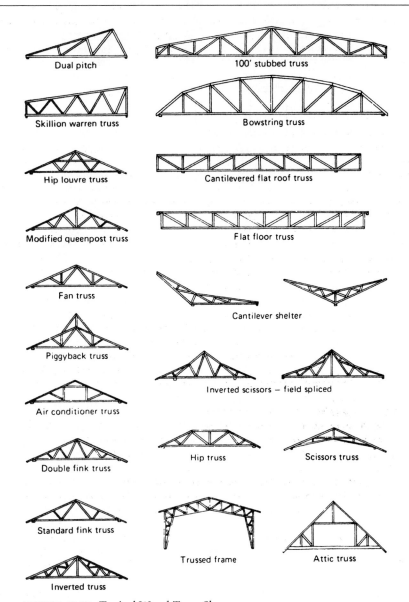

FIGURE 13.57. Typical Wood Truss Shapes.

shape (design) are possible. Ducts, piping, and conduit may be easily incorporated into the trusses. The trusses are only part of the system and must be used in conjunction with a deck of some type. Typical truss shapes are shown in Figure 13.57.

Specifications. Check the type of truss required. If any particular manufacturer is specified, note how the members are to be attached (to each other and to the building) and what stress-grade-marked lumber is required. Note also any requirements regarding the erection of the trusses and any finish requirements.

Estimating. The estimator needs to determine the number, type, and size required. If different sizes are

required, they are kept separate. Any special requirements for special shapes are noted, with a written proposal from the manufacturer or supplier. If the estimator is not familiar with the project, arrangements should be made to see the contract documents for the complete price. A check is made to see how the truss is attached to the wall or column supporting it.

If the trusses are to be installed by the general contractor, an allowance for the required equipment (booms, cranes) and workers must be made. The type of equipment and the number of workers will depend on the size and shape of the truss and the height of erection. Trusses 400 feet long may be placed at the rate of 1,000 to 1,500 sf of coverage per hour, using two mobile cranes mounted on trucks and seven workers.

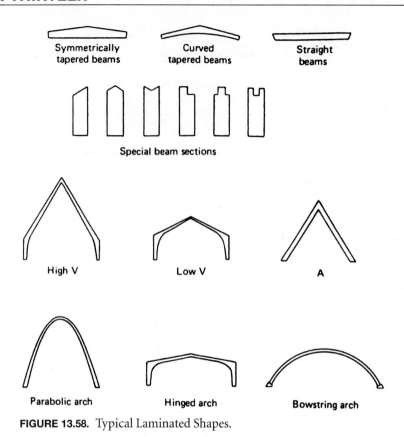

FIGURE 13.58. Typical Laminated Shapes.

13–11 LAMINATED BEAMS AND ARCHES

Laminated structural members are pieces of lumber glued under controlled temperature and pressure conditions. The glue used may be either interior or exterior. They may be rectangular beams or curved arches, such as parabolic, bowstring, V, or A (Figure 13.58). The wood used includes Douglas fir, southern pine, birch, maple, and redwood. Spans just over 300 feet have been built by using laminated arches in a cross vault. They are generally available in three grades: industrial appearance, architectural appearance, and premium appearance grades. The specifications should be carefully checked so that the proper grade is used. They are also available prefinished.

Specifications. Check for quality control requirements, types of adhesives, hardware, appearance grade, finish, protection, preservative, and erection requirements.

Estimating. The cost of materials will have to come from the manufacturer or supplier, but the estimator performs the takeoff. The takeoff should list the linear footage required for each type and the size or style of beam or arch. Note also the type of wood, appearance grade, and finishing requirements. If the laminated shapes are not prefinished, be certain that finishing is covered somewhere in the specifications.

If the laminated beams and arches are to be installed by the general contractor, consider who will deliver the material to the site, how it will be unloaded, where it will be stored, and how much equipment and how many workers will be required to erect the shapes. Erection time varies with the complexity of the project. In a simple erection job, beams may be set at the rate of six to eight lengths per hour, while large, more complex jobs may require four hours or more for a single arch. Carefully check the fastening details required— the total length of each piece, how pieces will be braced during construction, and whether an arch is one piece or segmental.

13–12 WOOD DECKING

Wood decking is available as a solid-timber decking, plank decking, and laminated decking. Solid-timber decking is available either in natural finish or prefinished; the most common sizes are 2×6, 3×6, and 4×6. The most commonly used woods for decking are southern pine, western red cedar, inland white fir, western white spruce, and redwood.

Plank decking is tongue-and-groove decking fabricated into panels. The most common panel size is 21 inches wide; lengths are up to 24 feet. Installation costs are reduced by using this type of deck. Note wood species, finish, size required, and appearance grade.

Block decking nominal dimensions		Laminated decking actual dimensions		
3 x 6	4 x 6	2 1/4 x 5 1/2	3 1/16 x 5 1/2	3 13/16 x 5 1/2
3.43	4.58	3.33	4.4	5.5

Multiply the square foot area to be covered by the factor listed to arrive at the required board measure. These factors include no waste.

FIGURE 13.59. Wood Deck Estimating Factors.

Laminated decks are available in a variety of thicknesses and widths. Installation costs are reduced substantially. Note the wood species, finish, and size required. Appearance grades are also available.

Estimating. No matter what type of wood deck is required, the takeoff is the square foot area to be covered. Because the decking is sold by the board measure, the square footage is multiplied by the appropriate factor shown in Figure 13.59. Take particular notice of fastening details and the size of decking required. (See board measure, Section 13–2.) Consider also whether the decking is priced delivered to the job site or must be picked up, where the decking will be stored, how it will be moved onto the building, and the number of workers required.

Solid decking requires between 12 and 16 labor hours for an mbm (thousand board feet), while plank decking may save 5 to 15 percent of installation cost on a typical project. Laminated decking saves 15 to 25 percent of the installation cost since it is attached by nailing instead of with heavy spikes.

Specifications. Check for wood species, adhesives, finishes, appearance grade, fastening, size requirements, and construction of the decking. The size of decking should also be noted.

13–13 PLYWOOD SYSTEMS

The strength and versatility of plywood have caused it to be recommended for use in structural systems. The systems presently in use include box beams, rigid frames, folded plates, and space planes for long-span systems. For short-span systems, stressed skin panels and curved panels are used. The availability, size, and ease of shaping of plywood make it an economical material to work with. Fire ratings of one hour can also be obtained. Technical information regarding design, construction, estimating, and cost comparison may be obtained through the American Plywood Association. Plywood structural shapes may be supplied by a local manufacturer, shipped in, or built by the contractor's workforce. The decision will depend on local conditions such as suppliers available, workers, and space requirements.

Specifications. Note the thickness and grade of plywood, whether interior or exterior glue is required, fastening requirements, and finish that will be applied. All glue used in

assembling the construction should be noted, and any special treatments, such as with fire or pressure preservatives, must also be taken off.

Estimating. Take off the size and shape of each member required. Total the linear footage of each different size and shape. If a manufacturer is supplying the material, it will probably be priced by the linear foot of each size or as a lump sum. If the members are to be built by the contractor's workforce, either on or off the job site, a complete takeoff of lumber and sheets of plywood will be required.

13–14 WOOD CHECKLIST

Wood:

species

finish required

solid or hollow

laminated

glues

appearance grade

special shapes

primers

Fastening:

bolts

nails

spikes

glue

dowels

Erection:

cranes

studs

wood joists

rafters

sheathing

insulation

trim

bracing

bridging

plates

sills

WEB RESOURCES

www.wwpa.org

www.apawood.org

www.wpma.org

www.bc.com

www.southernpine.com

www.calredwood.org

REVIEW QUESTIONS

1. What unit of measure is used for lumber?

2. Determine the number of board feet for the following order:

   $190 - 2 \times 4 - 8'0''$

   $1{,}120 - 2 \times 4 - 12'0''$

   $475 - 2 \times 6 - 16'0''$

   $475 - 2 \times 12 - 16'0''$

   $18 - 2 \times 8 - 18'0''$

3. How would you determine the number of studs required for a project?

4. How do you determine the number of joists required?

5. What is the difference between pitch and slope, as the terms pertain to roofing?

6. Determine the length of rafter required for each of the following conditions if the run is 16'0'':

   1/12 pitch

   1/6 pitch

   3 in 12 slope

   4 in 12 slope

7. What unit of measure is used for plywood when it is used for sheathing? How is plywood waste kept to a minimum?

8. What unit of measure is most likely to be used for laminated beams?

9. What unit of measure is most likely to be used for wood decking? How is it determined?

10. What unit of measure is used for wood trim? If it requires a finish, where is this information noted?

11. Determine the flooring material required for the residence in Appendix B.

12. Determine the wall materials required for the residence in Appendix B.

13. Determine the ceiling materials required for the residence in Appendix B.

14. Determine the roof materials required for the residence in Appendix B.

CHAPTER FOURTEEN

THERMAL AND MOISTURE PROTECTION

14–1 WATERPROOFING

Waterproofing is designed to resist the passage of water and, usually, to resist the hydrostatic pressures to which a wall or floor might be subjected. (Dampproofing resists dampness, but it is not designed to resist water pressure.) Waterproofing can be effected by the admixture mixed with the concrete, *membrane waterproofing;* by placing layers of waterproofing materials on the surface, the *integral method;* and by the *metallic method.*

14–2 MEMBRANE WATERPROOFING

Membrane waterproofing (Figure 14.1) consists of a buildup of tar or asphalt and membranes (plies) into a strong impermeable blanket. This is the only method of waterproofing that is dependable against hydrostatic head. In floors, the floor waterproofing must be protected against any expected upward thrust from hydrostatic pressure.

The actual waterproofing is provided by an amount of tar or asphalt applied between the plies of reinforcement. The purpose of the plies of reinforcement is to build up the amount of tar or asphalt that meets the waterproofing requirements and to provide strength and flexibility to the membrane.

Reinforcement plies are of several types, including a woven glass fabric with an open mesh, saturated cotton fiber, and tarred felts.

Newly applied waterproofing protection membranes should always be protected against rupture and puncture during construction and backfilling. Materials applied over the membrane include building board and rigid insulation for protection on exterior walls against damage during construction and backfilling.

Another type of waterproofing is the use of sprayed-on asphalt with chopped fiberglass sprayed on simultaneously. The fiberglass reinforcement helps the resultant film bridge hairline cracks that may occur in the wall. The sprayed-on waterproofing requires that one to two days of good weather pass before backfilling can begin. Take care during backfill so that the film is not damaged.

Estimating Membrane Waterproofing.   The unit of measurement is the square (1 square = 100 sf). To estimate the quantity of reinforcement felt required, the estimator determines the square footage of the walls and floors that require membrane waterproofing and keeps walls and floors separate.

The specifications must be checked to determine the number of membrane (plies) required, the kind and weight of reinforcement ply, the number of coatings, the type of coating (tar or asphalt), and the pounds of coating material required to complete each 100 sf of waterproofing. The estimator also checks to determine what industry requirements must be met and whether the specifications limit the manufacturer's materials that may be used.

If one layer of reinforcement is specified, the amount must allow for laps over the footings and at the top of the wall, as well as the lap required over each strip of reinforcement. A 4-inch side lap of layers plus the top and bottom laps will require about 20 percent additional. A 6-inch lap will require about 25 percent additional to the area being waterproofed. Lap is generally listed in the specifications. The number of plies used usually ranges from two to five, and the amount to be added for laps and extra material drops to about 10 percent of the actual wall or floor area.

Most felt and cotton fabric reinforcement is available in rolls of 432 sf. Glass yarn reinforcement is generally available in rolls of 450 sf. Different manufacturers may have various size rolls and sheets available of certain types of reinforcements.

The amount of tar required varies depending on the number of plies; the amount used per square (one square = 100 sf) should be in the specifications. Figure 14.2 shows the amount of reinforcing and coating required. Membrane waterproofing is similar to applying built-up roofing, but

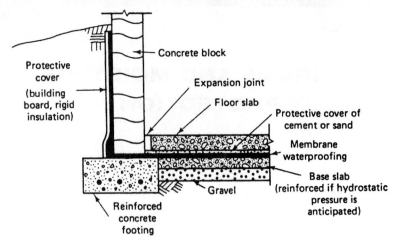

FIGURE 14.1. Membrane Waterproofing.

Material	Plies					
	1 Ply 4" Lap	1 Ply 6" Lap	2 Ply	3 Ply	4 Ply	5 Ply
Reinforcement s.f.	120	125	230	330	440	550
Mopping of Bituminous Coatings	2	2	3	4	5	6
Pounds of Bituminous Coatings	60 - 70	60 - 70	90 - 100	120 – 135	150 - 170	180 - 205

FIGURE 14.2. Materials Required for Membrane Waterproofing per Square.

the cost is greater for vertical surfaces than for horizontal surfaces, and there is generally less working space on the vertical.

The square footage of protective coating must also be determined; this is the same as the area to be waterproofed. The type of protective coating, thickness, and method of installation are included in the specifications.

The sprayed-on asphalt and chopped fiberglass require special equipment to apply, and most manufacturers specify that only if these are applied by approved applicators will they be responsible for the results. Materials required per square are 8 to 10 gallons of asphalt and 7 to 10 pounds of fiberglass. The surface that is being sprayed must be clean and dry.

Labor. Membrane waterproofing is almost always installed by a subcontractor (usually a roofing subcontractor). A typical subcontractor's bid (to the contractor) is shown in Figure 14.3. Typical labor productivity rates are shown in Figure 14.4.

Agrees to furnish and install 3-ply built-up membrane waterproofing in accordance with the contract documents for the sum of ----------------------$8,225.00

FIGURE 14.3. Subcontractor's Bid for Membrane Waterproofing.

14-3 INTEGRAL METHOD

The admixtures added to the cement, sand, and gravel generally are based on oil-/water-repellent preparations. Calcium chloride solutions and other chemical mixtures are used for liquid admixtures, and stearic acid is used for powdered admixtures. These admixtures are added in conformance with manufacturers' recommendations. Most of these admixtures enhance the water resistance of the concrete, but offer no additional resistance to water pressure.

Estimating. The amount to be added to the mix varies considerably, depending on the manufacturer and composition of the admixture. The powdered admixtures should be mixed with water before they are introduced into the mix.

Type of Work	Labor Hours per 100 s.f.
Cleaning	0.25 to 3.5
Painting Brush, per Coat	0.5 to 2.0
Parging, per Coat	1.0 to 2.5
Metallic	1.5 to 3.0
Waterproofing	
Asphalt of Tar, per Coat	0.7 to 2.0
Felt or Reinforced, per Layer	1.2 to 2.5

FIGURE 14.4. Labor Hours Required for Applying Waterproofing and Dampproofing.

Liquid admixtures may be measured and added directly, which is a little less trouble. The costs involved derive from delivering the admixture to the job site, moving it into a convenient place for use, and adding it to the mix. The containers may have to be protected from the weather, particularly the dry admixture containers.

14–4 METALLIC METHOD

Metallic waterproofing employs a compound that consists of graded fine-iron aggregate combined with oxidizing agents. The principle involved is that, when the compound is properly applied, it provides the surface with a coating of iron that fills the capillary pores of the concrete or masonry and, as the iron particles oxidize, the compound creates an expanding action that fills the voids in the surface and becomes an integral part of the mass. Always apply in accordance with the manufacturer's specifications, and always apply the compound to the inside of the wall.

Estimating. The estimator needs to check the specification for the type of compound to be used and the thickness and method of application specified. From the drawings, he or she can determine the square footage to be covered and, using the manufacturer's information regarding square feet coverage per gallon, determine the number of gallons required.

Labor will depend on the consistency of the compound, the thickness, and the method of application required as well as the convenience afforded by the working space. Equipment includes scaffolds, planks, ladders, mixing pails, and trowels.

14–5 DAMPPROOFING

Dampproofing is designed to resist dampness and is used on foundation walls below grade and exposed exterior walls above grade. It is not intended to resist water pressure. The methods used to dampproof include painting the wall with bituminous materials below grade and transparent coatings above grade. *Parging* with a rich cement base mixture is also used below grade, often with a bituminous coating over it.

14–6 PAINTING METHOD

Among the most popular paints or compounds applied by brush or spray are tar, asphalt, cement washes, and silicone-based products. This type of dampproofing may be applied with a brush or mop, or may be sprayed on. The type of application and the number and thickness of coats required will be determined by the specifications and manufacturer's recommendations.

Below grade on masonry or concrete, the dampproofing material is often black mastic coating that is applied to the exterior of the foundation. It is applied with a brush, mop, or spray. The walls must be thoroughly coated with the mastic, filling all voids or holes. The various types available require from one to four coats to do the job, and the square footage per gallon varies from 30 to 100. The more porous the surface, the more material it will require.

Transparent dampproofing of exterior masonry walls is used to make them water repellent. The colorless liquids must be applied in a quantity sufficient to completely seal the surface. Generally, one or two coats are required by the manufacturers, with the square footage per gallon ranging from 50 to 200.

Estimating Painting. To estimate these types of dampproofing, it is necessary first to determine the types of dampproofing required, the number of coats specified (or recommended by the manufacturer), and the approximate number of square feet that one gallon of material will cover so that the amount of material required may be determined. If two coats are required, twice the material is indicated.

The amount of labor required to do the work varies with the type of compound being used, the number of coats, the height of work, and the method of application. When the surface to be dampproofed is foundation walls below grade and the work is in close quarters, it will probably take a little longer. Sprayed-on applications require much fewer labor hours but more expensive equipment. High buildings require scaffolds that will represent an added cost factor.

Equipment that may be required includes ladders, planks, mixing cans, brushes, mops, spraying equipment, and scaffolding.

EXAMPLE 14-1 FOUNDATION WALLS

The below-grade foundation wall on the building in Appendix A must be brushed with a bituminous product on the 8′4″-high foundation walls to within 2 inches of the finished grade. Figure 14.5 shows the building layout, and Figure 14.6 is a tabulation of the linear feet of the foundation wall. The coverage rate is 60 sf per gallon.

$$\text{Assume the height to be sprayed is } 8'2''$$

$$\text{Area to be dampproofed} = 112'8'' \times 8'2''$$

$$\text{Area to be dampproofed} = 112.67' \times 8.167' = 920 \text{ sf}$$

$$\text{Gallons of asphalt} = 920 \text{ sf}/60 \text{ sf per gallon} = 15.3 \text{ gallons}$$

$$\text{Use 16 gallons}$$

Using the labor productivity rates from Figure 14.4 and a local prevailing wage rate of $13.25 per hour, the labor costs can be determined.

$$\text{Labor hours} = 1.25 \text{ labor hours per sq} \times 9.2 \text{ squares}$$
$$= 11.5 \text{ labor hours}$$

$$\text{Labor cost} = \text{Labor hours} \times \text{Wage rate}$$

$$\text{Labor cost} = 11.5 \text{ hour} \times \$13.25 \text{ per hour} = \$152$$

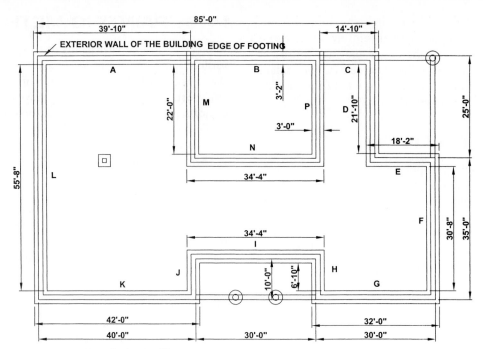

FIGURE 14.5. Foundation Walls.

Side	Length of 8'-4" High Foundation Wall
B	32'-4"
M	24'-0"
N	32'-4"
P	24'-0"
Total	112'-8"

FIGURE 14.6. Foundation Walls Requiring Dampproofing.

14–7 PARGING (PLASTERING)

The material used to parge on exterior portions of foundation walls below grades is a mixture of portland cement, fine aggregates (sand), and water. A water-repellent admixture is often added to the mix and is troweled or sprayed in place, the most common thickness being about 1/2 inch. Compounds with a cement base are also available from various manufacturers.

On some projects, the specifications require parging and the application of a coating of bituminous material in a liquid form. The bituminous type of coating is discussed in Section 14–6.

Estimating Parging. To determine the amount of material required for parging, the estimator first checks the specifications to determine exactly the type of materials required. The materials may be blended and mixed on the job, or they may be preblended compounds. Information concerning the specially formulated compounds must be obtained from the manufacturer. The job-mixed parging requires that the mix proportions be determined. The next step is to determine the number of square feet to be covered and the thickness. With this information, the cubic feet required can be calculated, or the number of gallons can be determined if the materials are preblended.

The amount of labor will depend on the amount of working space available and the application technique. Labor costs will be higher when the materials are mixed on the job, but the material cost will be considerably lower.

Equipment required includes a mixer, trowel or spray accessories, scaffolding, planks, shovels, and pails. Water must always be available for mixing purposes.

14–8 INSULATION

Insulation in light-frame construction may be placed between the framing members (studs or joists) or nailed to the rough sheathing. It is used in the exterior walls and the ceiling of most buildings.

Stud Spacing (Inches)	Insulation Width	To determine l.f. of Insulation required multiply s.f. of Area by	OR	l.f. of insulation required pre 100 s.f. of wall area
12	11	1.0		100
16	15	.75		75
20	19	.60		60
24	23	.50		50

FIGURE 14.7. Insulation Requirements.

Insulation placed between the framing members may be pumped in or laid in rolls or in sheets, while loose insulation is sometimes placed in the ceiling. Roll insulation is available in widths of 11, 15, 19, and 23 inches to fit snugly between the spacing of the framing materials. Sheets of the same widths and shorter lengths are also available. Rolls and sheets are available unfaced, faced on one side, faced on both sides, and foil faced. The insulating materials may be of glass fiber or mineral fiber. Nailing flanges, which project about 2 inches on each side, lap over the framing members and allow easy nailing or stapling. To determine the number of linear feet required, the square footage of the wall area to be insulated is easily determined by multiplying the distance around the exterior of the building by the height to which the insulation must be carried (often the gable ends of a building may not be insulated). Add any insulation required on interior walls for the gross area. This gross area should be divided by the factor given in Figure 14.7. If the studs are spaced 16 inches on center, the 15-inch-wide insulation plus the width of the stud equals 16 inches, so a batt 1 foot long will cover an area of 1.33 sf.

Ceiling insulation may be placed in the joists or in the rafters. The area to be covered should be calculated; and if the insulation is placed in the joists, the length of the building is multiplied by the width. For rafters (gabled roofs), the methods shown in Chapter 14 (for pitched roofs) should be used: the length of the building times the rafter length from the ridge to plate times two (for both sides). Divide the area to be covered by the factor given in Figure 14.7 to calculate the linear footage of a given width roll.

Ceiling insulation may also be poured between the joists; a material such as vermiculite is most commonly used. Such materials are available in bags and may be leveled easily to any desired thickness. The cubic feet of material required must be calculated. Loose mineral or fiberglass may also be "blown in" between the ceiling joists. Because this material compresses easily, it is typically not done until all construction foot traffic is completed.

Insulation board that is nailed to the sheathing or framing is estimated by the square foot. Sheets in various thicknesses and sizes are available in wood, mineral, and cane fibers. The area to be covered must be calculated, and the number of sheets must be determined.

For all insulation, add 5 percent waste when net areas are used. Net areas will give the most accurate takeoff. Check the contract documents to determine the type of insulation, thickness or R-value required, and any required methods of fastening. When compiling the quantities, estimators keep each different thickness or type of insulation separate.

EXAMPLE 14-2 RESIDENTIAL BUILDING INSULATION

Determine the insulation required for the residential building introduced in Chapter 13 and shown in Figure 14.8.

Floor: 5½"-thick R 19 roll insulation

$$\text{sf of floor} = 50' \times 24' = 1,200 \text{ sf}$$

$$1 \text{ roll} = 15'' \times 56'$$

Effective coverage per roll $= 16'' \times 56' = 1.33' \times 56' = 75 \text{ sf}$

Rolls of insulation $= 1,200 \text{ sf of floor}/75 \text{ sf of coverage per roll}$
$$= 16 \text{ rolls}$$

Use 16 rolls

Ceiling: 7½" R 22 roll insulation

$$\text{sf of ceiling} = \text{sf of floor}$$

$$\text{sf of ceiling} = 1,200 \text{ sf}$$

$$1 \text{ roll} = 15'' \times 32'$$

Effective coverage per roll $= 16'' \times 32' = 1.33' \times 32' = 43 \text{ sf}$

Rolls of insulation $= 1,200 \text{ sf of ceiling}/43 \text{ sf of coverage per roll}$
$$= 27.9 \text{ rolls}$$

Use 28 rolls

Walls: 3½" R 11 roll insulation

$$\text{sf of wall} = 7'8'' \text{ high} \times 148 \text{ lf of exterior walls} = 1,135 \text{ sf}$$

$$1 \text{ roll} = 15'' \times 56'$$

Effective coverage per roll $= 16'' \times 56' = 1.33' \times 56' = 75 \text{ sf}$

Rolls of insulation $= 1,135 \text{ sf of wall}/75 \text{ sf of coverage per roll}$
$$= 15.1 \text{ rolls}$$

Use 16 rolls

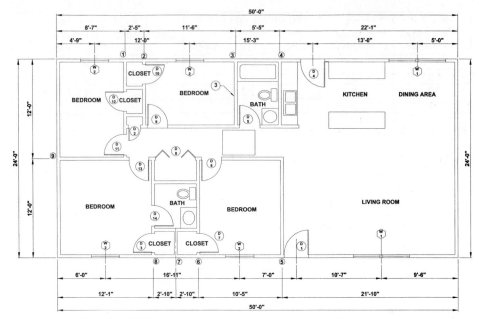

FIGURE 14.8. Residential Floor Plan.

14–9 ROOFING

Roofing is considered to include all the material that actually covers the roof deck. It includes any felt (or papers) as well as bituminous materials that may be placed over the deck. These are commonly installed by the roofing subcontractor and included in that portion of the work. Flashing required on the roof, at wall intersections, joints around protrusions through the roof, and the like are also included in the roofing takeoff. Miscellaneous items such as fiber blocking, cant strips, curbs, and expansion joints should also be included. The sections and details, as well as the specifications, must be studied carefully to determine what is required and how it must be installed.

14–10 ROOF AREAS

In estimating the square footage of the area to be roofed, estimators must consider the shape of the roof (Figure 14.9). When measuring a flat roof with no overhang or with parapet walls, they take the measurements of the building from the outside of the walls. This method is used for parapet walls because it allows for the turning up of the roofing on the sides of the walls. If the roof projects beyond the building walls, the dimensions used must be the overall outside dimensions of the roof and must include the overhang. The drawings should be carefully checked to determine the roofline around the building, as well as where and how much overhang there may be. The floor plans, wall sections,

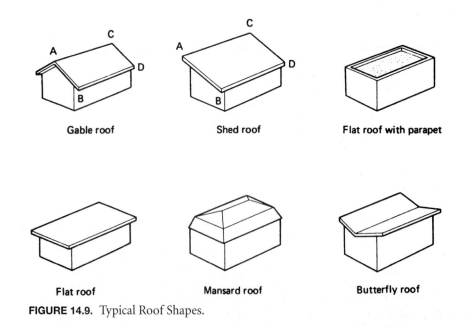

FIGURE 14.9. Typical Roof Shapes.

and details should be checked to determine the amount of overhang and type of finishing required at the overhang. Openings of less than 30 sf should not be deducted from the area being roofed.

To determine the area of a gable roof, multiply the length of the ridge (A to C) by the length of the rafter (A to B) by two (for the total roof surface). The area of a shed roof is the length of the ridge (A to C) times the length of rafter (A to B). The area of a regular hip roof is equal to the area of a gable roof that has the same span, pitch, and length. The area of a hip roof may be estimated the same as the area of a gable roof (the length of roof times the rafter length times two). However, gable roofs require more materials because they generate more waste.

The length of rafter is easily determined from the span of the roof, the overhang, and slope. Refer to Figure 13.49 (Example 13-21) for information required to determine the lengths of rafters for varying pitches and slopes.

14–11 SHINGLES

Asphalt Shingles

Available in a variety of colors, styles, and exposures, strip asphalt shingles are 12 and 15 inches wide and 36 inches long. They are packed in bundles that contain enough shingles to cover 20, 25, or 33.33 sf of roof area. The *exposure* (amount of shingle exposed to the weather) generally is 4, 4.5, or 5 inches. Sometimes individual shingles 12 to 16 inches long are also used. Asphalt shingles may be specified by the weight per square, which may vary from 180 to 350 pounds. Shingles may be fire-rated and wind resistant, depending on the type specified.

In determining the area to be covered, always allow one extra course of shingles at the eaves, as the first course must always be doubled. Hips and ridges are taken off by the linear foot and considered 1 foot wide to determine the square footage of shingles required. Waste averages 5 to 8 percent. Galvanized, large-headed nails, 1/2 to 1¾ inches long, are used on asphalt shingles. From 1½ to 3 pounds of nails are required per square.

Asphalt shingles are generally placed over an underlayment of building paper or roofing felt. The felt is specified by the type of material and weight per square. The felt should have a minimum top lap of 2 inches and end lap of 4 inches; to determine the square footage of felt required, multiply the roof area by a lap and waste factor of 5 to 8 percent.

EXAMPLE 14-3   ASPHALT SHINGLE ROOFING

Using the residence that was introduced in Chapter 13, determine the felt and shingle requirements. In Example 13-21, the rafter length of 13.9 feet and ridge length of 52 feet were determined. The first step is to determine the quantity of 15-pound roofing felt that is required.

$$\text{Roof area} = 13.9' \text{ rafter length} \times 52' \text{ ridge length} \times 2 \text{ sides}$$
$$= 1,446 \text{ sf}$$

Slope: 3 on 12

Felt requirements: 2 layers—lapped 19 inches

Square feet of felt in a roll = 36″ wide × 144′ long = 432 sf

$$\text{Roll coverage} = \text{Square feet in roll} \times \frac{\text{Exposure}}{\text{Width}} \quad \textbf{Formula 14-1}$$

Felt Coverage

Exposure = Roll width − Lap

Exposure = 36″ − 19″ = 17″

Roll coverage = 432 sf × (17″/36″) = 204 sf

Rolls of felt required = 1446 sf/204 sf per roll = 7.08 rolls

Use 8 rolls

Use 235-pound shingles

Starter course = 52 lf of eaves per side × 2 sides
= 104 lf (Assume 1′ wide)

Starter course = 1.04 squares

Squares of roofing = 14.46 squares + 1.04 squares = 15.5 squares

Order 16 squares

Ridge shingles = 5-inch exposure

Ridge length = 52 lf = 624″

Pieces of ridge shingles = 624″/5″ per shingle = 125 pieces ■

Labor.   Subcontractors who specialize in roofing will do this work, and they may price it on a unit basis (per square) or a lump sum. The time required to install shingles is shown in Figure 14.10.

Wood Shingles

Available in various woods—the best of which are cypress, cedar, and redwood—wood shingles come hand-split (rough texture) or sawed. The hand-split variety is commonly

Roofing Material	Labor Hours per Square
Shingles	
Asphalt (strip)	0.8 to 4.0
Asphalt (single)	1.5 to 6.0
Wood (single)	3.0 to 5.0
Metal (single)	3.0 to 6.0
Heavyweight Asphalt	Add 50%
Tile	
Clay	3.0 to 5.0
Metal	3.5 to 6.0
Built-up	
2 Ply	0.8 to 1.4
3 Ply	1.0 to 1.6
4 Ply	1.2 to 2.0
Aggregate Surface	0.3 to 0.5

FIGURE 14.10. Roofing Labor Productivity Rates.

referred to as *shakes*. Lengths are 16, 18, and 24 to as long as 32 inches, but random widths of 4 to 12 inches are common. Wood shingles are usually sold by the square based on a 10-inch weather exposure. A double-starter course is usually required; in some installations, roofing felt is also needed. Valleys will require some type of flashing, and hips and ridges require extra material to cover the joint; thus extra-long nails become requisite. Nails should be corrosion resistant, and 2 to 4 pounds of 1¾- to 2-inch long nails are required per square.

14–12 BUILT-UP ROOFING

Built-up roofing consists of layers of overlapping roof felt with each layer set in a mopping of hot tar or asphalt. Such roofing is usually designated by the number of plies of felt that are used; for example, a three-ply roof has four coats of bituminous material (tar or asphalt) and three layers of felt. Although built-up roofing is used primarily for flat and near-flat roofs, it can be used on inclines of up to 9 inches per foot, providing certain special bituminous products are used.

The specifications must be read carefully, because no one type of system exists for all situations. There may be vapor barrier requirements and varying amounts of lapping of felt. Different weights and types of felts and bituminous materials may be used. The deck that is to receive the roofing must be considered, as must the service it is intended to sustain.

Many specifications require a bond on the roof. This bond is furnished by the material manufacturer and supplier through the roofing contractor. It generally guarantees the water tightness of the installation for a period of 10, 15, or 20 years. During this period of time, the manufacturer will make repairs on the roof that become necessary due to the normal wear and tear of the elements to maintain the roof in a watertight condition. Each manufacturer has its own specifications or limitations in regard to these bonds, and the roofing manufacturer's representative should be called in for consultation. Special requirements may pertain to approved roofers, approved flashing, and deck inspection; photographs may also be necessary. If it appears that the installation may not be approved for bonding, the architect/engineer should be so informed during the bid period. The manufacturer's representative should call on the architect/engineer and explain the situation so that it can be worked out before the proposals are due.

The takeoff follows the same general procedure as any other roofing:

1. The number of squares to be covered must be noted.
2. Base sheets—a base sheet or vapor barrier is used, when required; note the number of plies, lap (usually 4 inches), weight per ply per square.
3. Felt—determine the number of plies, weight of felt per square, and type of bituminous impregnation (tar or asphalt).
4. Bituminous material—note the type, number of coats, and the pounds per coat per square (some specifications call for an extra-heavy topcoat or pour).

5. Aggregate surface (if required)—the type and size of slag, gravel, or other aggregate, and the pounds required per square should all be noted.
6. Insulation (if required)—note type, thickness, and any special requirements (refer also to Section 14–8).
7. Flashing—note type of material, thickness, width, and linear feet required.
8. Trim—determine the type of material, size, shape and color, method of attachment, and the linear footage required.
9. Miscellaneous materials—blocking, cant strips, curbs, roofing cements, nails and fasteners, caulking, and taping are some examples.

Bituminous materials are used to cement the layers of felt into a continuous skin over the entire roof deck. Types of bituminous material used are coal tar pitch and asphalts. The specifications should indicate the amount of bituminous material to be used for each mopping so that each ply of felt is fully cemented to the next. In no instance should felt be allowed to touch felt. The mopping between felts averages 25 to 30 pounds per square, while the top pour (poured, not mopped), which is often used, may be from 65 to 75 pounds per square. It is this last pour into which any required aggregate material will become embedded.

Felts are available in 15- and 30-pound weights, 36-inch widths, and rolls 432 or 216 sf. With a 2-inch lap, a 432-sf roll will cover 400 sf, while the 216-sf roll will cover 200 sf. In built-up roofing, a starter strip 12 to 18 inches wide is applied; then over that one, a strip 36 inches wide is placed. The felts that are subsequently laid overlap the preceding felts by 19, 24⅔, or 33 inches, depending on the number of plies required. Special applications sometimes require other layouts of felts. The specifications should be carefully checked to determine exactly what is required as to weight, starter courses, lapping, and plies. Waste averages about 8 to 10 percent of the required felt, which allows for the material used for starter courses.

Aggregate surfaces such as slag or gravel are often embedded in the extra-heavy top pouring on a built-up roof. This aggregate acts to protect the membrane against the elements, such as hail, sleet, snow, and driving rain. It also provides weight against wind uplift. To ensure embedment in the bituminous material, the aggregate should be applied while the bituminous material is hot. The amount of aggregate used varies from 250 to 500 pounds per square. The amount required should be listed in the specifications. The amount of aggregate required is estimated by the ton.

Since various types of materials may be used as aggregates, the specifications must be checked for the size, type, and gradation requirements. It is not unusual for aggregates such as marble chips to be specified. This type of aggregate will result in much higher material costs than gravel or slag. Always read the specifications thoroughly.

The joints between certain types of roof deck materials—such as precast concrete, gypsum, and wood fiber—may be required to have caulking and taping of all joints. This application of flashing cement and a 6-inch-wide felt strip will minimize the deleterious effects of uneven joints on the roofing membrane, movement of the units, and moisture transmission through the joint; it also reduces the possibility of bitumen seepage into the building. The linear footage of joints to be covered must be determined for pricing purposes.

14–13 CORRUGATED SHEETS (INCLUDING RIBBED, V-BEAM)

Corrugated sheets may be made of aluminum, galvanized steel, and various combinations of materials such as zinc-coated steel. They are available in various sizes, thicknesses, and corrugation shapes and sizes, in a wide range of finishes. Common steel gauges of from 12 to 29 are available with lengths of 5 to 12 feet and widths of 22 to 36 inches. Aluminum thicknesses of 0.024, 0.032, 0.040, and 0.050 inches are most commonly found, with sheet sizes in widths of 35 and 48⅓ inches and lengths of 3 through 30 feet.

Estimate the corrugated sheets by the square; note the type of fastening required. Corrugated sheets require that corrugated closures seal them at the ends. The enclosures may be metal or rubber and are estimated by the linear foot.

The amount of end and side lap must also be considered when estimating the quantities involved. Because of the variety of allowances, it is not possible to include them here. Given the information, the supplier or manufacturer's representative will supply the required information. Fasteners should be noncorrosive and may consist of self-taping screws, weldable studs, cleats, and clips. The specifications and manufacturers' recommendations should be checked to determine spacing and any other special requirements; only then can the number of fasteners be estimated. Flashing and required trim should be estimated by the linear foot.

14–14 METAL ROOFING

Materials used in metal roofing include steel (painted, tinned, and galvanized), copper, and terne. Basically, in this portion we are considering flat and sheet roofing. The basic approach to estimating this type is the same as that used for other types of roofing: first the number of squares to be covered must be determined; then consideration must be given to available sheet sizes (often 14 × 20 inches and 20 × 28 inches for tin) and quantity in a box. Items that require special attention are the fastening methods and the type of ridge and seam treatment.

Copper roofing comes in various sizes. The weights range from 14 to 20 ounces per square foot, and the copper selected is designated by its weight (16-ounce copper). The copper may have various types of joining methods; among the most popular are standing seam and flat seam. The estimator must study the drawings and specifications carefully, as the type of seam will affect the coverage of a copper sheet. When using a standing seam, 2¼ inches in width are lost from the sheet size (in actual coverage) to make the seam. This means that when standing seams are used, 12 percent must be added to the area being covered to allow for the forming of the seam; if end seams are required, an allowance must also be added for them. Each different roofing condition must be planned to ensure a proper allowance for seams, laps, and waste.

14–15 SLATE

Slate is available in widths of 6 to 16 inches and lengths of 12 to 26 inches. Not all sizes are readily available, and it may be necessary to check the manufacturer's current inventory. The basic colors are blue-black, gray, and green. Slate may be purchased smooth or rough textured. Slate shingles are usually priced by the amount required to cover a square if the manufacturer's recommended exposure is used. If the exposure is different from the manufacturer's, Formulas 14-2 and 14-3 (Section 14–17) may be used. To the squares required for the roof area, add 1 sf per foot of length for hips and rafters. Slate thicknesses and corresponding weights are shown in Figure 14.11, assuming a 3-inch lap. Each shingle is fastened with two large-head, solid copper nails, 1¼ or 1½ inches long. The felt required under the slate shingles must also be included in the estimate. Waste for slate shingles varies from 8 to 20 percent, depending on the shape, number of irregularities, and number of intersections on the roof.

14–16 TILE

The materials used for roofing tiles are cement, metals, and clay. To estimate the quantities required, the roof areas are obtained in the manner described earlier. The number, or linear footage, of all special pieces and shapes required for ridges, hips, hip starters, terminals, and any other special

Thickness of slate (inches)	$\frac{3}{16}$	$\frac{1}{4}$	$\frac{3}{8}$	$\frac{1}{2}$	$\frac{3}{4}$	1
Weight per square (3″ lap)	700–800	900–1000	1300–1400	1700–1800	2500–2800	3400–3600

FIGURE 14.11. Approximate Weight of Slate per Square.

pieces must be carefully taken off. If felt is used under the tile, its cost must also be included in the estimate of the job, and furring strips must be installed for certain applications.

As unfamiliar installations arise, the estimators may call the manufacturer, the manufacturer's representative, or a local dealer. These people can review the project with the estimators and help them arrive at material takeoffs, and perhaps even suggest a local subcontractor with experience in installing the tile.

14–17 SHEETS, TILE, AND SHINGLES FORMULA

Formula 14-2 may be used to determine the number of sheets, tiles, or shingles required to cover a square (100 sf) of roof area for any required lap or seam.

$$N = \frac{(100 \text{ ft}^2) \times (144 \text{ in}^2 \text{ per ft}^2)}{(W - S) \times (L - E)}$$ **Formula 14-2**

Number of Sheets, Tiles, or Shingles per sq.

Where

    N = number of sheets (tiles or shingles)

    W = width of sheet (inches)

    L = length of sheet (inches)

    S = side lap or seam lap requirement (for some types, there may be none)

    E = end lap or seam requirements

To determine the square feet of any given roofing sheet, tile, or shingles required to cover one square, Formula 14-3 may be used:

$$A = \frac{100 \times W \times L}{(W - S) \times (L - E)}$$ **Formula 14-3**

Square Feet of Material per Square

Where

    A = square feet of material per square

14–18 LIQUID ROOFING

Liquid roofing materials were developed primarily for free-form roofs. Composition varies among the manufacturers, but the application sequences are similar. First, primer is applied liberally over the entire surface. Then all major imperfections are caulked, joints taped, and flashing is applied at all intersecting surfaces. Next, three coats of liquid roofing are applied; depending on the deck and slope, they may be either base and finish coats or only the finish coat. Finish coats are available in a variety of colors. Equipment is simple, consisting only of rollers, hand tools, brushes, and a joint-tape dispenser. Labor will be affected primarily by the shape of the roof, ease of moving about, and so forth. On some buildings, it is necessary to erect scaffolding to apply the roofing; in such cases, the cost obviously will increase accordingly. Before bidding this type of roofing, estimators

should discuss the project with the manufacturer's representative so that the latest technical advice may be incorporated in the bid.

14–19 FLASHING

Flashing is used to help keep water from getting under the roof covering and from entering the building wherever the roof surface meets a vertical wall. It usually consists of strips of metal or fabric shaped and bent to a particular form. Depending on the type of flashing required, it is estimated by the piece, linear footage with the width noted, or square feet. Materials commonly used include copper, asphalt, plastic, rubber, composition, and combinations of these materials. Bid the gauge of thickness and width specified. Expansion joints are estimated by the linear foot, and it may be necessary that curbs be built up and the joint cover either prefabricated or job assembled. Particular attention to the details on the drawings is required so that the installation is understood.

14–20 INSULATION

Insulation included as a part of roofing is of the type that is installed on top of the deck material. This type of insulation is rigid and is in a sheet (or panel). Rigid insulation may be made of urethane, fiberboards, or perlite and is generally available in lengths of 3 to 12 feet in 1-foot increments and widths of 12, 16, 24, and 48 inches. Thicknesses of 1/2, 3/4, 1, 1½, 2, 2½, 3, and 4 inches are available. Insulation is estimated by the square with a waste allowance of 5 percent, provided there is proper planning and utilization of the various sizes.

When including the insulation, keep in mind that its installation will require extra materials, either in the line of additional sheathing paper, moisture barriers, mopping, or a combination of these. Also, the specifications often require two layers of insulation, usually with staggered joints, which require twice the square footage of insulation (to make up two layers), an extra mopping of bituminous material, and extra labor. Estimators must read the specifications carefully and never bid on a project they do not fully understand.

NOTE: Some roofing manufacturers will not bond the performance of the roof unless the insulation meets their specifications. This item should be checked, and if the manufacturer or the representative sees any problem, the architect/engineer should be notified so that the problem may be cleared up during the bid period.

14–21 ROOFING TRIM

Trim—such as gravel stops, fascia, coping, ridge strips, gutters, downspouts, and soffits—is taken off by the linear foot. All special pieces used in conjunction with the trim (e.g., elbows, shoes, ridge ends, cutoffs, corners and brackets) are estimated by the number of pieces required.

The usual wide variety of materials and finishes is available in trim. Not all trim is standard stock; much of it must be specially formed and fabricated, which adds considerably to the cost of the materials and to the delivery time. Estimators must be thorough and never assume that the trim needed is standard stock.

14–22 LABOR

The labor cost of roofing will depend on the hourly output and hourly wages of the workers. The output will be governed by the incline of the roof, size, irregularities in the plan, openings (skylights, etc.), and the elevation of the roof above the ground since the higher the roof, the higher the materials have to be hoisted to the work area.

Costs can be controlled by the use of crews familiar with the type of work and experienced in working together. For this reason, much of the roofing done on projects is handled by a specialty contractor with equipment and trained personnel.

14–23 EQUIPMENT

Equipment requirements vary considerably, depending on the type of roofing used and the particular job being estimated. Most jobs will require hand tools for the workers, ladders, some scaffolding, and some type of hoist, regardless of the type of roofing being applied. Specialty equipment for each particular installation may also be required. Built-up roofing may demand that mops, buggies, and heaters (to heat the bituminous materials) be used; some firms have either rotary or stationary felt layers. Metal roofing requires shears, bending tools, and soldering outfits.

Equipment costs are estimated either by the square or by the job, with the cost including such items as transportation, setup, depreciation, and replacement of miscellaneous items.

14–24 CHECKLIST

Paper
Felt
Composition (roll)
Composition (built-up)
Tile (clay, metal, concrete)
Shingles (wood, asphalt, slate)
Metal (copper, aluminum, corrugated, steel)
Insulation
Base
Solder:
  paints
  plaster
  foundation walls
  slabs
  sump pits
  protective materials
  exterior
  interior
  admixtures
  drains
  pumps

Flashing
Ridges
Valleys
Fasteners
Trim
Battens
Blocking (curbs)
Cant strips
Waterproofing:
  integral
  membrane
Dampproofing:
  integral
  parge
  vapor barriers
  bituminous materials
  drains
  foundation walls
  slabs

WEB RESOURCES

www.nrca.net
www.asphaltroofing.org
www.roofhelp.com

REVIEW QUESTIONS

1. What is the difference between waterproofing and dampproofing?

2. What is membrane waterproofing, and how is it estimated?

3. What is parging, and what unit of measure is used?

4. What is the unit of measure for roll batt insulation? What type of information should be noted on the estimate?

5. What is the unit of measure for shingle roofing? What type of information should be noted on the estimate?

6. What is the unit of measure for built-up roofing? What type of information should be noted on the estimate?

7. Determine the amount of dampproofing required for the building in Appendix C.

8. Assume that membrane waterproofing is required under the slab for the building in Appendix C. Determine the material required if a three-ply system is specified under the floor slab and up the walls within 2 inches of grade.

9. Determine the amount of insulation required for the walls and ceiling for the building in Appendix C.

10. Determine the amount of cant strips and other accessories required on the building in Appendix C.

11. Determine the amount of built-up roofing, slag, and flashing required for the building in Appendix C.

CHAPTER FIFTEEN

DOORS AND WINDOWS

15–1 WINDOW AND CURTAIN WALL FRAMES

Window and curtain wall frames may be made of wood, steel, aluminum, bronze, stainless steel, or plastic. Each material has its particular types of installation and finishes, but from the estimator's viewpoint, there are two basic types of windows: stock and custom-made.

Stock windows are more readily available and, to the estimator, more easily priced as to the cost per unit. The estimator can count the number of units required and list the accessories to work up the material cost.

Custom-made frames cannot be accurately estimated. Approximate figures can be worked up based on the square footage and type of window, but exact figures can be obtained only from the manufacturer. In this case, the estimator will call either the local supplier or the manufacturer's representatives to be certain that they are bidding the job. Often, copies of the drawings and specifications are sent to them, which they may use to prepare a proposal.

When checking proposals for the windows on a project, the estimator needs to note whether the glass or other glazing is included, where delivery will be made, and whether installation is done by the supplier, a subcontractor, or the general contractor. The proposal must include all the accessories that may be required, including mullions, screens, and sills, and the material being bid must conform to the specifications.

If the contractor is going to install the windows, the estimator needs to check whether they will be delivered preassembled or whether they must be assembled on the job. The job may be bid by the square foot or linear foot of frame, but the most common method is to bid in a lump sum.

Shop drawings should always be required for windows and curtain wall frames, because even stock sizes vary slightly in terms of masonry and rough openings required for their proper installation. Custom-made windows always require shop drawings so that the manufactured sizes will be coordinated with the actual job conditions.

If the estimator decides to do a complete takeoff of materials required, she should (1) determine the linear footage of each different shape required of each type of material, (2) determine the type and thickness of glazing required, and (3) calculate the sheet sizes required.

Wood Windows. Wood windows are commonly made of ponderosa pine, southern pine, and Douglas fir. Custom-made frames may be of any of these species or of some of the more exotic woods such as redwood and walnut. Finishes may be plain, primed, preservative-treated, or even of wood that is clad with vinyl.

Vinyl Windows. The frames for vinyl windows are made of extruded polyvinyl chloride (PVC). Vinyl windows come in a variety of colors, with the color being added to the PVC before extrusion. Vinyl windows are a common choice for residences, because they are economical, energy efficient, require little maintenance, and are available in custom sizes.

The same general estimating procedure is followed for wood and vinyl windows as for other frames since these frames may be stock or custom-made. Shop drawings should be made and carefully checked so that all items required are covered by the proposal. If painting is required, it must be covered in that portion of the specifications.

EXAMPLE 15-1   RESIDENTIAL WINDOWS

Determine the quantity of windows for the residence found in Appendix B. The window schedule (Figure 15.1) typically denotes all window sizes and types. The estimator simply must verify the list and count the number of each size and type (Figure 15.2).

WINDOW SCHEDULE		
MARK	SIZE	TYPE / DESCRIPTION
1	6'-0" X 5'-0"	WOOD CASEMENT
2	4'-0" X 3'-0"	WOOD CASEMENT
3	3'-0" X 4'-0"	WOOD CASEMENT
4	4'-0" X 4'-0"	WOOD CASEMENT

FIGURE 15.1. Sample Window Schedule.

Size	Type / Description	Count
6'-0" X 5'-0"	Wood Casement	1
4'-0" X 3'-0"	Wood Casement	1
3'-0" X 4'-0"	Wood Casement	2
4'-0" X 4'-0"	Wood Casement	2
Total Window Count		6

FIGURE 15.2. Window Takeoff.

Installation:

6 wood windows in masonry

Productivity rate (Figure 15.3) 3 labor hours / window

Labor hours = 6 windows × 3 labor hours per window
= 18 labor hours

Labor cost ($) = 18 hours × $12.25 per hour = $220.50

Windows	Labor Hours per Window
Wood	
In Wood Frames	2.0 to 3.5
In Masonry	2.5 to 5.0
Metal	
In Wood Frames	2.0 to 3.5
In Masonry	2.0 to 4.0
Over 12 s.f.	Add 20%
Bow and Bay Windows	2.5 to 4.0

FIGURE 15.3. Labor Hour Productivity Rates.

Aluminum Windows and Curtain Wall. The shapes for aluminum mullions used in curtain walls or storefronts are made by an extrusion process and roll forming. Finishes include mill finish (natural silvery sheen), anodizing (protective film; it may be clear or a variety of color tones), paint, lacquer, and colored or opaque coatings. Obviously, the finish required must be noted, because it will affect the price of the materials. The specifications will require certain thicknesses of metal throughout the construction of the frames.

Care on the job is important, because exposed aluminum surfaces may be subject to splashing with plaster, mortar, or concrete masonry cleaning solutions that often cause corrosion of the aluminum. Exposed aluminum should be protected during construction with either a clear lacquer or a strippable coating. Concealed aluminum that is in contact with concrete, masonry, or absorbent material (such as wood or insulation) and may become intermittently wet must be protected permanently by coating either the aluminum or the adjacent materials. Coatings commonly used include bituminous paint and zincchromate primer.

The costs of such protection must be included in and are part of the cost of the project. If the frames are to be installed by a subcontractor, the estimator must check to see whether the subcontractor has included the required protection in the proposal; if not, the estimator will have to make an allowance for it. Also, who will apply the protection? If a strippable coating is used on exposed aluminum, the estimator must note who will remove the coating, whether the subcontractor has included this in the price, or whether the general contractor will handle it. Any piece of work that must be done costs money, so estimators ensure that all items are included.

Before the job is bid, the estimator should check the details showing how the windows are to be installed. They may have to be installed as the construction progresses (where the window is built into the wall), or they may be slipped into openings after the building has been constructed. The method is important as there is always a time lag for approvals, and manufacture (of custom units) and coordination will be important.

EXAMPLE 15-2   COMMERCIAL BUILDING

Determine the quantity of windows in the small commercial building found in Appendix A. Figure 15.4 is an excerpt of the window schedule for that project.

Single hung 3'0" × 7'0" count 2
Fixed glass 3'0" × 7'0" count 17

Productivity rate (Figure 15.3) 3 labor hours / window

Labor hours = 19 windows × 3 labor hours per window
= 57 labor hours

Labor cost ($) = 57 hours × $12.25 per hour = $698.25

WINDOW SCHEDULE

MARK	TYPE	SIZE	MATERIAL	DESCRIPTION	REMARKS
1	A	30X70	ALUMINUM	FIXED GLASS	
2	A	30X70	ALUMINUM	FIXED GLASS	
3	A	30X70	ALUMINUM	FIXED GLASS	
4	A	30X70	ALUMINUM	FIXED GLASS	
5	A	30X70	ALUMINUM	FIXED GLASS	
8	A	30X70	ALUMINUM	FIXED GLASS	
9	A	30X70	ALUMINUM	FIXED GLASS	
11	B	30X70	ALUMINUM	SINGLE HUNG	
12	B	30X70	ALUMINUM	SINGLE HUNG	
13	A	30X70	ALUMINUM	FIXED GLASS	
14	A	30X70	ALUMINUM	FIXED GLASS	
15	A	30X70	ALUMINUM	FIXED GLASS	
16	A	30X70	ALUMINUM	FIXED GLASS	
17	A	30X70	ALUMINUM	FIXED GLASS	
18	A	30X70	ALUMINUM	FIXED GLASS	
19	A	30X70	ALUMINUM	FIXED GLASS	

FIGURE 15.4. Window Schedule.

EXAMPLE 15-3   STOREFRONT GLASS

Perform a takeoff of the materials for the storefront glass required for the building entrance from Appendix A. The north window wall is shown in Figure 15.5. The required tubing is shown in Figure 15.6, and the required glass is shown in Figure 15.7.

Aluminum curtain walls are almost always installed by a sub-contractor, usually the curtain wall supplier. The time required for curtain wall installation is shown in Figure 15.8.

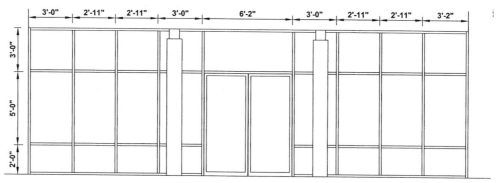

FIGURE 15.5. Elevation of North Storefront.

Length	Pcs.	Linear Ft.
10'-0"	10	100
3'-0"	12	36
2'-11"	16	47
3'-2"	4	13
6'-2"	2	13
Total l.f.		209

FIGURE 15.6. Aluminum Tubing Takeoff.

Size	Pcs.	s.f. Ea.	Total s.f.
3'-0" X 3'-0"	3	9	27
3'-0" X 5'-0"	3	15	45
3'-0" X 2'-0"	3	6	18
2'-11" X 3'-0"	4	8.75	35
2'-11" X 5'-0"	4	14.5	58
2'-11" X 2'-0"	4	5.8	23
3'-2" X 3'-0"	1	9.5	9.5
3'-2" X 5'-0"	1	15.8	15.8
3'-2" X 2'-0"	1	6.3	6.3
6'-2" X 3'-0"	1	18.5	18.5

FIGURE 15.7. Glass Takeoff.

Metal Curtain Walls	S.F per Labor Hours
Up to 1 Story High	8 to 18
1 to 3 Stories	6 to 12
Over 3 Stories	6 to 10

Including glazing. The larger the opening
for the glass or panels the more work
which can be done per labor hour.

FIGURE 15.8. Curtain Wall Productivity Rates.   ■

15-2 ACCESSORIES

The items that may be required for a complete job include glass and glazing, screening, weatherstripping, hardware, grilles, mullions, sills, and stools. The specifications and details of the project must be checked to find out what accessories must be included, so the estimator can make a list of each accessory and what restrictions apply for each.

Glass. *Glass* is discussed in Section 15–10. At this point, estimators note whether it is required, what thickness, type, and quality, and whether it is to be a part of the unit or installed at the job site. The square footage of each type of glass must be known as well.

Glazing. *Glazing* is the setting of whatever material is installed in the frames. Often the material is glass, but porcelain, other metal panels, exposed aggregate plywood, plastic laminates, and even stone veneer and precast concrete have been set in the frames. Many stock frames are glazed at the factory, but it is not uncommon for stock windows, especially steel frames, to be glazed on the job. Custom-made frames are almost always glazed on the job. The subcontractor and supplier will help to determine who glazes them. Glazing costs will depend on the total amount involved, size and quality of the material to be glazed, and the type of glazing method used. If the frame is designed so that the material can be installed from inside the building, no scaffolding will be required, and work will proceed faster. On wood frames glazing compounds are usually employed, whereas on metal frames either glazing compounds or neoprene gaskets are used.

Screens. Screening mesh may be painted or galvanized steel, plastic-coated glass fiber, aluminum, or bronze. If specified, be certain that the screens are included in the proposals received. If the selected subcontractor or supplier for the frames does not include the screens, a source for those items must be obtained elsewhere. If so, be certain that the screens and frames are compatible and that a method of attaching the screen to the frame is determined. The various sizes required will have to be noted so that an accurate price for the screening can be determined.

Hardware. Most types of frames require *hardware,* which may consist only of locking and operating hardware with the material of which the frame is made. Because the hardware may be sent unattached, it must be applied at the job site. Various types of locking devices, handles, hinges, and cylinder locks on sliding doors are often needed. Materials used may be stainless steel, zinc-plated steel, aluminum, or bronze, depending on the type of hardware and where it is being used.

Weatherstripping. Most specifications require some type of *weatherstripping,* and many stock frames come with weatherstripping factory installed—in that case, the factory-installed weatherstripping must be the type specified. Some of the more common types of weatherstripping are vinyl, polypropylene woven pile, neoprene, metal flanges and clips, polyvinyl, and adhesive-backed foam. It is usually sold by the linear foot for the rigid type and by the roll for the flexible type. The estimator will do the takeoff in linear feet and work up a cost.

Mullions. *Mullions* are the vertical bars that connect adjoining sections of frames. The mullions may be of the same material as the frame, or a different material, color, or finish. Mullions may be small T-shaped sections that are barely noticeable or large elaborately designed shapes used to accent and decorate. The mullions should be taken off by the linear footage with a note as to thicknesses, finish, and color.

Sills. The sill is the bottom member of the frame. The member on the exterior of the building, just below the bottom member of the frame, is another sill. This exterior sill serves to direct water away from the window itself. These exterior sills may be made of stone, brick, precast concrete, tile, metal, or wood. The details of the frame and its installation must be studied to determine the type of material, its size, and how it fits in. Basically, there are two types of sills: the slip sill and the engaged sill. The slip sill is slightly smaller than the opening for the frame and can be slipped in place after the construction of the walls is complete or either just before or just after the frames have been installed (depending on the exact design). The engaged sill must be installed as the walls go up, since it is wider than the opening and extends into the wall construction. Sills are taken off by the linear foot; the takeoff should be accompanied by notes and sketches that show exactly what is required.

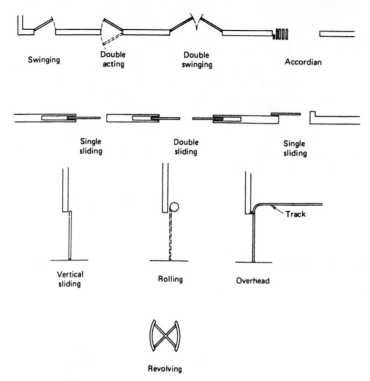

FIGURE 15.9. Typical Door Operation.

Stools.   The interior member at the bottom of the frame (sill) is called the *stool*. The stool may be made of stone, brick, precast concrete, terrazzo, tile, metal, or wood. It may be of the slip or engaged variety. The quantity should be taken off in linear feet with notes and sketches, showing material size and installation requirements.

Flashing.   Flashing may be required at the head and sill of the frame. The specifications and details will state whether it is required and the type. Usually the flashing is installed when the building is being constructed, but the installation details diagram how it is to be installed. Also, the estimator checks on who buys the flashing and who is supposed to install it. These seemingly small items should not be neglected, because they amount to a good deal of time and expense.

Lintels.   The horizontal supporting member of the opening is called a *lintel*. It may be wood, concrete, steel, or block with steel reinforcing. The lintels are not installed as part of the window, but rather they are installed as the building progresses. At this time, however, the estimator double-checks whether the lintels have been included, both as material and installation costs.

15-3 DOORS

Doors are generally classified as interior or exterior, although exterior doors are often used in interior spaces. The list of materials of which doors are made includes wood, aluminum, steel, glass, stainless steel, bronze, copper, plastics, fiberglass, and hardboard. Doors are also grouped according to the mode of their operation. Some different types of operation

are illustrated in Figure 15.9. Accessories required include glazing, grilles and louvers, weatherstripping (for sound, light, and weather), molding, trim, mullions, transoms, and more. The frames and hardware must also be included.

Many specially constructed doors serve a particular need: Some examples are fire-rated, sound-reduction, and lead-lined doors. The frames may be of as many different materials as the doors themselves, and the doors sometimes come prehung in the frame. The doors may be prefinished at the factory or job finished.

Wood Doors.   Wood doors are basically available in two types: solid and hollow core. Solid-core doors may have a core of wood block or low-density fibers and are generally available in thicknesses of 1⅛ and 1¼ inches for fiber core and 1¼ inches through 2½ inches in the wood block core. Widths of 5 feet are available in both types with a height of 12 feet as maximum for wood block core. The fiber core door is available in 6′8″, 7′0″, and 8′0″ heights, but different manufacturers offer different sizes. In checking the specifications, the estimator notes the type of core required, any other special requirements such as the number of plies of construction, fire rating, and the type of face veneer.

Hollow-core doors are any doors in which the cores are not solid. The core may have interlocking wood grid strips, ribs, struts, or corrugated honeycomb. The specifications should spell out exactly what type is required. Thicknesses usually range from 1⅛ inches to 1¼ inches, with widths of 1 to 4 feet and heights up to 8 feet.

The face veneer required in the specifications should be checked carefully, as there is a considerable difference in prices. Refer to Figure 15.10 for a quick reference cost guide to various

Veneer	Index Value
Rotary sound ph. mahogany	72
Rotary sound natural birch	92
Rotary good birch	98
Rotary premium grad birch	100
Rotary premium red oak	100
Rotary premium select red birch	116
Plain sliced premium red oak	133
Plain sliced premium African mahogany	150
Plain sliced premium natural birch	165
Plain sliced premium cherry	170
Plain sliced premium walnut	180
Sliced quartered premium walnut	250
Plain sliced premium teak	250
Laminated plastic faces (premachined)	250

Each veneer has an index value comparing it with the price of Rotary cut premium grade birch. Fluctuation in veneer costs will cause variations in the index values.

FIGURE 15.10. Veneer Cost Comparison.

face veneers. From the specifications, the estimator determines whether the doors are to be prefinished at the factory or job finished. He or she decides who will finish the doors if they must be job finished, and what type of finish is required. Doors that are prefinished at the factory may require some touch-up on the job, and damaged doors will have to be replaced. No matter what type of door is required, the items to be checked are the same. The specifications must be read carefully.

Metal Doors.   The materials used in a metal door may be aluminum, steel, stainless steel, bronze, or copper. The doors may be constructed of metal frames with large glazed areas, hollow metal, tin clad, or a variety of other designs. The

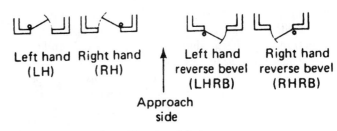

Left hand Right hand   Left hand   Right hand
(LH)   (RH)   reverse bevel reverse bevel
                       (LHRB)   (RHRB)

Approach
side

FIGURE 15.11. Door Hand and Swing.

important thing is to read the specifications, study the drawings, and bid what is required. Often, special doors, frames, and hardware are specified, and the material is custom-made. Finishes may range from lacquered and anodized aluminum and primed or prefinished steel to natural and satin-finish bronze. Bid only that which is specified.

Door Swings.   The swing of the door is important to the proper coordination of door, frame, and hardware. The names are often confused, so they are shown in Figure 15.11 to act as a reminder and learning aid. When checking door, frame, and hardware schedules, estimators must check the swing as well.

Fire-Rated Doors.   *Fire-rated* doors are produced under the factory inspection and labeling service program of Underwriters Laboratories, Inc. (U.L.) and can be identified by the labels on the door. The label states the rating given to that particular door. The ratings given are also related to the temperatures at which the doors are rated. Doors labeled in this manner are commonly called "labeled doors." The letters used, hourly ratings, temperatures, and common locations are given in Figure 15.12. Fire doors

Underwriters label classification	Heat transmission rating	Most commonly recommended locations in building
A 3 hr. situations	Temperature rise 30 minutes 250 degrees max.	Opening in fire walls. Areas of high hazard contents. Curtain and division walls.
A* 3 hr. situations		Openings in fire walls. Areas of high hazard contents. Curtain and division walls.
B 1 ½ hr. situations	Temperature rise 30 minutes 250 degrees max.	Openings in stairwells, elevators shafts, vertical shafts.
B* 1 ½ hr. situations		Openings in stairwells, elevator shafts, vertical shafts.
C* ¾ hr. situations		Openings in room partitions and corridors
D* 1 ½ hr. situations		Openings in exterior walls where exposure to fire is severe.
E* ¾ hr. situations		Openings in exterior walls where exposure to fire is moderate.

* Not rated for heat transmission

FIGURE 15.12. Fire Door Requirements.

may be made of hollow metal, sheet metal, composite wood with incombustible cores, and other composite constructions. Special hardware will be required, and the frame may also have to be labeled.

Fire-rated doors may be conventionally operated, operated by the fire alarm system, or be horizontal or vertical slide doors that close when the fusible link releases.

Acoustical Doors.   *Acoustical doors* are doors specially designed and constructed for use in all situations when sound control is desired. The doors are generally metal or wood, with a variety of cores used. Because the doors available offer a wide range of sound control, the estimator should use caution in selecting the right door to meet the specifications. When high sound control is required, double doors may be desirable. Acoustical doors alone will not solve the problem of noise transmission around the door. An automatic drop seal for the bottom of the door and adjustable doorstops (or other types of sound control devices) are used at the doorjamb and head where the door meets the frame to reduce sound transmission. Often, the seals and devices are sold separately from the door. In these cases, estimators determine the linear footage of the seal needed and include its cost and labor charges in the estimate.

Overhead Doors.   *Overhead doors* are available in all sizes. They are most commonly made of wood and steel, although aluminum and stainless steel are sometimes used. Overhead doors are first designated by the type of operation of the door: rolling, sectional, canopy, and others. The estimator must determine the size of door, size of opening, type and style of door, finishes required, door hanger type, installation details, and whether it is hand, electric, or chain operated. The overhead door is priced as a unit with all the required hardware. Often, the supplier will act as a subcontractor and install the doors.

Miscellaneous Doors.   There are a wide variety of specialized doors available to fill a particular need. Among these are rolling grilles, which roll horizontally or vertically to provide protection or control traffic. Other specialty doors are revolving, dumbwaiter, rubber shock-absorbing doors, cooler and freezer doors as well as blast- and bulletproof doors. The estimating procedures are basically the same, regardless of the type of door required.

Folding Doors and Partitions.   *Folding (accordion) doors and partitions* offer the advantage of increased flexibility and more efficient use of the floor space in a building. They are available in fiberglass, vinyl, and wood, or in combinations of these materials. They can be made to form a radius and have concealed pockets and overhead track; a variety of hardware is also available. Depending on the type of construction, the maximum opening height ranges from 8 to 21 feet and the width from 8 to 30 feet. The doors may be of steel construction with a covering, rigid fiberboard panels, laminated wood, or solid wood

panels. Specially constructed dividers are also available with higher sound-control ratings. The exact type required will be found in the specifications, as will the hardware requirements. Both the specifications and drawings must be checked for installation details and the opening sizes required.

15-4   PREFITTING AND MACHINING (PREMACHINING) DOORS

The doors may be machined at the factory to receive various types of hardware. Factory machining can prepare the door to receive cylindrical, tubular, mortise, unit, and rim locks. Other hardware such as finger pulls, door closers, flush catchers, and hinges (butts) are also provided for. Bevels are put on the doors, and any special requirements are handled.

Premachining is popular as it cuts job labor costs to a minimum, but coordination is important because the work is done at the factory from the hardware manufacturer's templates. This means that approved shop drawings, hardware and door schedules, and the hardware manufacturer's templates must be supplied to the door manufacturer.

From the estimator's point of view, premachining takes an item that is difficult to estimate and simplifies it considerably. Except where skilled door hangers are available, premachined doors offer cost control with maximum results. For this reason, they are being used more and more frequently.

Residential doors are often prehung. A prehung, interior door includes the door, frame, and hinges (butts) as a single unit. An exterior, prehung door includes the door, frame, exterior (brick) mold, hinges, threshold, and weather stripping. Prehung doors are often machined for the lockset.

15-5   PREFINISHING DOORS

Prefinishing of doors is the process of applying the desired finish on the door at the factory instead of finishing the work on the job. Doors that are premachined are often prefinished as well. Various coatings are available, including varnishes, lacquer, vinyl, and polyester films (for wood doors). Pigments and tints are sometimes added to achieve the desired visual effect. Metal doors may be prefinished with baked-on enamel or vinyl-clad finishes. Prefinishing can save considerable job-finishing time and generally yields a better result than job finishing. Doors that are prefinished should also be premachined so that they will not have to be "worked on" on the job. The prefinished door must be handled carefully and installed at the end of the job so that the finish will not be damaged; it is often difficult to repair a damaged finish. Care during handling and storage is also requisite.

EXAMPLE 15-4   RESIDENTIAL DOORS

Determine the doors required for the residence in Appendix B. The first place to start is with the door schedule (Figure 15.13). Just as with the window schedule, this table should list all of the unique types of doors. The task for the estimator is to find and count the different types of doors.

DOOR SCHEDULE		
MARK	**SIZE**	**TYPE / DESCRIPTION**
1	6'-0" X 6'-8"	SOLID W/ SIDELITES
2	6'-0" X 6'-8"	ALUM & GLASS
3	2'-8" X 6'-8"	WOOD HOLLOW CORE
4	3'-0" X 6'-8"	WOOD SOLID CORE
5	2'-4" X 6'-8"	WOOD HOLLOW CORE
6	7'-0" X 8'-0"	O/H DOOR
7	2 - 2'-6" X 6'-8"	BYPASS HOLLOW CORE

FIGURE 15.13. Residence Door Schedule.

1–3/0 × 6/8 panel door with glass sidelites

2–6/0 × 6/8 aluminum and glass sliding doors

5–2/8 × 6/8 hollow core (h.c.), wood, prefinished

2–3/0 × 6/8 solid core, wood, prefinished

3–2/4 × 6/8 hollow core, wood, prefinished

1–7/0 × 8/0 overhead garage door, 4-panel, aluminum

4–5/0 × 6/8 hollow core, bipass, prefinished

The labor costs are typically expressed as so many labor hours per door. Figure 15.14 shows productivity rates for installing doors.

2 aluminum and glass doors at 2.0 labor hours per door
$$= 4 \text{ labor hours}$$

8 prehung hollow core at 2.0 labor hours per door = 16 labor hours

2 prehung solid core at 2.5 labor hours per door = 5 labor hours

1 overhead door at 5 labor hours = 5 labor hours

4 bipass doors at 3.0 labor hours per door = 12 labor hours

1 exterior door with sidelites at 6 labor hours = 6 labor hours

Labor Cost ($) = 48 hours × $13.50 per hour = $648

Doors and Frames	Labor Hours per Unit
Residential, Wood	
Prehung	2.0 to 3.5
Pocket	1.0 to 2.5
Not prehung	3.0 to 5.0
Overhead	4.0 to 6.0
Heavy duty	Add 20%
Commercial	
Aluminum Entrance, per door	4.0 to 6.0
Wood	3.0 to 5.0
Metal, prefitted	1.0 to 2.5

FIGURE 15.14. Door Installation Productivity Rate. ■

EXAMPLE 15-5   COMMERCIAL BUILDING

The commercial building in Appendix A has a door schedule found on sheet A9.1. This door schedule not only shows all of the doors but also shows their location, fire rating, and required hardware.

Figure 15.15 is that schedule. If the drawings are that detailed, the takeoff simply becomes an accuracy check.

DOOR SCHEDULE						
MARK	TYPE	SIZE	RATING (IN MINUTES)	MATERIAL	HW #	REMARKS
101	B	60X70X1-3/4		ALUMINUM	2	
102	A	30X70X1-3/4	60 MIN	LAM PLAST	1	
103	A	30X70X1-3/4	60 MIN	LAM PLAST	1	
104	A	30X70X1-3/4	60 MIN	LAM PLAST	1	
105	A	30X70X1-3/4	60 MIN	LAM PLAST	1	
106	A	30X70X1-3/4		METAL	3	
107	D	40X70X1-3/4	60 MIN	LAM PLAST	4	
108	D	40X70X1-3/4	60 MIN	LAM PLAST	4	
109	A	30X70X1-3/4	60 MIN	LAM PLAST	1	
110	A	30X70X1-3/4	60 MIN	LAM PLAST	1	
111	A	30X70X1-3/4	60 MIN	LAM PLAST	5	
112	A	30X70X1-3/4	60 MIN	LAM PLAST	5	
113	A	30X70X1-3/4	60 MIN	LAM PLAST	5	

FIGURE 15.15. Commercial Building Door Schedule. ■

15–6   DOOR FRAMES

The door frames are made of the same type of materials as the doors. In commercial work, the two most common types are steel and aluminum. The steel frames (also called *door bucks*) are available in 14-, 16-, and 18-gauge steel and come primed. Steel frames are available knocked down (KD), setup and spot-welded (SUS), or setup and arc-welded (SUA).

Many different styles and shapes are available, and the installation of the frame in the wall varies considerably. Usually steel frames are installed during the building of the surrounding construction. For this reason, it is important that the door frames are ordered quickly; slow delivery will hinder the progress of the job. Also, the frames must be anchored to the surrounding construction.

The sides of the door frame are called the *jambs;* the horizontal pieces at the top are called the *heads.* Features available on the frames include a head 4 inches wide (usually about 2 inches), lead lining for X-ray frames, anchors for existing walls, base anchors, a sound-retardant strip placed against the stop, weatherstripping, and various anchors. Fire-rated frames are usually required with fire-rated doors.

When frames are ordered, the size and type of the door, the hardware to be used, and the swing of the door all must be known. Standard frames may be acceptable on some jobs, but often special frames must be made.

Steel frames are primed coated at the factory and must be finished on the job. Aluminum, bronze, brass, and stainless steel have factory-applied finishes, but have to be protected from damage on the job site. Wood frames are available primed and prefinished.

When steel frames are used in masonry construction, it is critical that the frames be installed as the masonry is laid, which means that the frames will be required very early in the construction process. Because door frames for most commercial buildings must be specially made from approved shop drawings and coordinated with the doors and hardware used, the estimator must be keenly aware of this item and how it may affect the flow of progress on the project.

15–7   HARDWARE

The hardware required on a project is divided into two categories: rough and finished. *Rough hardware* comprises the bolts, screws, nails, small anchors, and any other miscellaneous fasteners. This type of hardware is not included in the hardware schedule, but it is often required for installation of the doors and frames. *Finished hardware* is the hardware that is exposed in the finished building and includes items such as hinges (butts), hinge pins, door-closing devices, locks and latches, locking bolts, kickplates, and other miscellaneous articles. Special hardware, such as panic hardware, is required on exit doors.

Finished hardware for doors is either completely scheduled in the specifications, or a cash allowance is made for the purchase of the hardware. If a cash allowance is made, this amount is included in the estimate (plus sales taxes, etc.) only for the purchase of materials. When hardware is completely scheduled in the specification or on the drawings, this schedule should be sent to a hardware supplier for a price. Only on small projects will the estimator figure a price on hardware, unless the firm is experienced in this type of estimating. The cost for installation of finished hardware will vary depending on what type and how much hardware is required on each door, and whether the door has been premachined.

15–8   ACCESSORIES

Items that may be required to complete the job include weatherstripping, sound control, light control, and saddles. The specifications and details will spell out what is required. A list containing each item must be made. The takeoff for accessories should be made in linear feet or the number of each size piece, for example, "five saddles, 3'0" long."

Weatherstripping for the jambs and head may be metal springs, interlocking shapes, felt or sponge, neoprene in a metal frame, and woven pile. At the bottom of the door (sill), the weatherstripping may be part of the saddle, attached to the door, or both. It is available in the same basic types as used for jambs and heads, but the attachment may be different. Metals used for weatherstripping may be aluminum, bronze, or stainless steel.

Saddles (thresholds) are most commonly wood, aluminum, or bronze. Various shapes, heights, and widths may be specified. Sound control and light control usually employ felt or sponge neoprene in aluminum, stainless steel, or bronze housing. The sill protection is usually automatic, closing at the sill; but at the jambs and head, it is usually adjustable.

15–9   CHECKLIST FOR DOORS AND FRAMES

1. Sizes and number required
2. Frame and core types specified
3. Face veneer specified (wood and veneer doors)
4. Prefinished or job finished (if so, specify the finish)
5. Prehung or job-hung (if so, specific installer)
6. Special requirements
   (a) louvers
   (b) windows
   (c) fire rating
   (d) lead lining
   (e) sound control
7. Type, size, style of frame, and the number of each required
8. Method of attachment of the frame to the surrounding construction
9. Finish required on the frame and who will apply it
10. Hardware—types required and installer
11. Accessories—types required and time to install them

Everything takes time and costs money. At the construction site or in the factory, someone must do every job so that all requirements, materials, and labor hours will be included in the estimate.

15–10   GLASS

Glass is the most common material to be glazed into the frames for windows, curtain walls, storefronts, and doors. The most commonly used types are plate glass, clear window glass, wire glass, and patterned glass. Clear window glass is available in thicknesses of 0.085 to 2.30 inches; the maximum size varies with the thickness and type. Generally available as single and double strength, heavy sheet and picture glass with various qualities are available in each classification. Clear window glass has a characteristic surface wave that is more apparent in the larger sizes.

Plate glass is available in thicknesses of 3/16 to 1¼ inches and as polished plate glass, heavy-duty polished plate glass, rough, or polished plate glass. The more common types are regular, grey, bronze, heat absorbing, and tempered.

Wire glass is available with patterned and polished finishes, and with various designs of the wire itself. The most common thickness is 1/4 inch, and it is also available in colors. This type of glass is used when fire-retardant and safety characteristics (breakage) may be required.

Patterned glass, used primarily for decoration, is available primarily in 7/32- and 1/8-inch thicknesses. Pattern glass provides a degree of privacy, yet allows diffused light into the space.

Other types of glass available include a structural-strength glass shaped like a channel and tempered, sound control, laminated, insulating, heat- and glare-reducing, colored, and bullet-resisting glass.

The frame may be single glazed (one sheet of glass) or double glazed (two sheets of glass) for increased sound and heat insulation. If the specifications call for double glazed, twice as many square feet of glass will be required.

Glass is estimated by the square foot with sizes taken from the working drawings. Because different frames may require various types of glass throughout the project, special care must be taken to keep each type separate. Also to be carefully checked is which frames need glazing, since many windows and doors come with the glazing work already completed. The types of setting blocks and glazing compound required should be noted as well.

REVIEW QUESTIONS

1. Determine the window and/or curtain wall and door materials required for the building in Appendix C. Use workup sheets and sketches.

2. What accessories should be checked for when taking off windows and curtain wall?

3. Define glazing. Why must the estimator determine who will perform the required glazing?

4. What information is required to price a door?

5. Describe the advantages in prefitting and prefinishing doors.

6. Determine the door and hardware requirements for the building in Appendix C.

7. Why should the type of finish required on the door and door frames be noted on the workup sheet?

8. Describe briefly the ways hardware may be handled on a project.

9. What precautions must an estimator take when using an allowance, from the specifications, in the estimate?

10. What is the unit of measure for glass, and why should the various types and sizes required be listed separately?

11. Determine the window and door materials required for the residence in Appendix B.

CHAPTER SIXTEEN

FINISHES

16–1 DRYWALL AND WETWALL CONSTRUCTION

Although drywall construction utilizes wallboards and wetwall utilizes plaster and stucco, many components of the two systems of construction are interchangeable. Both require that supporting construction be applied under them, and the same types are used for both. Many of the fasteners, attachments, and accessories are the same or very similar.

All supporting systems and furring should be installed in accordance with the specifications, and the manufacturer's recommendations for spacing, accessories, and installation should be consulted. If the specifications by the architect/engineer are stricter than the manufacturer's, it is those specifications that must be followed. If the project specifications are less stringent than those of the manufacturer, it is advisable to call the architect/engineer's office to be certain what must be bid. Many manufacturers will not guarantee the performance of their materials on the job unless those materials are installed in accordance with their recommendations.

16–2 SUPPORTING CONSTRUCTION

The wallboard can be applied directly to wood, metal, concrete, or masonry that is capable of supporting the design loads and provides a firm, level, plumb, and true base.

Wood and metal supporting construction often consists of self-supporting framing members, including wall studs, ceiling joists, and roof trusses. Wood and metal furring members such as wood strips and metal channels are used over the supporting construction to plumb and align the framing, concrete, or masonry.

Concrete and Masonry.   Concrete and masonry often have wallboard applied to them. When used, either exterior and below grade, furring should be applied over the concrete or masonry to protect the wallboard from damage due to moisture in the wall; this is not required for interior walls. Furring may also be required to plumb and align the walls.

The actual thickness of the wall should be checked so that the mechanical and electrical equipment will fit within the wall thickness allowed. Any recessed items such as fire extinguishers and medicine cabinets should be carefully considered.

Wood Studs.   The most common sizes used are 2×4 and 2×3, but larger sizes may be required on any particular project: Spacing may vary from 12 to 24 inches on center, again depending on job requirements. Openings must be framed around, and backup members should be provided at all corners. The most common method of attachment of the wallboard to the wood studs is by nailing, but screws and adhesives are also used. Many estimators take off the wood studs under rough carpentry, particularly if they are load bearing. The number of board feet required can be determined from Figure 13.1, using the length of partition, in conjunction with Figure 13.7, which takes into consideration the various spacings. Special care should be taken with staggered and double walls so that the proper amount of material is estimated.

Wood Joists.   The joists themselves are estimated under rough carpentry (Chapter 13). When the plans require the wallboard to be applied directly to the joists, the bottom faces of the joists should be aligned in a level plane. Joists with a slight crown should be installed with the crown up, and if slightly crooked or bowed joists are used, it may be necessary to straighten and level the surface with the use of nailing stringers or furring strips. The wallboard may be applied by nailing or by screwing.

Wood Trusses.   When used for the direct application of wallboard, trusses sometimes require cross-furring to provide a level surface for attachment. Stringers attached at

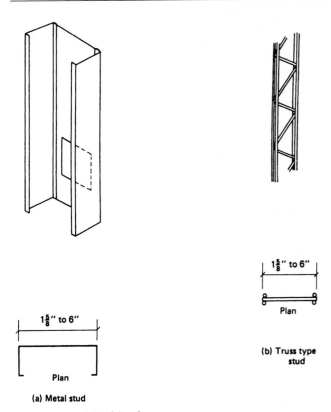

FIGURE 16.1. Metal Studs.

third points will also help align the bottom chord of wood trusses, and a built-in chamber is suggested to compensate for further deflection. Because the trusses are made up of relatively small members spanning large distances, they have a tendency to be more difficult to align and level for the application of wallboard.

Metal Studs. The metal studs (Figure 16.1) most commonly used are made of 25-gauge, cold-formed steel, electrogalvanized to resist corrosion. Most metal studs have notches at each end and knockouts located about 24 inches on center to facilitate pipe and conduit installation. The size of the knockout, not the size of the stud, will determine the maximum size of pipe, or other material, that can be passed through. Often, when large pipes, ducts, or other items must pass vertically or horizontally in the wall, double stud walls are used, spaced the

required distance apart. Studs are generally available in thicknesses of $1\frac{5}{8}$, $2\frac{1}{2}$, $3\frac{5}{8}$, 4, and 6 inches. Metal runner track is used at the top and bottom of the wall to connect the studs together, just as the top and bottom plates connect a wood wall together. In nonbearing walls, metal runner track is often used as headers over openings. The metal runners used are also 25-gauge steel and sized to complement the studs.

A variety of systems have been developed by the manufacturers to meet various requirements of attachment, sound control, and fire resistance. Many of the systems have been designed for ease in erection and yet are still demountable for revising room arrangements. The estimator must carefully determine exactly what is required for a particular project before beginning the takeoff.

The wallboard is typically attached with screws and, in certain applications, with adhesives or nails. Different shapes of studs are available to accommodate either the screws or the nails. Metal studs and runners are sold by the pound or by the linear foot. Once the linear footage of studs has been determined, it is easy to calculate the weight required. The linear footage of each different type of wall must be determined. The walls must be separated according to thickness or the type of stud, backing board and wallboard, as well as according to any variations in ceiling height, application techniques, and stud spacing.

Once again, special care should be taken with double walls so that the proper amount of material may be estimated. If the studs and runners are sold by the pound, the weight per linear foot should be determined from the manufacturers' brochures and multiplied by the number of linear feet to determine the total weight.

Open-Web Joists. The joists themselves are estimated under metals (Chapter 12). A joist, however, may be used as a base for wallboard. Because the bottom chords of the joists are seldom well aligned and the spacing between joists is often excessive, the most common methods of attachment are with the use of furring and with a suspension system. Each of these methods is discussed in this chapter.

Metal Furring. Metal furring (Figure 16.2) is used with all types of supporting construction. It is particularly advantageous where sound control or noncombustible assemblies

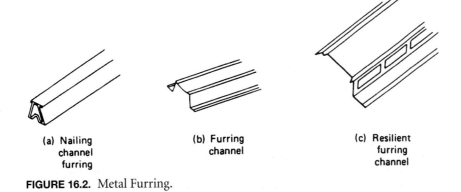

(a) Nailing channel furring   (b) Furring channel   (c) Resilient furring channel

FIGURE 16.2. Metal Furring.

are required. Various types of channels are available. Cold-rolled channels, used for drywall or wetwall construction, are made of 16-gauge steel, 1/4 to 2 inches wide and available in lengths of up to 20 feet. These channels must be wire-tied to supporting construction. They are used primarily as a supporting grid for the lighter drywall channels to which the wallboard may be screw attached.

Drywall channels are 25-gauge electrogalvanized steel and are designed for screw attachment of wallboard; nailable channels are also available. The channels may be used in conjunction with the cold-rolled channels or installed over the wood, steel, masonry, or concrete supporting the construction. These drywall channels may be plain or resilient. The resilient channels are often used over wood and metal framing to improve sound isolation and to help isolate the wallboard from structural movement.

Metal furring is sold by the linear foot. The estimator needs to determine the size and types of furring required, the square footage to be covered, and then the linear footage of each type. Also note the type and spacing of fasteners. Labor and equipment will depend on the type of supporting construction, height and length of walls, shape of walls (straight or irregular), and fastening.

Wood Furring.   Strips are often used with wood frame, masonry, and concrete to provide a suitably plumb, true, or properly spaced supporting construction. These furring strips may be 1 × 2 inches spaced 16 inches on center, or 2 × 2 inches spaced 24 inches on center. Occasionally, larger strips are used to meet special requirements. They may be attached to masonry and concrete with cut nails, threaded concrete nails, and powder or air-actuated fasteners.

When the spacing of the framing is too great for the intended wallboard thickness, cross furring is applied perpendicular to the framing members. If the wallboard is to be nailed to the cross-furring, the furring should be a minimum of 2 × 2 inches in order to provide sufficient stiffness to eliminate excessive hammer rebound. The furring (1 × 2 or 1 × 3 inches) is often used for screw and adhesive attached wallboard.

The furring is attached by nailing with the spacing of the nails 16 or 24 inches on center. The estimator will have to determine the linear footage of furring required, the nailing requirements, equipment, and labor hours. The labor will vary depending on the height of the wall or ceiling, whether straight or irregular walls are present, and the type of framing to which it is being attached.

EXAMPLE 16-1   STEEL STUDS

Using the small commercial building in Appendix A for an example, all of the interior walls are framed with 3⅝ inches wide metal studs, 16 inches on center (1.33 ft). The walls behind the plumbing fixtures are framed with 6-inch-wide studs. The quantification of steel studs is performed in virtually the same manner, as are wood studs. First, identify the linear feet of the wall by stud thickness. In this example (Figure 16.3), the following linear footages are found:

3⅝-inch-wide wall—325′4″

6-inch-wide wall — 20′8″ (Devising wall between restrooms)

Number of stud spaces = Linear feet of wall/Spacing

Number of stud spaces = 325.33′/1.33′ = 245 spaces

Add 1 to get the number of studs

Use 246 studs — 12′ long (3⅝″ wide)

6-inch-wide wall — 20′8″ (Dividing wall between restrooms)

Number of stud spaces = 20.67′/1.33′ = 16 spaces

Use 17 studs — 12′ long (6″ wide)

3 5/8" Wide Studs at 12' Long

Room	Wall Length			Room	Wall Length		
	Ft	In	Ft.		Ft	In	Ft.
107	12	0	12.00	102	16	8	16.67
	14	8	14.67		14	8	14.67
106	14	0	14.00	103	16	4	16.33
	14	8	14.67	109	12	7	12.58
105	12	4	12.33		10	8	10.67
104	10	7	10.58		12	7	12.58
	13	6	13.50	112	12	3	12.25
114	13	6	13.50		10	8	10.67
	21	1	21.08		12	3	12.25
	13	6	13.50	111	20	8	20.67
101	2	0	2.00		3	0	3.00
	13	4	13.33		12	3	12.25
				110	12	7	12.58
					3	0	3.00
Sub Total - Linear Feet		155.17					170.17
TOTAL LINEAR FEET					325	4	325.33

FIGURE 16.3. Length of Walls.

Just as with the wood studs, the number of openings, intersections, and corners needs to be counted.

13 openings — Add 3 studs per opening — 39 studs

11 interior intersections — Add 2 studs per intersection
— 22 studs

4 interior corners — Add 2 studs per corner — 8 studs

Total studs (3⅝″) = 246 + 39 + 22 + 8
= 315 studs (3⅝″)

Add 5 percent for waste — order 331 (3⅝″ × 12′) studs

16–3 SUSPENDED CEILING SYSTEMS

When the plaster, wallboard, or tiles cannot be placed directly on the supporting construction, the wallboard is suspended below the structural system. This may be required if the supporting construction is not properly aligned and true, or if lower ceiling heights are required.

A large variety of systems are available for use in drywall construction, but basically they can be divided into two classes (Figure 16.4): exposed grid systems and concealed grid systems. Within each group, many different shapes of pieces are used to secure the plaster wallboard or tile, but basically the systems consist of hangers, main tees (runners), cross tees (hangers), and furring channels. No matter which type is used, accessories such as wall moldings, splines, and angles must be considered. For wetwall construction, a lath of some type is required.

The suspension system and wallboard may also be used to provide recessed lighting, acoustical control (by varying the type of wallboard and panel), fire ratings, and air distribution (special tile and suspension system).

The suspension system itself is available in steel with an electro-zinc coating as well as prepainted and aluminum, and with plain, anodized, or baked enamel finishes. Special shapes, for example, steel shaped like a wood beam that is left exposed in the room, are also available.

The suspension system may be hung from the supporting construction with 9- or 10-gauge hanger wire spaced about 48 inches on center, or it may be attached by the use of furring strips and clips.

From the specifications and drawings, the type of system can be determined. The pieces required must be

Add 5 percent for waste — order 18 (6″ × 12′) studs

Top and bottom runner track

Track (3⅝″) = 325′4″ × 2 = 650′8″

Track (6″) = 20′8″ × 2 = 41′4″

Since runner track comes in 20-foot pieces, the following would be required (8 percent waste factor):

Order 36 pieces of 3⅝″ × 20′ runner track
Order 3 pieces of 6″ × 20′ runner track   ■

listed, and a complete breakdown of the number of linear feet of each piece is required. Estimators should check the drawings for a reflected ceiling plan that will show the layout for the rooms, as this will save considerable time in the estimate. They should also take note of the size of the tile to be used and how the entire system will be attached to the supporting construction. Later, this information should be broken down into the average amount of material required per square to serve as a reference for future estimates.

16–4 TYPES OF ASSEMBLIES

Basically, drywall construction may be divided into two basic types of construction: single-ply and multi-ply. *Single-ply construction* consists of a single layer of wallboard on each side of the construction, whereas *multi-ply construction* uses two or more layers of wallboard (and often, different types of board) in the various layers (Figure 16.5). The multi-ply construction may be semi-solid or solid, or may have various combinations of materials. Estimators analyze what is required carefully so that the takeoff and pricing may be as complete and accurate as possible.

Various types of demountable and reusable assemblies are also available. Estimators must take time to analyze the assembly specified, break it down into each piece required, and study its pieces and how they are assembled; then an estimate can be made. As questions arise about an unfamiliar assembly, they should not hesitate to call the supplier or manufacturer for clarification.

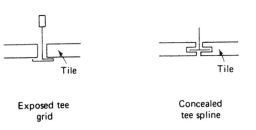
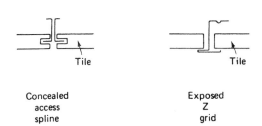

FIGURE 16.4. Typical Grid Systems.

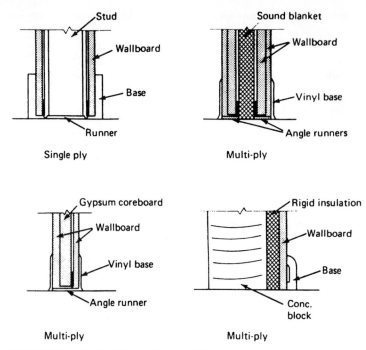

FIGURE 16.5. Typical Drywall Assemblies.

16–5 WALLBOARD TYPES

There are various types of wallboards available for use in drywall construction; among them are gypsum wallboard, tiles, wood panels, and other miscellaneous types. The various types and the special requirements of each are included in this section.

Gypsum Wallboard. Gypsum wallboard is composed of a gypsum core encased in a heavy manila-finished paper on the face side and a strong liner paper on the backside. It is available in a 4-foot width and lengths ranging from 8 to 16 feet, with thicknesses of 1/4, 3/8, 1/2, and 5/8 inches. It is available in regular, fire-resistant (Type X), and moisture-resistant.

Tiles. Tiles used in drywall construction are most commonly available in gypsum, wood and mineral fibers, vinyl, plastic, and metal. Thicknesses vary from 3/16 to 2¹³/₁₆ inches. The most common thicknesses are 1/2, 5/8, and 3/4 inch, with the tile sizes being 12 × 12, 12 × 24, 12 × 36, 16 × 16, 16 × 32, 24 × 24, and 24 × 48 inches. Not all sizes are available in each type, and all manufacturers must be checked to determine what they stock.

In addition to different materials, tiles come in a variety of surface patterns and finishes. They may be acoustical or nonacoustical, and may possess varying edge conditions, light reflection, sound absorption, flame resistance, and flame spread. Tile is even available with small slots that can be opened to provide ventilation in the space below. Plastic louvers and translucent panels are also available.

Prices may be quoted by the piece, square foot, square yard, or particular size package. A square-foot takeoff will provide the estimating information required.

Tile may be applied to various supporting constructions and is often used in conjunction with suspended ceilings.

Wood Panels. Available in many different wood veneers and a variety of finishes, panels are usually 4 feet wide, with lengths varying from 6 to 16 feet. The panels may be constructed of a solid piece of wood or laminated plywood with cores of veneer, flakeboard, or lumber. Either hardwood or softwood may be used.

The estimator must read the specifications carefully to determine exactly what type of paneling is required. Although inexpensive paneling is available at $6 to $15 per sheet and moderately priced panels from $25 to $50 per sheet, some paneling sells for $150 or more per sheet when special face patterns are required. The estimator should never guess at a price for materials, but rather always get a written quote from suppliers and manufacturers.

Paneling is taken off by the square foot, square yard, panels required, or square. The fastening device and trim to be used must be noted in the estimate.

Miscellaneous Panels. Many types of panels are used on walls and ceilings. The majority of them will be priced by the square foot. Among the types available are vinyl-coated plywood, plastic-coated plywood, hardboards, and metal-coated panels. Each separate type has its own method of fastening, accessories, and requirements. The estimator should always check with the manufacturer for installation recommendations.

Specifications. Items that should be checked are the type, thickness, sheet size, and method of attachment required.

The spacing of fasteners must also be determined either from the specifications or, if it is not given there, from the manufacturer's recommendations. The estimator should make a list of accessories, the materials from which they are fabricated, and the finish required.

Estimating.   All wallboard should be taken off by the square foot, with the estimator double-checking for panel layout of the job. The first step is to determine the linear footage of each type of wall, carefully separating any wall with different sizes or types of material, fasteners, or any other variations. There are two sides to most walls, and each side requires a finish. After a complete listing has been made, the square footage of wallboard may be determined. Different varieties may be encountered, and each must be kept separate.

Equipment required may simply be wood or metal horses, planks, platforms, and scaffolding as well as small electric tools and staplers. On projects with high ceilings, scaffolds on wheels are often used so that the workers can work more conveniently, and the scaffold may be moved easily from place to place.

Labor for drywall construction will vary depending on the type of wallboard trim, fasteners, whether the walls are straight or jogged, height of walls or ceiling, and the presence of other construction underway at the time. Many subcontractors are available with skilled workers specially trained for this type of work.

16-6 DRYWALL PARTITIONS

Drywall partitions consist of wallboard over supporting construction; backing board may also be required. The types of materials used for this construction will depend on the requirements of the job with regard to appearance, sound control, fire ratings, strength requirements, and cost. Materials may be easily interchanged to meet all requirements.

Each component of the drywall construction assembly must be taken off and estimated separately. Plain gypsum wallboard will require a finish of some type, such as painting or wallpaper.

Gypsum wallboard with a rugged vinyl film that is factory laminated to the panel is also available. The vinyl-finished panel is generally fastened with adhesives or clips and matching vinyl-covered trim; it is also available in fire-resistant gypsum wallboard where fire-rated construction is desired or required.

Fire-resistant gypsum wallboard is available generally in 1/2- and 5/8-inch thicknesses. These panels have cores, containing special mineral materials, and can be used in assemblies that provide up to two-hour fire ratings in walls and three-hour fire ratings in ceilings and columns.

Other commonly used gypsum wallboards include insulating panels (aluminum foil on the back), water-resistant panels (for use in damp areas; they have special paper and core materials), and backing board, which may be used as a base for multi-ply construction and acoustical tile application, and which may be specially formulated for a fire-resistant base for acoustical tile application.

The supporting construction for ceilings generally utilizes wood joists and trusses, steel joists and suspended ceilings, and concrete and masonry, sometimes in conjunction with the various types of furring materials available.

Drywall construction is generally estimated by the square foot, square yard, and square, each estimator using whichever seems most comfortable. The most common approach to estimating drywall partitions is to take the linear footage of each different type and thickness of the wall from the plan and list them on the workup sheet. Walls that are exactly the same should be grouped together; any variation in the construction of the wall will require that it be considered separately. If the ceiling heights vary throughout the project, the lengths of walls of each height must also be kept separately.

Once the estimator makes a listing of wall lengths and heights, multiply the length times height and determine the square footage of the partition. With this information, the amounts of material required may be estimated. Deduct all openings from the square footage, and add 8 to 10 percent for waste.

16-7 COLUMN FIREPROOFING

Columns may be fireproofed by using drywall construction consisting of layers of fire-resistant gypsum wallboards held in place by a combination of wire, steel studs, screws, and metal angles. Up to a three-hour fire rating may be obtained by using gypsum board. A four-hour fire rating is available when the gypsum board is used in conjunction with gypsum tile (usually 2 or 3 inches thick). To receive the fire ratings, all materials must be installed in accordance with U.L. designs. A complete takeoff of materials is required. No adhesives may be used.

16-8 ACCESSORIES

Accessories for the application and installation of drywall construction include mechanical fasteners and adhesives, tape and compound for joints, fastener treatment, trim to protect exposed edges and exterior corners, and base plates and edge moldings.

Mechanical Fasteners.   Clips and staples may be used to attach the base ply in multi-ply construction. The clip spacing may vary from 16 to 24 inches on center and may also vary depending on the support spacing. Staples should be 16-gauge galvanized wire with a minimum of a 7/16-inch-wide crown with legs having divergent points. Staples should be selected to provide a minimum of 5/8-inch penetration into the supporting structure. They are spaced about 7 inches on center for ceilings and 8 inches on center for walls.

Nails used to fasten wallboard may be bright, coated, or chemically treated; the shanks may be smooth or angularly threaded with a nail head that is generally flat or slightly concave. The angularly threaded nails are most commonly used since they provide more withdrawal resistance, require less penetration, and minimize nail popping. For a fire rating, it is usually required to have 1 inch or more of penetration; and in this case, the smooth shank nails are most often used. Nails with small heads should never be used for attaching wallboard to the supporting construction. The spacing of nails generally varies from 6 to 8 inches on center, depending on the size and type of nail and the type of wallboard being used. Nails are bought by the pound. The approximate weight of the nails that would be required per 1,000 sf (msf) of gypsum board varies between 5 and 7 pounds.

Screws may be used to fasten both wood and metal supporting construction and furring strips. In commercial work, the drywall screws have virtually eliminated the use of nails. Typically, these screws have self-drilling, self-tapping threads with flat recessed Phillips heads for use with a power screwdriver. The drywall screws are usually spaced about 12 inches on center except when a fire rating is required, when the spacing is usually 8 inches on center at vertical joists. There are three types of drywall screws: one for fastening to wood, one for sheet metal, and one for gypsum board. The approximate number of screws required per 1,000 sf of gypsum board based on vertical board application and spacing of 12 inches on center is 1,000. If 8-inch on-center spacing is used on the vertical joints, about 1,200 screws are required. If the boards are applied horizontally with screws spaced 12 inches on center, only about 820 screws are required.

Adhesives may be used to attach single-ply wallboard directly to the framing, concrete, or masonry, or to laminate the wallboard to a base layer. The base layer may be gypsum board, sound-deadening board, or rigid-foam insulation. Often the adhesives are used in conjunction with screws and nails that provide either temporary or permanent supplemental support. Basically, the three classes of adhesives used are stud adhesives, laminating adhesives, and contact adhesives. There are also various modifications within each class. Information regarding the exact adhesives required should be obtained from the specifications of the project and cross-checked with the manufacturer. Also, determine the special preparation, application, and equipment requirements from the manufacturer. Information concerning coverage per gallon and curing requirements should also be obtained.

Trim.   A wide variety of trims are available in wood and metal for use on drywall construction. The trim is generally used to provide maximum protection and neat, finished edges throughout the building. The wood trim is available unfinished and prefinished in an endless selection of sizes, shapes, and costs. The metal trim is available in an almost equal amount of sizes and shapes. Finishes range from plain steel, galvanized steel, and prefinished painted to trim with permanently bonded finishes that match the wallboard; even aluminum molding, plain and anodized, is available. Most trim is sold by the linear foot, so the takeoff should also be made in linear feet. Also to be determined is the manner in which the moldings are to be attached to the construction.

Joint Tape and Compounds.   Joint tape and compounds are employed when a gypsum wallboard is used, and it is necessary to reinforce and conceal the joints between the wallboard panels and to cover the fastener heads. These items provide a smooth, continuous surface in interior walls and ceilings.

The tape used for joint reinforcement is usually a fiber tape designed with chamfered edges feathered thin, with a cross-fiber design.

Joint compounds are classified as follows: (1) embedding compound, used to embed and bond the joint tape; (2) topping compound, used for finishing over the embedding compound (it provides final smoothing and leveling over fasteners and joints); and (3) all-purpose compound, which combines the features of both of the other two, providing embedding and bonding of the joint tape and giving a final smooth finish. The compounds are available premixed by the manufacturer or in a powdered form to be job-mixed.

The amount of tape and compound required for any particular job will vary depending on the number of panels used with the least number of joints and the method of fastening specified. To finish 1,000 sf of surface area, about 380 lf of tape and 50 pounds of powder joint compound (or 5 gallons of ready-mixed compound for the average job) will be required.

Blankets.   Various types of blankets are used in conjunction with the drywall construction. The blankets are most commonly placed in the center of the construction, between studs, or on top of the suspended ceiling assembly. The two basic types of blankets are heat insulating and sound control. The heat-insulating blankets are used to help control heat loss (winter) and heat gain (summer), while the sound-control blankets are used to improve sound transmission classification (STC) ratings of the assembly. Both types are available in a variety of thicknesses and widths. Once the square footage of the wall or ceiling that requires the blanket has been determined, the amount of blanket required is virtually the same for metal stud walls, but the stud spacing should be noted so that the proper width blanket will be ordered. The estimator needs to check the specifications to determine any special requirements for the blankets, such as aluminum foil on one or both sides, paper on one or both sides, and method of attachment. The most common method of attachment is with staples.

EXAMPLE 16-2   COMMERCIAL BUILDING DRYWALL

For the commercial building in Appendix A, determine the required quantity of 5/8-inch-thick drywall. The information to perform this takeoff comes from the stud quantification (Example 16-1). In that takeoff, it was found that there were roughly (325.33 feet + 20.67 feet) 346 lf of interior partitions. The floor to ceiling height is 10 feet. In this example, a 10-foot drywall will be used, but the installers will have to be careful to lift the drywall roughly 3/4 inches off the floor. If there is a concern about the hangers being able to do this, then a 12-foot drywall should be purchased and cut. However, the latter adds material and labor costs, and increases the amount of construction waste that must be hauled away from the job.

If 10-foot sheets are to be ordered and installed vertically, it would require 183 sheets.

Drywall (sheets) = 346 lf/4′ per sheet = 87 sheets per side

Double the quantity for both sides

Add 5 percent for waste — Use 183 sheets

Taping joints is also required. Enough tape must be provided to cover the vertical joints between the gypsum board sheets. From the information above, 1,000 sf of surface area will require 380 lf of tape and 5 gallons of ready-mix joint compound.

sf of surface = 346 lf × 10′ ceiling height × 2 sides = 6,920 sf

Use 7,000 sf

Linear feet of tape = 7 (thousands of sf) × 380 lf per thousand

Linear feet of tape = 2,660 lf of tape

Add 10 percent for waste and assume 50-foot rolls

Order 59 rolls

Gallons of joint compound = 7 (thousands of sf)

× 5 gallons per thousand sf

Gallons of joint compound = 35 gallons of joint compound

Drywall screws = 7 (thousands of sf)

× 1,000 screws per thousand sf

Drywall screws = 7,000 drywall screws ■

EXAMPLE 16-3   RESIDENTIAL BUILDING

Exterior walls: gypsum board, 1/2″ × 4′ × 8′
Ceiling height: 8 feet
148 lf (from Section 13–4, Example 13-10)

sf of wall = 148 lf × 8′ ceiling height = 1,184 sf

Drywall (sheets) = 1,184 sf/32 sf per sheet = 37 sheets

Interior walls: 1/2-inch gypsum board each side
Ceiling height: 8 feet
149 lf (from Section 13–4, Example 13-18)

sf of wall = 149 lf of wall × 8′ ceiling height × 2 sides = 2,384 sf

Drywall (sheets) = 2,384 sf of wall/32 sf per sheet = 75 sheets

Ceiling area: 24′0″ × 50′0″ (5/8-inch gypsum wall board)

Ceiling area (sf) = 24 × 50 = 1,200 sf

Drywall (sheets) = 1,200 sf/32 sf per sheet = 38 sheets

Using a 5 percent waste factor, the following would be ordered:

40 sheets 5/8″-thick 4′ × 8″ gypsum wallboard

118 sheets — 1/2″-thick 4′ × 8′ gypsum wallboard

Metal Trim: corners bead: 3 pieces at 8′ ■

Drywall	Labor Hours per 100 s.f.
Gypsum Board	
Nailed to Studs:	
Walls	1.0 to 2.2
Ceilings	1.5 to 2.8
Jointing	1.1 to 1.8
Glued:	
Walls	0.8 to 2.4
Ceilings	1.5 to 3.0
Jointing	0.5 to 1.2
Ceilings over 8′	Add 10% to 15%
Screwed to Metal Studs	Add 10%
Rigid Insulation, Glued to Walls	1.2 to 3.0

FIGURE 16.6. Labor Hours Required for Drywall Installation.

Labor.   Subcontractors who specialize in drywall installation generally do this type of work and may price it on a unit basis (per square foot) or as a lump sum. The productivity rates for hanging drywall are shown in Figure 16.6. The hourly wage rates need to come from the local market conditions.

EXAMPLE 16-4   RESIDENTIAL BUILDING

Gypsum wallboard, nailed to studs:

Wall area = 1,184 sf + 2,384 sf = 3,568 sf

Hanger (walls), 1.2 labor hours per 100 sf

Taper (walls), 1.4 labor hours per 100 sf

Total 2.6 labor hours per 100 sf

Hang and tape (labor hours) = 35.68 (hundreds of sf)
× 2.6 labor hours per 100 sf

Hang and tape = 92.8 labor hours

Labor cost ($) = 92.8 hours × $10.00 per hour

Labor cost ($) = $928.00 (walls)

Hanger (ceiling), 1.8 labor hours per 100 sf

Taper (ceiling), 1.6 labor hours per 100 sf

Total 3.4 labor hours per 100 sf

Hang and tape (labor hours) = 12.00 (hundreds of sf)
× 3.4 labor hours per 100 sf

Hang and tape = 40.8 labor hours

Labor cost ($) = 40.8 hours × $10.00 per hour
= $408.00 (ceiling)

Total cost = $408.00 + 928.00 = $1,336.00 ■

16–9   WETWALL CONSTRUCTION

Wetwall construction consists of supporting construction, lath, and plaster. The exact types and methods of assembly used for this construction will depend on the requirements of the particular job regarding appearance, sound control, fire ratings, strength requirements, and cost.

The supporting construction may be wood, steel, concrete, gypsum, tile, masonry, or lath. Certain types of lath used with the plaster are self-supporting. The plaster itself may be two or three coats, with a variety of materials available for each coat.

Proper use of plasters and bases provides the secure bond necessary to develop the required strength. A mechanical bond is formed when the plaster is pressed through the holes of the lath or mesh and forms keys on the backside. A suction or chemical bond is formed when the plaster is applied over masonry and gypsum bases, with the tiny needle-like plaster crystals penetrating into the surface of the base. Both mechanical and suction bonds are developed with perforated gypsum lath.

16–10   PLASTER

In its plastic state, plaster can be troweled to form. When set, it provides a hard covering for interior surfaces such as walls and ceilings. Plaster is the final step in wetwall construction (although other finishes may be applied over it). Together with the supporting construction and some type of lath, the plaster will complete the assembly. The type and thickness of the plaster used will depend on the type of supporting construction, the lath, and the intended use. Plaster is available in one-coat, two-coat, and three-coat work, and it is generally classified according to the number of coats required. The last and final coat applied is called the *finish coat,* while the coat, or combination of coats, applied before the finish coat is referred to as the *base coat.*

Base Coats.   Base coat plasters provide a plastic working material that conforms to the required design and serves as a base over which the finish coats are applied. Base coats are available mill-mixed and job-mixed. Mill-mixed base coats are available with an aggregate added to the gypsum at the mill. Aggregates used include wood fibers, sand, perlite, and vermiculite. For high-moisture conditions, a portland cement and lime plaster base coat is available.

Three-coat plaster must be used on metal lath. The first coat (scratch coat) must be of a thickness sufficient to form keys on the back of the lath, fill it in completely, and cover the front of the lath. The thickness may vary from 1/8 to 1/4 inch. The second coat (brown coat) ranges from 1/4 to 3/8 inch thick, and the finish coat ranges from 1/16 to 1/8 inch thick.

Two-coat plaster may be used over gypsum lath and masonry. The first coat is the base coat (scratch or brown), and the second coat is the finish coat. Base coats range from 1/4 to 1/8 inch in thickness, and the finish coat ranges from 1/16 to 1/8 inch. Perforated gypsum lath will require enough material to form the mechanical keys on the back of the sheets.

Finish Coats.   Finish coats serve as leveling coats and provide either a base for decorations or the required resistance to abrasion. Several types of gypsum-finish plasters are available, including those that require the addition of only water

and those that blend gypsum, lime, and water (or gypsum, lime, sand, and water). The finish coat used must be compatible with the base coat. Finishing materials may be classified as prepared finishes, smooth trowel finishes, or sand float finishes. Finish coat thickness ranges from 1/16 to 1/8 inch.

Specialty finish coats are also available. One such specialty coat is radiant heat plaster for use with electric cable ceilings. It is a high-density plaster that allows a higher operator temperature for the heating system, as it provides more efficient heat transmission and greater resistance to heat deterioration. Applied in two coats—the first to embed the cable, the second a finish coat over the top—its total thickness is about 1/4 to 1/8 inch. It is usually mill-prepared and requires only the addition of water.

One-coat plaster is a thin-coat, interior product used over large sheets of gypsum plaster lath in conjunction with a glass fiber tape to finish the joints. The plaster coat is 1/16- to 1/32-inch thick.

Keene cement plaster is used where a greater resistance to moisture and surface abrasions is required. It is available in a smooth and sand-float finish. It is a dead-burned gypsum mixed with lime putty and is difficult to apply unless sand is added to the mixture; with sand as an additive, it is less resistant to abrasion.

Acoustical plasters, which absorb sound, are also available. Depending on the type used, they may be troweled or machine-sprayed onto the wall. Trowel applications are usually stippled, floated, or darbied to a finish. Some plasters may even be tinted various colors. Thickness ranges from 3/8 to 1/2 inch.

Special plasters for ornamental plastering work, such as moldings and cornices, are also available. Molding and casting plasters are most commonly used for such work.

Stucco.   Stucco is used in its plastic state. It can be troweled to form. When set, it provides a hard covering for exterior walls and surfaces of a building or structure. Stucco is generally manufactured with portland cement as its base ingredient and with clean sand and sometimes lime added. Generally applied as three-coat work, the base coats are mixed about one part portland cement to three parts clean sand. If lime is added, no more than 6 to 8 pounds per 100 pounds of portland cement should be used in the mix. The lime tends to allow the mix to spread more easily. The finish coat is usually mixed 1:2 (cement to sand), and no coat should be less than 1/4 inch thick.

Stucco is usually applied to galvanized metal lath that is furred out slightly from the wall, but it can also be applied directly to masonry. Flashing is often required and must be included in the estimate.

Various special finishes may also be required and will affect the cost accordingly. Finishes may be stippled, broomed, pebbled, swirled, or configured in other designs.

Synthetic stuccos are also available. They may be applied by installing reinforcing wire over a wood wall and by adding one or two base coats and a top coat. Alternately, they may be applied over rigid insulating board by installing a

Per 1/16" Thickness of Plaster Allow	Perforated Plaster Board Lath - Add	Metal Lath Add
5.2 C.F.	4.0 C.F.	8.0 C.F.

FIGURE 16.7. Approximate Plaster Quantities.

base coat reinforced with a mesh sheet and by applying a top coat to form an exterior insulation and finish system (EIFS).

Specifications. The specifications should state exactly what type of plaster is required and where. The types often will vary throughout the project, and each must be kept separate. The number of coats, thickness of each coat, materials used, and the proportions of the mix must all be noted. The estimator needs to check also the type of finish required, what trueness of the finish coat is required, the room finish schedule on the drawings—since often finishes and the finish coat required are spelled out—and determine the accessories, grounds, trim, and anything else that may be required for a complete job.

Estimating. Gypsum plasters are usually packed in 100-pound sacks and priced by the ton. The estimator makes the wetwall takeoff in square feet, converts it to square yards, and must then consider the number of coats, thickness of coats, mixes to be used, and the thickness and type of lath required. The amounts of materials required may be determined with the use of Figure 16.7. This table gives the cubic feet of plaster required per 100 sy of surface.

EXAMPLE 16-5   PLASTER QUANTITY

For 100 sy of wall area, a 1/4-inch thickness over metal lath will require the following:

5.2 cf per 1/16" × 4 (for 1/4") = 20.8 cf of plaster

Add 8 cf for over metal lath

Use 29 cf of plaster   ∎

Depending on the type of plaster being used, the approximate quantities of materials can be determined. The mix design varies from project to project and must be carefully checked. Figure 16.8 shows some typical quantities of materials that may be required. Many projects have mixes

designed for a particular use included in the specifications. Read them carefully and use the specified mix to determine the quantities of materials required. Once the quantities of materials have been determined, the cost for materials may be determined and the cost per square yard (or per 100 sy) may be calculated.

Labor time and costs for plastering are subject to variations in materials, finishes, local customs, type of job, and heights and shapes of walls and ceilings. The ability of workers to perform this type of work varies considerably from area to area. In many areas, skilled plasterers are scarce, meaning that the labor cost will be high and problems may occur with the quality of the work done. It is advisable to contact the local unions and subcontractors to determine the availability of skilled workers. In most cases, one helper will be required to work with two plasterers.

If the plastering is bid on a unit basis, the estimator needs to be certain that there is an understanding of how the yardage will be computed, as the methods of measuring vary in different localities. The yardage may be taken as the gross area, the net area, or the gross area minus the openings that are over a certain size. In addition, curved and irregular work may be charged and counted extra; it will not be done for the same costs as the flat work.

Equipment required includes a small power mixer, planks, scaffolds, mixing tools, mixing boxes, and miscellaneous hand tools. Machine-applied plaster will require special equipment and accessories, depending on the type used.

Labor. Subcontractors who specialize in wetwall installations do this type of work, and they may price it on a unit basis (per square yard) or lump sum.

16-11 LATH

Lath is used as a base; the plaster is bonded to the lath. Types of lath include gypsum tile, gypsum plaster, metal, and wood.

The type of lath required will be specified and will vary depending on the requirements of the project. The estimator must read the specifications carefully and note on their workup sheet the type of lath required and where it is required. It is not unusual for more than one type to be used on any one job.

Mix	Maximum Amount of Aggregate, in C.F. per 100 Pounds of Gypsum Plaster	Volume Obtained from Mixes Shown (C.F.)
100 : 2	2	2
100 : 2 ½	2 ½	2 ½
100 : 3	3	3
Aggregate weights vary.	Sand 95–100 pounds per cubic foot Perlite 40–50 pounds per cubic foot Vermiculite 40–50 pounds per cubic foot	

FIGURE 16.8. Plaster Materials.

Gypsum Plaster Lath.   In sheet form, gypsum plaster lath provides a rigid base for the application of gypsum plasters. Special gypsum cores are faced with multilayered laminated paper. The different lath types available are plain gypsum, perforated gypsum, fire-resistant, insulating, and radiant heat. Depending on the supporting construction, the lath may be nailed, stapled, or glued. The type of spacing of the attachments depends on the type of construction and the thickness of the lath. Gypsum lath may be attached to the supporting construction by the use of nails, screws, staples, or clips.

Plain gypsum lath is available in thicknesses of 3/8 and 1/2 inch with a face size of 16 × 48 inches. The 3/8-inch thickness is also available in 16 × 96-inch sizes. (A 16 3/16-inch width is available in certain areas only.) When the plaster is applied to this base, a chemical bond holds the base to the gypsum lath. Gypsum lath 24 × 144 inches is also available in certain areas.

Perforated gypsum lath is available in a 3/8 inch thickness with a face size of 16 × 48 inches. Holes 3/4 inches in diameter are punched in the lath spaced 16 inches on center. The perforated lath permits higher fire ratings, because the plaster is held by mechanical as well as chemical bonding.

Fire-resistant gypsum lath has a specially formulated core of special mineral materials. It has no holes, but it provides additional resistance to fire exposure. It is available in a 3/8-inch thickness with a face size of 16 × 48 inches.

Insulating gypsum lath is plain gypsum lath with aluminum foil laminated to the back face. It serves as a plaster base, an insulator against heat and cold, and a vapor barrier. It is available in 3/8- and 1/2-inch thicknesses with a face size of 16 × 48 inches.

Radiant heat lath is a large gypsum lath for use with plaster in electric cable ceilings. It improves the heat emission of the electric cables and increases their resistance to heat deterioration. It is available 48 inches wide, and in 1/2- and 5/8-inch thicknesses and lengths of 8 to 12 feet. This type of lath is used with plaster that is formulated for use with electric cable heating systems.

Estimating.   Gypsum lath is sold by the sheet or 1,000 sf. The estimator will calculate the number of square feet required (the square yards of plaster times nine equals square feet), and divide by the number of square feet in a sheet. Note the type and thickness required. Depending on the number of jogs and openings, about 6 percent should be allowed for waste. The materials used for attachment must be estimated, and a list of accessories must be made.

Metal Lath.   Metal lath is sheet steel that has been slit and expanded to form a multitude of small mesh openings. Ordinary, expanded metal lath (such as diamond mesh, flat-rib lath) is used in conjunction with other supporting construction. There are also metal laths that are self-supporting (such as 3/8-inch rib lath), requiring no supporting construction.

Metal lath is available painted, galvanized, or asphalt dipped; sheet sizes are generally 24 × 96 inches (packed 16 sy per bundle) or 27 × 96 inches (20 sy per bundle). Basically, the three types of metal lath available for wetwall construction are diamond, flat-rib lath, and 3/8-inch rib lath. Variations in the designs are available through different manufacturers.

The metal lath should be lapped not less than 1/2 inch at the sides and 1 inch at the ends. The sheets should be secured to the supports at a maximum of 6 inches on center. The metal lath is secured to the steel studs or channels by the use of 18-gauge tie wires about 6 inches on center. For attachment to wood supporting construction, nails with a large head (about 1/2 inch) should be used.

Diamond Lath.   Diamond lath is an all-purpose lath that is ideal as a plaster base, as a reinforcement for walls and ceilings, and as fireproofing for steel columns and beams. It is easily cut, bent, and formed for curved surfaces. It is available in weights of 2.5 and 3.4 pounds per square yard; both sizes are available in copper alloy steel either painted or asphalt-coated. Galvanized diamond lath is available only in 3.4 pounds per square yard.

Flat-rib lath is a 1/8-inch lath with "flat ribs," which make a stiff type of lath. This increased stiffness generally permits wider spacing between supports than diamond lath, and the design of the mesh allows the saving of plaster. The main longitudinal ribs are spaced 1½ inches apart, with the mesh set at an angle to the plane of the sheet. Available in copper alloy and steel in weights of 2.75 and 3.4 pounds per square yard, and in galvanized steel in a weight of 3.4 pounds per square yard, it is used with wood or steel supporting construction on walls and ceilings, and for fireproofing.

The 3/8-inch rib lath combines a small mesh with heavy reinforcing ribs. The ribs are 3/8-inch deep, 4½ inches on center. Used as a plaster base, it may be employed in studless wall construction and in suspended and attached ceilings. Rib lath permits wider spacing of supports than flat-rib and diamond lath. This type is also used as a combination form and reinforcement for concrete floor and roof slabs. Copper alloy steel lath is available in 3.4 and 4.0 pounds per square yard, and the galvanized is available in 3.4 pounds per square yard.

Specifications.   The type of lath, its weight, and finish must be checked. The spacing of the supporting construction will affect the amount of material and labor required to attach the lath. The type and spacing of attachment devices should be checked as well as a list of accessories.

Estimating.   The metal lath is taken off by the square yard in the same manner as plaster. It is usually quoted at a cost per 100 sy with the weight and finish noted. For plain surfaces, add 6 to 10 percent for waste and lapping; for beams, pilasters, and columns, add 12 to 18 percent. When furring is required, it is estimated separately from the lath. Determine what accessories will be required and the quantity of each.

Wood Lath. Although largely displaced by the other types of laths available, on occasion wood lath is encountered. The most commonly used size is 3/8 inch thick, 1⅜ inches wide (actual), and 48 inches long spaced 3/8 inch apart; therefore, one lath takes up the space of 1¾ × 48 inches. The wood lath would be taken off by the square yard, with about 14.3 pieces required per square yard. They are attached by nails or staples.

Gypsum Tile. Gypsum tile is a precast, kiln-dried tile used for non-load-bearing construction and fireproofing columns. Thicknesses available are 2 inches (solid), 3 inches (solid or hollow), and 4 and 6 inches (hollow). The 2-inch tile is used for fireproofing only, not for partitions. A face size of 12 × 30 inches (2.5 sf) is available. Used as a plaster base, it provides excellent fire and sound resistance. Gypsum tile may be taken off as part of wetwall construction or under masonry.

Specifications. The estimator determines the type and thickness required from the specifications, makes a list of all clips and accessories, and decides how the gypsum tile is to be installed.

Estimating. The number of units required must be determined. If the square feet, squares, or square yards have been determined, their area can be easily converted to the number of units required. The thickness required must be noted as well as the accessories and the amount of each required. Resilient clips may also be used. The lath required for the building will be included in the subcontractor's bid, but the estimator should check the subcontractor's proposal to be certain that it calls for the same lath as the contract documents.

16–12 ACCESSORIES

The accessories available for use with wetwall construction include various types of corner beads, control and expansion joints, screeds, partition terminals, casing beads, and a variety of metal trim to provide neat-edged cased openings. Metal ceiling and floor runners are also available, as are metal bases. Resilient channels may also be used. These accessories are sold by the linear foot, so the estimator makes the takeoff accordingly.

A complete selection of steel clips, nails, staples, and self-drilling screws is available to provide positive attachment of the lath. Special attachment devices are available for each particular wetwall assembly. The estimator will have to determine the number of clips or screws required on the project. The specifications will state the type of attachment required and may also give fastener spacing. The estimator may check the manufacturer's recommended fastener spacing to help determine the number of fasteners required.

Accessories required should be included in the subcontractor's bid, but the estimator should check the sub-

contractor proposal against the contract documents to be certain that they are the same size, thickness of metal, and finish.

16–13 DRYWALL AND WETWALL CHECKLIST

Wetwall	Drywall
lath, metal	studs
furring	wallboards
studs	furring
channels	channels
lath, gypsum	tape
gypsum block	paste (mud)
corner beads	adhesives
accessories	staples
number of coats	clips
type of plaster	nails
tie wire	screws
molding	
stucco	

16–14 FLOORING

Flooring may be made of wood, resilient tile or sheets, carpeting, clay and ceramic tiles, stone, and terrazzo. Each type has its own requirements as to types of installation, depending upon job conditions, subfloor requirements, methods of installation, and moisture conditions.

Wood Flooring

The basic wood flooring types are strip, plank, and block. The most widely used wood for flooring is oak. Other popular species are maple, southern pine and Douglas fir—with beech, ash, cherry, cedar, mahogany, walnut, bamboo, and teak also available. The flooring is available unfinished or factory finished.

Strip flooring is flooring up to 3¼ inches wide and comes in various lengths. *Plank flooring* is from 3¼ inches to 8 inches wide with various thicknesses and lengths. The most common thickness is 25/32 inch, but other thicknesses are available. It may be tongue and grooved, square edged, or splined. Flooring may be installed with nails, screws, or mastic. When a mastic is used, the flooring used should have a mastic recess so that the excess mastic will not be forced to the face of the flooring. Nailed wood flooring should be blind nailed (concealed); nail just above the tongue with the nail at a 45 degree angle. Waste on strip flooring may range from 15 to 40 percent, depending on the size of the flooring used. This estimate is based on laying the flooring straight in a rectangular room, without any pattern involved.

Strip and plank flooring may be sold either by the square foot or by the board foot measure. The estimator

Size of Wood Flooring (Actual)	To Change s.f. to Board Feet, Multiply the s.f. amount by
25/32" x 1 ½"	1.55
25/32" x 2 ¼"	1.383
25/32" x 3 1/4	1.29
3/8" x 1 ½"	1.383
½" x 2"	1.30

Values allow 5 percent waste.

FIGURE 16.9. Wood Flooring, Board Measure.

should figure the square footage required on the drawings, noting the size of flooring required and the type of installation, and then, if required, should convert square feet to board feet (Figure 16.9), not forgetting to add waste.

Block flooring is available as parquet (pattern) floors, which consists of individual strips of wood or larger units that may be installed in decorative geometric patterns. Block sizes range from 6 × 6 inches to 30 × 30 inches, thickness from 5/16 to 3/4 inch. They are available tongue and grooved or square edged. Construction of the block varies considerably: It may be pieces of strip flooring held with metal or wood splines in the lower surface; *laminated blocks,* which are cross-laminated piles of wood; or *slat blocks,* which are slats of hardwood assembled in basic squares and factory assembled into various designs. Block flooring is estimated by the square foot, with an allowance added for waste (2 to 5 percent). The type of flooring, pattern required, and method of installation must be noted.

Wood flooring may be unfinished or factory finished. Unfinished floors must be sanded with a sanding machine on the job and then finished with a penetrating sealer, which leaves virtually no film on the surface, or with a heavy solid type finish, which provides a high luster and protective film. The penetrating sealer also will usually require a coat of wax. The sanding of the floors will require from three to five passes with the machine. On especially fine work, hand sanding may be required. The labor required will vary, depending on the size of the space and the number of sanding operations required. The surface finish may require two or three coats to complete the finishing process. Factory-finished wood flooring requires no finishing on the job, but care must be taken during and after the installation to avoid damaging the finish.

In connection with the wood floor, various types of supporting systems may be used. Among the more common are treated wood sleepers, a combination of 1/8-inch hot asphalt fill and treated sleepers, steel splines, and cork underlayment.

Laminate Flooring

Laminate flooring is made to look like wood flooring or tile. Laminate flooring consists of four layers of materials fused together. The top or wear layer consists of a clear layer containing aluminum oxide and protects the flooring against wear, staining, and fading. The second layer provides the pattern for the flooring. The third layer is made from high-density fiberboard and may be impregnated with plastic resin (melamine) for added strength and moisture resistance. The final layer is the backing layer that provides a moisture barrier. Laminate floors are available with warranties ranging from 10 to 30 years and provide a low-maintenance alternative to wood floors.

Laminate floors are installed over an approved padding material. Be sure that the selected pad meets the manufacturer's recommendation. Some laminate floors are glued together, while others snap together. Laminate floors float over the pad and are free to expand and contract with changes in the humidity and temperature. To allow for this expansion and construction, trim must be provided at the edges of the floor. A wide variety of trim is available in matching colors. Laminate floors are estimated in the same manner as wood floors.

Resilient Flooring

Resilient flooring may be made of asphalt, vinyl, rubber, or cork. Resilient sheets are available in vinyl and linoleum. The flooring may be placed over wood or concrete subfloors by the use of the appropriate adhesive. The location of the subfloor (below grade, on grade, or suspended above grade) will affect the selection of resilient flooring, since moisture will adversely affect some types. All types may also be used on suspended wood subfloors and concrete as long as it is sufficiently cured. Where moisture is present below grade and on grade, the materials may be used *except* cork and rag felt-backed vinyl.

Tile sizes range from 9 × 9 inches to 12 × 12 inches, except for rubber tile, which is available up to 36 × 36 inches and vinyl accent strips, which come in various sizes. Thicknesses range from 0.050 to 1/8 inch except for rubber and cork tiles, which are available in greater thicknesses. Sheet sizes most commonly used are 6 and 12 feet wide, with 4' to 6' also available.

The subfloor may require an underlayment on it to provide a smooth, level, hard, and clean surface for the placement of the tile or sheets. Over wood subfloors, panel underlayments of plywood, hardboard, or particleboard may be used, while the concrete subfloors generally receive an underlayment of mastic. The panel underlayment may be nailed or stapled to the subfloor. The mastic underlayments may be latex, asphalt, polyvinyl-acetate resins, or portland and gypsum cements.

Adhesives used for the installation of resilient flooring may be troweled on with a notched trowel or brushed on. Because there are so many types of adhesives, it is important that the proper adhesive be selected for each application. Check the project specifications and the manufacturers' recommendations to be certain that they are compatible.

The wide range of colors and design variations are in part responsible for the wide use of resilient flooring. For the estimator, this means taking care to bid the color, design, thickness (gauge), size, and finish that is specified.

Accessories include wall base, stair tread, stair nosings, thresholds, feature strips, and reducing strips. The color and design variations are more limited in the accessories than in the flooring materials.

Specifications. From the specifications, the color, design, gauge (thickness), size of tile or sheet, adhesives, subfloor preparation, and any particular pattern requirements may be found. Fancy patterns may be shown on the drawings and in most cases will have a higher percentage of waste.

Note which areas will require the various types of resilient flooring, since it is unusual for one type, color, size, design, and so forth to be used throughout the project. The wall base may vary in height from area to area; this can be determined from the specifications, room finish schedule, and details.

Estimating. Resilient tile is estimated by the square foot: the actual square footage of the surface to be covered plus an allowance for waste. The allowance will depend on the area and shape of the room. When designs and patterns are made of tile or a combination of feature strips and tile, a sketch of the floor and an itemized breakdown of required materials should be made. The cost of laying tile will vary with the size and shape of the floor, size of the tile, type of subfloor and underlayment, and the design. Allowable waste percentages are shown in Figure 16.10.

Feature strips must be taken off in linear feet, and if they are to be used as part of the floor pattern, the square footage of feature strips must be subtracted from the floor area of tile or sheets required.

For floors that need sheet flooring, the estimator must do a rough layout of the floors involved to determine the widths required, the location where the roll will be cut, and ways to keep waste to a minimum. Waste can amount to between 30 and 40 percent if the flooring is not well laid out or if small amounts of different types are required.

Each area requiring different sizes, designs, patterns, types of adhesives, or anything else that may be different must be kept separately if the differences will affect the cost

of material and amount of labor required for installation (including subfloor preparation).

Wall base (also referred to as *cove base*) is taken off by the linear foot. It is available in vinyl and rubber, with heights of 2½, 4, and 6 inches and in lengths of 42 inches, and 50 and 100 feet. Corners are preformed, so the number of interior and exterior corners must be noted.

Adhesives are estimated by the number of gallons required to install the flooring. To determine the number of gallons, divide the total square footage of flooring by the coverage of the adhesive per gallon (in square feet). The coverage usually ranges from 150 to 200 sf per gallon but will vary depending on the type used and the subfloor conditions.

EXAMPLE 16-6   COMMERCIAL BUILDING

Assume that the entire commercial building in Appendix A was to be covered with resilient flooring (main level only).

$$\text{Net sf of floor area} = 5,893 - 375 - 300 = 5,218 \text{ sf}$$

$$\text{Area taken by interior walls (Example 16-1)}$$

$$\text{Area (2} \times \text{4 walls)} = 325' \times 4\tfrac{1}{2}'' = 325' \times 0.375' = 122 \text{ sf}$$

$$\text{Area (2} \times \text{6 walls)} = 21' \times 7'' = 21' \times 0.583' = 12 \text{ sf}$$

$$\text{Area under walls} = 122 \text{ sf} + 12 \text{ sf} = 134 \text{ sf}$$

$$\text{Net area of floor (sf)} = 5,218 \text{ sf} - 134 \text{ sf} = 5,084 \text{ sf}$$

$$\text{Add 5 percent for waste}$$

$$\text{Resilient flooring required 5,338 sf}$$

Gallons of Adhesive

$$\text{Required adhesive: coverage 150 sf per gallon}$$

$$\text{Gallons of adhesive} = 5,084 \text{ sf resilient flooring/150 sf per gallon}$$

$$\text{Use 34 gallons}$$

Linear Feet of Vinyl Base
From Example 16-1, the interior walls measure 346 lf. From the drawings in Appendix A, there are 375 lf of exterior walls.

$$\text{Gross feet of base} = (2 \times 346) + 375 = 1,067 \text{ lf}$$

$$\text{Deductions for doors:}$$

$$\text{1 at } 6'0'' \text{ (1 side)} = 6 \text{ ft}$$

$$\text{2 at } 4'0'' \text{ (2 sides)} = 8 \text{ ft}$$

$$\text{10 at } 3'0'' \text{ (1 at 1 side, 9 at 2 sides)} = 57 \text{ ft}$$

$$\text{Net base} = 1,067 \text{ lf} - 71 \text{ lf} = 996 \text{ lf}$$

$$\text{Add 5 percent for waste—use 1,046 lf}$$

$$\text{sf of base} = 1,046' \times 4'' = 1,046' \times 0.33' = 346 \text{ sf}$$

With an adhesive coverage of 150 sf per gallon—order 3 gallons

■

Area s.f.	Percent Waste
Up to 75	10 – 12
75 – 150	7 – 10
150 – 300	6 – 7
300 – 1,000	4 – 6
1,000 – 5,000	3 – 4
5,000 and up	2 - 3

FIGURE 16.10. Approximate Waste for Resilient Tile.

Labor. Subcontractors who specialize in resilient floor installation generally do this type of work, and they may price it on a unit basis (per square foot) or on a lump sum. The time required for resilient floor installation is shown in Figure 16.11.

Tile	Labor Hours per 100 s.f.
Resilient Squares	
9″ × 9″	1.5 to 2.5
12″ × 12″	1.0 to 2.2
Seamless Sheets	0.8 to 2.4
Add for felt underlayment	Add 10%
Less than 500 s.f.	Add 15%

FIGURE 16.11. Labor Hours Required for Resilient Floor Installation.

Carpeting

Carpeting is selected and specified by the type of construction and the type of pile fibers. The types of construction (how they are made) are tufted, woven, and knitted; punched and flocked have become available more recently. In comparing carpeting of similar construction, factors such as pile yard weight, pile thickness, and the number of tufts per square inch are evaluated in different carpets. Carpeting is generally available in widths of 9, 12, 15, and 18 feet, but not all types come in all widths, with 12-foot widths being the most common.

Pile fibers used include wool, nylon, acrylics, modacrylics, and polypropylene for long-term use. Acetate, rayon, and polyester fibers are also used. The type of pile used will depend on the type of use intended and the service required. The installed performance of the carpet is not dependent on any single factor, but on all the variables involved in the construction of the carpet and the pile characteristics.

The cushion over which the carpeting is installed may be manufactured of animal hair, rubberized fibers, or cellular rubber. The cushion increases the resilience and durability of the installation. The cushion may be bonded to the underside of the carpet, but it is more common to have separate cushions. The type of cushion used will depend on the intended usage of the space, and a variety of designs are available for each type of material. The various cushions within each group are rated by weight in ounces per square yard. The heavier the cushion is, the better and more expensive it will be. Cushioning is generally available in widths of 27, 36, and 54 inches, and 6, 9, and 12 feet.

Specifications. The specifications should state the type of carpeting required, pile yarn weight and thickness, number of tufts per inch, construction, backing, rows, and other factors relating to the manufacture of the carpet. Many specifications will state a particular product or "equal," which means the estimator will either use the product mentioned or ask other suppliers (or manufacturers) to price a carpet that is equal in quality. In the latter case, the estimator should compare the construction specifications of the carpeting to be certain that the one chosen is equal. Different types of carpeting may be used throughout the project. Take note of what types are used and where they are used.

The cushion type required and its material, design, and weight must be noted. If variations in the type of cushion required throughout the project are evident, they should be noted.

Estimating. Carpeting is estimated by the square yard with special attention given to the layout of the space for the most economical use of the materials. Waste and excess material may be large without sufficient planning. Each space requiring different types of carpeting, cushion, or color must be figured separately. Most carpets must be installed with all pieces running in the same direction. If the specifications call for the color to be selected by the architect/engineer at a later date, it may be necessary to call and try to determine how many different colors may be required. In this manner, a more accurate estimate of waste may be made.

Certain types of carpeting can be bought by the roll only, and it may be necessary to purchase an entire roll for a small space. In this case, waste may be high, since the cost of the entire roll must be charged to the project.

The cushion required is also taken off in square yards, with the type of material, design, and weight noted. Since cushions are available in a wider range of widths, it may be possible to reduce the amount of waste and excess material.

EXAMPLE 16-7 RESIDENTIAL BUILDING

Using 12-foot-wide carpet, the following linear feet of carpet will be required.

Bedroom 1	8′0″
Bedroom 2	8′2″
Bedroom 3	11′5″
Bedroom 4	10′0″
Living/Dining Room	30′0″
Hall (3′ × 30′ = 90 sf)	9′0″
Closets	6′0″
Total Carpet	8′7″

Square feet of carpet = 82′7″ × 12′ = 991 sf of carpet

Square yards of carpet = 991 sf/9 sf per sy = 111 sy ∎

Labor. Subcontractors who specialize in carpet installations do this type of work, and they may price it on a unit basis (per square yard) or lump sum. Additional charges are common for stairs. The time required for the installation is shown in Figure 16.12.

Carpet	Squares per Labor Hour
Carpet and Pad, Wall to Wall	8 to 20
Carpet, Pad Backing, Wall to Wall	10 to 22
Deduct for Gluing to Concrete Slab	10%
Less than 10 Squares	Add 15% to 20%

FIGURE 16.12. Labor Hours Required for Carpet Installation.

Tile

Tile may be used on floors and walls. The tile used for floors is usually ceramic or quarry tile, whereas the tile used for walls and wainscots may be ceramic, plastic, or metal.

Ceramic tile is available in exterior or interior grades, glazed or unglazed. Individual tile size may range from 3/8 × 3/8 inches to 16 × 16 inches. Tile may come in individual pieces or sheets of 1/2 to 2¼ sf per sheet. Tile mounted in sheets will be much less expensive to install than unmounted tile. Ceramic tiles come in various shapes and a wide range of sizes and colors. The tile may be installed by use of portland cement mortar, dry-set mortar, organic adhesives, and epoxy mortars. The portland cement mortar is used where leveling or slopes are required in the subfloor; the thickness of this mortar ranges from 3/4 to 1¼ inches, and it requires damp curing. The mortar will receive a coat of neat grout cement coating and the tile will be installed over the neat cement. The other methods are primarily thin-set (1/16 to 1/4 inch) one-coat operations. After the tile has been installed, the joints must be grouted. The grouts may be portland cement-based, epoxies, resins, and latex.

Specifications. The type of tile (material) should be determined for each space for which it is specified. The type of tile and the finish often vary considerably throughout the job. The specifications usually provide a group from which a tile will be selected and a price range (e.g., American Olean, price range A). The groupings and price ranges vary among manufacturers, so care must be taken in the use of specifications written in this manner. Other specifications will spell out precisely what is required in each area, which makes it easier to make an accurate estimate. Each area requiring different types, sizes, or shapes of tile must be taken off separately.

The methods of installation must be noted, and if the methods vary throughout the job, they must be kept separate. The type of grout required in each area must also be noted.

The types of trimmers are also included in the specifications. The number of trimmers required is kept separate from the rest of the tile takeoff because it is more expensive. Note exactly what is required, because some trimmer shapes are much more expensive than others. The contractor should bid what is specified, and if the specifications are not clear, they should contact the architect/engineer.

Estimating. Floor and wall tiles are estimated by the square foot. Each area must be kept separate, according to the size and type being used. It is common to have one type of tile on the floor and a different type on the walls. The different colors also vary in cost even if the size of the tile is the same, so caution is advised. The trim pieces should be taken off by the linear feet of each type required. Because of the large variety of sizes and shapes at varying costs, the specifications must be checked carefully, and the bid must reflect what is required. If portland cement mortar is used as a base, it is installed by the tile contractor. This requires the purchase of cement, sand, and sometimes wire mesh. Tile

available in sheets is much more quickly installed than individual tiles. Adhesives are sold by the gallon or sack, and approximate coverage is obtained from the manufacturer. The amount of grout used depends on the size of the tile.

When figuring wall tile, estimators note the size of the room, number of internal and external corners, height of wainscot, and types of trim. Small rooms require more labor than large ones. A tile setter can set more tile in a large room than in several smaller rooms in a given time period.

Accessories are also available and, if specified, should be included in the estimate. The type and style are in the specifications. They may include soap holders, tumbler holders, toothbrush holders, grab rails, paper holders, towel bars and posts, doorstops, hooks, shelf supports, or combinations of these. These accessories are sold individually, so the number required of each type must be taken off. The accessories may be recessed, flush, or flanged, and this also must be noted.

16–15 PAINTING

The variables that affect the cost of painting include the material painted, the shape and location of the surface painted, the type of paint used, and the number of coats required. Each of the variables must be considered, and the takeoff must list the different conditions separately.

Although painting is one of the items commonly subcontracted, the estimator should still take off the quantities so that the subcontractor's proposal can be checked. In taking off the quantities, the square feet of the surface are taken off the drawings, and all surfaces that have different variables must be listed separately. With this information, the amount of materials can be determined by the use of the manufacturer's information on coverage per gallon.

The following methods for taking off the painting areas are suggested, with interior and exterior work listed separately.

Interior

Walls—actual area in square feet

Ceiling—actual area in square feet

Floor—actual area in square feet

Trim—linear footage (note width); amount of door and window trim

Stairs—square footage

Windows—size and number of each type, square feet

Doors—size and number of each type, square feet

Baseboard radiation covers—linear feet (note height)

Columns, beams—square feet

Exterior

Siding—actual area in square feet

Trim—linear footage (note width)

Doors—square feet of each type

Windows—square feet of each type

Masonry—square footage (deduct openings over 50 sf)

The specifications should list the type of coating, number of coats, and finish required on the various surfaces throughout the project. Interiors receive different treatment than exteriors; different material surfaces require different applications and coatings—all of this should be in the specifications. Paints may be applied by brush, roller, or spray gun. The method to be used is also included in the specifications.

Sometimes, the specifications call for prefinished and factory-finished materials to be job finished also. Except for possible touch-ups, this is usually due to an oversight in the architect/engineer's office and the estimator should seek clarification. The most common items to be factory finished are doors, floorings, windows, baseboard, radiation covers, and grilles. The estimator should keep a sharp eye out to see that each item of work is figured only once.

Structural steel work often requires painting also. It usually comes to the job primed, with only touch-up of the prime coat required. Sometimes, it is delivered unprimed with the priming done on the job. Touch-up painting is impossible to figure accurately and depends on the type of structural system being used, but an average of 5 to 10 percent of the area is usually calculated as the touch-up required.

The structural steel work is taken off by the tonnage of steel required with the types and sizes of the various members required. It must be noted which type of steel is to be painted: steel joists, rectangular or round tubes, H sections, or any other type. The shape of the member will influence the cost considerably. The square footage to be painted per ton of steel may vary from 150 for large members to 500 sf for trusses and other light framing methods.

Labor. Subcontractors who specialize in painting and staining often do this type of work, and they almost always price it on a lump-sum basis. The time required for painting and staining is shown in Figure 16.13.

Panting, Brushes	s.f. or l.f. per Labor Hour
Interior	
Primer and 1 Coat	200 to 260 s.f.
Primer and 2 Coats	150 to 200 s.f.
Stain, 2 Coats	150 to 200 s.f.
Trim	
Primer and 1 Coat	120 to 180 l.f.
Primer and 2 Coats	100 to 160 l.f.
Stain, 2 Coats	100 to 160 l.f.
Exterior	
Primer and 1 Coat	160 to 220 s.f.
Primer and 2 Coats	120 to 180 s.f.
Stain, 2 Coats	200 to 280 s.f.

FIGURE 16.13. Labor Hours Required for Painting and Staining.

16–16 FLOORS AND PAINTING CHECKLIST

Floors:

type of material
type of fastener
spikes
nails
adhesives
screws
finish
thickness
size, shape
accent strips
pattern
cushion
base
Corners

Painting:

filler
primer
paint, type
number of coats
shellac
varnish
stain
check specifications for all areas requiring paint

WEB RESOURCES

www.4specs.com
www.awci.org
www.gypsum.org
www.americangypsum.com
www.usg.com
www.gp.com
www.tlpca.org
www.woodfloors.org
www.forestdirectory.com
www.nofma.org
www.maplefloor.org
www.wfca.org
www.flooryou.com
www.floorfacts.com
www.paint.org
www.armstrong.com

REVIEW QUESTIONS

1. What is the difference between drywall and wetwall?

2. Why should walls of various heights, thicknesses, and finishes be listed separately?

3. What procedure is used to estimate the steel studs and runners used?

4. List the types of lath used for wetwall and the unit of measure for each.

5. What are the advantages and disadvantages of using subcontractors for drywall and wetwall construction?

6. What unit of measure is used for wood block flooring, and what information should be noted on the workup sheets?

7. What unit of measure is used for resilient flooring, and what information should be noted on the workup sheets?

8. What unit of measure is used to estimate carpet, and what can be done to minimize waste?

9. Determine the amounts of materials required for the drywall and/or wetwall for the building in Appendix B.

10. Determine the amounts of materials required for the drywall for the building in Appendix C.

11. Determine the amount of flooring required for the building in Appendix C.

12. Determine the interior and exterior areas of the residence to be painted in Appendix C.

13. Determine the interior finishes of all areas of the building in Appendix C.

14. Determine the floor finishes required for the building in Appendix C.

15. Determine the floor finishes required for the residence in Appendix B.

CHAPTER SEVENTEEN

ELECTRICAL

17-1 ELECTRICAL WORK

Under single contracts, the electrical work is the responsibility of a single prime contractor. In most cases, this means that electrical contractors will submit prices on the work to be completed to the prime contractor. The general contractor will include an electrical contractor price plus overhead and profit in the bid price on the project. In this case, the prime contractor is directly responsible for the work to the owner and must coordinate all of the parties involved in the project.

With separate contracts, the electrical contractor will bid the electrical work directly to the owner; the owner will select the contractor and sign a contract. In this case, the electrical contractors are responsible for the electrical work; and although there will be certain mutual responsibilities and coordination between electrical contractors and general contractors, according to the general conditions, Article 6, the general contractor's responsibilities are not as great as they would be under single contracts.

All electrical work must be installed in accordance with the code regulations of a given area. Throughout the United States, the National Electric Code (NEC) is used extensively. State and local regulations must also be considered. Before beginning the takeoff, the contractor should review the plans and carefully read the specifications. Often the specifications will require that "all work and installations be in conformance with all applicable national, state, and local codes." This statement means that contractors are responsible for compliance with the laws; if they are responsible for them, then they better be familiar with them. The codes contain information regarding wiring design and protection, wiring methods and materials, equipment, special occupancies, and other information.

Field experience in construction will be helpful in understanding the problems involved in electrical work and how the electrical aspect should be integrated into the rest of the construction. Without field experience, an under-

standing of the fundamentals of electrical work, and an ability to read and understand the drawings and specifications, it will be difficult to do a meaningful takeoff on these items.

17-2 SINGLE CONTRACTS

Because an electrical subcontractor will undoubtedly do the electrical work, that contractor will do the bidding. Learning to estimate electrical work is a special skill requiring extensive knowledge of the properties and behavior of electricity. The complexity of this topic would fill a book by itself. What is presented here are some issues that contractors face when dealing with electrical subcontractors. However, by using some common sense and complete files from past jobs, it is possible to obtain approximate estimates for checking whether the bids submitted are reasonable. With experience and complete files, it is possible to figure an estimate close to the low bid.

The wiring is considered the *rough* work, and the fixtures are considered the *finish* work. The wiring will usually be concealed in a conduit, which is installed throughout the building as it is erected. The wiring is pulled through the conduit much later in the job. Cable is also used extensively (almost exclusively on residential projects). Cables are installed when the building is being erected. The fixtures are usually the last items to go into the building, often after the interior finish work is complete.

To work up an estimate, estimators go through the plans and specifications in a systematic manner, taking each different item and counting the number of each. Every item must be kept separate. For example, floor outlets are different from wall outlets. The estimator must not hesitate to check off (lightly) each item as they count it to reduce the possibility of estimating the same item twice. Included in the list are all outlets, floor plugs, distribution panels, junction boxes, lighting panels, telephone boxes, switches, television receptacles, fixtures, and any other items, such as snow-melting mats.

It is the estimator's job to determine exactly where the responsibility begins for the wiring. Does it begin at the property line, at the structure, or 10 feet from the structure? If a transformer is required, who pays for it, who installs it, and who provides the base on which it will be set?

Different types of construction affect the installation of rough and finish work. When using steel joists, there is usually ample space through which to run conduits easily. Cast-in-place concrete requires that there be closer cooperation between the general contractor and the electrical contractor, because the conduit (as well as sleeves) and fixture hangers often must be cast in the concrete. The use of hollow-core, prestressed, precast concrete causes other problems, such as where to run the conduit and how to hang fixtures properly. The conduit can be run in the holes (or joints) that are in the direction of the span, but care must be taken to run them in other directions unless the conduit can be exposed in the room. These problems greatly increase the amount of conduit required as well as the cost of the installation. If installation is difficult, it also becomes more expensive. Similar problems occur when using precast double tees, except in that case no holes are available in the spanning direction. The problem can be alleviated to some extent by pouring a 2½ to 3-inch concrete floor over the slabs in which to run the conduit.

One last point: Estimators must not guess the price of fixtures. Prices vary considerably. What seems to be an inexpensive fixture may be very expensive. They should never trust guesswork, but rather check prices.

Light fixture manufacturers often prefer not to give anyone but the actual electrical subcontractor a firm price; however, if estimators are insistent, the manufacturer will cooperate. This is one reason why it is important to be on friendly terms with as many people as possible. Others often provide the key to success in whatever job is being bid.

The selection of an electrical subcontractor should not be based on price alone, although price is an important consideration. Other factors, such as the speed with which the subcontractors complete their work and the cooperation they show in dealing with the prime contractor and other subcontractors, are also important. Nothing causes hard feelings faster than subcontractors who are uncooperative. Because prime contractors are responsible to the owners for all work, it is in their own best interest to consider all factors while selecting an electrical subcontractor.

The major areas of coordination required between the electrical contractor and the general contractor are outlined in Section 17–4, while Figure 17.1 shows typical electrical symbols and abbreviations.

When electrical subcontractors bid a project, they are often asked to include required temporary wiring and lighting. By having the electrical subcontractor include these costs in their bid, the contractor has a negotiating advantage.

17–3 SEPARATE CONTRACTS

The electrical contractor does the takeoff and bidding, but this does not mean that the estimator for general construction should not review the drawings and specifications for this work. Often, when projects are being bid under separate contracts, the contractors for each phase receive the drawings only for that phase (or portion of work) on which they will be bidding. (The bidders for general construction may receive no electrical or mechanical plans.) In this case, a trip to the plan room or to the office of the electrical, HVAC, and plumbing contractors who are bidding the project should be made so that the drawings and specifications may be investigated.

Even under separate contracts, there are many areas of mutual responsibility and coordination. Keep in mind that the entire building must fit together and operate as one unit. The major areas of coordination for single and separate contracts are outlined in Section 17–4. If separate contracts are being used, the contractor needs to be aware of who is providing the temporary wiring and lighting.

17–4 COORDINATION REQUIREMENTS

Figure 17.2 lists the major areas of coordination required between the electrical and general contractors.

Coordination of work among the electrical, HVAC, and plumbing contractors themselves is also important, since the electrical, HVAC, and plumbing contractors may all have work to do on a particular piece of equipment. For example, the HVAC contractor may install the boiler unit in place, the electrical contractor may make all power connections, and the plumbing contractor may connect the water lines. There are many instances of several trades connecting to one item. Coordination and an understanding of the work to be performed by each contractor are important to a smooth-running job.

17–5 CHECKLIST

Rough:

conduit (sizes and lengths)

wire (type, sizes, and lengths)

outlets (floor, wall, overhead)

switches (2-, 3-, and 4-way)

panel boards

breakers (size, number of each)

outlets, weatherproof

control panels

power requirements

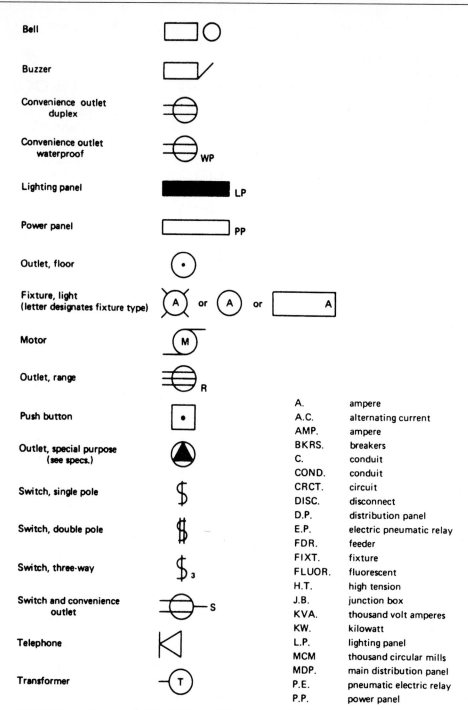

Bell	
Buzzer	
Convenience outlet duplex	
Convenience outlet waterproof	WP
Lighting panel	LP
Power panel	PP
Outlet, floor	
Fixture, light (letter designates fixture type)	A or A or A
Motor	M
Outlet, range	R
Push button	
Outlet, special purpose (see specs.)	
Switch, single pole	
Switch, double pole	
Switch, three-way	
Switch and convenience outlet	S
Telephone	
Transformer	T

A.	ampere
A.C.	alternating current
AMP.	ampere
BKRS.	breakers
C.	conduit
COND.	conduit
CRCT.	circuit
DISC.	disconnect
D.P.	distribution panel
E.P.	electric pneumatic relay
FDR.	feeder
FIXT.	fixture
FLUOR.	fluorescent
H.T.	high tension
J.B.	junction box
KVA.	thousand volt amperes
KW.	kilowatt
L.P.	lighting panel
MCM	thousand circular mills
MDP.	main distribution panel
P.E.	pneumatic electric relay
P.P.	power panel

FIGURE 17.1. Commonly Used Electrical Symbols and Abbreviations.

Item	Coordination Requirements
1. Underground utilities	Location, size, excavation by whom, from where?
2. Equipment	Recessed depth, size of access openings, method of feed, supports, by whom size limitations.
3. Distribution	Outlet locations, materials, method of feed, chases, in walls, under floor, overhead, special considerations.
4. Terminal fixtures and devices	Location, method of support, finish, color, and material.
5. Mounting surfaces	Mounting surface? Can it work?
6. Specialty equipment	Field provisions, storage.
7. Scheduling	Work to be done? When required? Job to be completed on time—who, why, when?

FIGURE 17.2. Coordination Requirements.

Piping symbols:

Vent – – – – – – – – – – –

Cold water —— · —— · —— · ——

Hot water —— ·· —— ·· —— ·· ——

Hot water return —— — —— — —— —

Gas I——— G ——— G ———

Soil, waste or leader ———————————
(above grade)

Soil, waste or leader — — — — —
(below grade)

Fixture symbols:

Baths

Water closet (with tank)

Water closet (flush valve)

Shower

Lavatory

DW   Dishwasher

SS   Service sink

HWH   Hot water heater

HWT
HWT   Hot water tank

DF   Drinking fountain

M   Meter

HB   Hose bib

C/O   CO   Cleanouts

FD   Floor drain

RD   Roof drain

A.F.D.	area floor drain
B.W.V.	backwater valve
CODP.	deck plate cleanout
C.W.	cold water
C.W.R.	cold water return
DEG.	degree
D.F.	drinking fountain
D.H.W.	domestic hot water
DR.	drain
D.W.	dishwasher
F.	fahrenheit
FDR.	feeder
FIXT.	fixture
F.D.	floor drain
F.H.	fire hose
F.E.	fire extinguisher unit
H.W.	hot water
H.W.C.	hot water circulating line
H.W.R.	hot water reserve
H.W.S.	hot water supply
H.W.P.	hot water pump
I.D.	inside diameter
LAV.	lavatory
LDR.	leader
O.D.	outside diameter
(R)	roughing only
R.D.	roof drain
S.C.	sill cock
S.S.	service sink
TOIL.	toilet
UR.	urinal
V.	vent
W.C.	water closet
W.H.	wall hydrant

FIGURE 18.1. Plumbing Symbols and Abbreviations.

out a bid that is too low and call it to the subcontractor's attention. If they suspect that the subcontractor's bid is too high, they should not hesitate to call and discuss it, or they should call if they suspect the bid is too low and a reasonable profit cannot be made. Job performance will probably suffer if the bid is too low. Cooperation and respect are two key ingredients to success. Contractors and estimators must treat their subcontractors well, and they will reciprocate.

Item	Coordination Requirements
1. Underground utilities	Location, size, excavation by whom, from where?
2. Building Entrance	Floor sleeves, supports.
3. Mechanical room equipment	Supports required, location, anchors, by whom?
4. Distribution	Wall sleeves, hangers, chases, in wall roof, vents, access doors.
5. Fixtures	Method of support, feed, outlets, built-in, floor drains, vents.
6. Finishes	Factory or field.
7. Specialty equipment	Field provisions, storage.
8. Scheduling	Work to be done? When required? Job to be completed on time—who, why, when?

FIGURE 18.2. Coordination Issues.

18–3 SEPARATE CONTRACTS

The plumbing contractors do the takeoff and bidding. The estimator for general construction must still check to determine the areas of mutual responsibility and coordination—they are similar to those required for the electrical work. There is always considerable work done in coordination, wall chases, anchoring supports, sleeves, and many other items outlined in Section 18–4.

18–4 COORDINATION REQUIREMENTS

Figure 18.2 lists the major areas of coordination required between the plumbing and general contractors.

Once again, coordination among the HVAC and plumbing contractors is also important in the understanding of who is responsible for what, why, when!

18–5 CHECKLIST

Rough:

permits
excavation and backfill
water, gas, and sewage lines
required pipes and fittings
cleanouts
valves
tanks
sleeves

Finish:

water closets
bath tubs
lavatories
drinking fountains
showers
tubs
service sinks
water heater
urinals
washers and dryers
dishwashers

Miscellaneous:

hookup to equipment
supplied by owner or other contractors may be required

REVIEW QUESTIONS

1. How do the various codes affect the installation of the plumbing portions of the project?
2. What type of work is most generally included under plumbing?
3. Why would an estimator call a subcontractor if it is suspected that the subcontractor's bid is too low?
4. Why should the estimator review the plumbing portions of the project whether it involves single or separate contracts?
5. How do the various types of construction affect the cost of the plumbing work?

CHAPTER NINETEEN

HEATING, VENTILATING, AND AIR-CONDITIONING

19-1 HVAC WORK (HEATING, VENTILATING, AND AIR-CONDITIONING)

As in electrical and plumbing work, this portion of the work may be bid under single or separate contracts. Bidding under separate contracts for HVAC work is similar to bidding under separate contracts for electrical and plumbing work. Understanding the mutual responsibilities and coordination required is a must (Section 19–4).

Like all work, this portion must be designed and installed in accordance with all federal, state, and local codes. Many different codes may have control, depending on the types of systems used.

Again, the field experience, understanding of design principles, and ability to understand drawings and specifications are key points. There is a tremendous variety of installations that may be used in this portion in dealing with electrical heat, ventilating ceilings, and various methods and fuels used for heating and air-conditioning.

19-2 SINGLE CONTRACTS

Subcontractors who specialize in this type of work will be submitting bids to estimators, so your area of responsibility falls within the preparation of an approximate bid for comparison.

Referring to the working drawings and specifications, estimators prepare a complete list of major items required. They determine exactly where the responsibility for each portion rests. If responsibility boundaries are unclear, estimators should request that the architect/engineer clarify them, as it is unwise to make assumptions about this or any portion of the project.

The takeoff includes piping, ductwork, drains, equipment, fixtures, and accessories. The specifications will state who is responsible for trenching, both from the road to the building and within the building. The takeoff list should include all equipment separately as to types and sizes.

Estimators need a general knowledge of heating, ventilating, and air-conditioning to understand the equipment involved. Figure 19.1 contains commonly used HVAC symbols and abbreviations. There are many different systems that could be used on a building. If the estimator is unfamiliar with a system, it is wise to call the manufacturer's representative to get a full explanation. In this way, the estimator will understand what is required of the general contractor and the subcontractor to guarantee a successful installation.

Estimators check the specifications carefully to be sure that the materials they (or their subcontractor) bid are those specified. To improve accuracy from job to job, good estimators keep a complete file of all costs and a breakdown of costs from subcontractors—and they check job against job. It is suggested that they go through the specifications item by item to be sure that all are included in the takeoff.

Another method used for approximate estimates is the use of square feet of building. Also, prices per btu, cfm, and tons of air-conditioning are sometimes quoted. If used with caution, these procedures may be effectively employed for comparison of prices.

19-3 SEPARATE CONTRACTS

The same general conditions prevail with HVAC work under separate contracts as when the electrical and plumbing are bid as separate contracts. Estimators must understand the mutual responsibilities involved and what the coordination requirements are. The work still must be integrated into the building construction. They will review the drawings and check that there is space for the ducts, pipes, and other mechanical lines.

19-4 COORDINATION REQUIREMENTS

Listed in Figure 19.2 are the major areas of coordination required between the HVAC and general contractors.

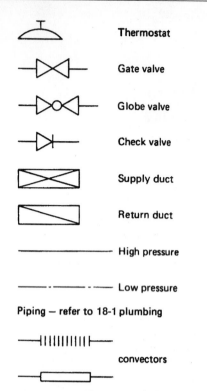

BLR.	boiler
CFM.	cubic feet per minute
CONV.	convector.
CUH	cabinet unit heater
HTR.	heater
HTG.	heating
H & V.	heating and ventilating
HVAC.	heating, ventilating, air conditioning
IBR.	Institute of Boilers and Radiators
R.A.	return air
T.	thermostat
T.R.	top register
U.H.	unit heater
VENT.	ventilate, ventilation

FIGURE 19.1. HVAC Symbols and Abbreviations.

Item	Coordination Requirements
1. Underground utilities	Location, size, excavation by whom, from where?
2. Equipment	Method of support, location, by whom, anchors, access for receiving and installing, size limitations, flues, roof curbs.
3. Piping	Wall sleeves, size limitations, chases, in walls, under floors, floor sleeves, expansion compensators, access doors.
4. Ductwork	Sizes, support, access doors, drops, outlet sizes, chases, outside air louvers, roof curbs, lintels.
5. Terminal equipment	Recesses for CUH, RC, FC, etc., size of radiation, method of concealing, grille fastenings.
6. Mechanical-electrical responsibility	Who is doing what?
7. Finishes	Field or factory?
8. Scheduling	Work to be done? When required? Job to be completed on time—who, why, when?

FIGURE 19.2. Coordination Issues.

19–5 CHECKLIST

Heating (hot water):

boiler

stoker

oil tanks

gauges

fuel

piping

insulation

circulating pumps

piping accessories

radiators

fin tubes

enclosures

clocks

hangers

unit heaters

Heating (warm air):

boiler

fuel

oil tanks

gauges

accessories

thermostats

wiring

chimney

ducts

diffusers

fans

valves

filters

humidifiers

dehumidifiers

insulation

baffles

Air-Conditioning:

central

units

coolant

fans

piping

diffusers

registers

wiring

ducts

filters

humidifiers

dehumidifiers

inlets

returns

fresh air

louvers

thermostats

REVIEW QUESTIONS

1. Would electric heat most likely be placed under electrical, plumbing, or HVAC?

2. How do the various types of construction affect the cost of the heating work?

3. Why are subcontractors hired under single contracts to perform the HVAC work?

CHAPTER TWENTY

PROFIT

20–1 PROFIT

Profit is not included in a chapter with other topics because it is the last thing considered and must be taken separately. Keep in mind that construction is among the tops in the percentage of business failures. Would you believe that some people even forget to add profit?

First, let us understand that by *profit* we mean the amount of money added to the total estimated cost of the project; this amount of money should be clear profit. All costs relating to the project, including project and office overhead and salaries, are included in the estimated cost of the building.

There are probably more approaches to determining how much profit should be included than could be listed. Each contractor and estimator seems to have a different approach. A few typical approaches are listed as follows:

1. Add a percentage of profit to each item as it is estimated, allowing varying amounts for the different items; for example, 8 to 15 percent for concrete work, but only 3 to 5 percent for work subcontracted out.

2. Add a percentage of profit to the total price tabulated for materials, labor, overhead, and equipment. The percentage would vary from small jobs to larger jobs (perhaps 20 to 25 percent on a small job and 5 to 10 percent on a larger one), taking into account the accuracy of the takeoff and pricing procedures used in the estimate.

3. Various methods of selecting a figure are employed that will make a bid low while not being too low, by trying to analyze all the variables and other contractors who are bidding.

4. There are "strategies of bidding" that some contractors (and estimators) apply to bidding. Most of the strategies require bidding experience to be accumulated and competitive patterns from past biddings to be used as patterns for future biddings. This will also lend itself nicely to computer operations.

5. Superstition sometimes plays a part. And why not, since superstition is prevalent in our lives? Many contractors and estimators will use only certain numbers to end their bids; for example, some always end with a 7 or take a million-dollar bid all the way down to 50 cents.

One approach is to include all costs of the project before profit is considered. Then make a review of the documents to find whether the drawings and specifications were clear, whether you understood the project you were bidding, and how accurate a takeoff was made (it should always be as accurate as possible). The other factors to be considered are the architect/engineer, that person's reputation, and how the work is handled.

After reviewing the factors, the contractors must decide how much money (profit, over and above salary) they want to make on this project. This amount should be added to the cost of construction to give the amount of the bid (after it is adjusted slightly to take into account superstitions and strategy types of bids). Exactly what is done at this point, slightly up or down, is an individual matter, but you should definitely know your competitors, keep track of their past bidding practices, and use those against them whenever possible.

Since profit is added at the end of the estimate, the estimator has a pretty good idea of the risks and problems that may be encountered. Discuss these risks thoroughly with other members of the firm. It is far better to bid what you feel is high enough to cover the risks than to neglect the risks, bid low, and lose money. There is sometimes a tendency to "need" or "want" a job so badly that risks are completely ignored. Try to avoid this sort of foolishness—it only invites disaster. If a project entails substantial risk and it is questionable that a profit can be made, consider not even bidding it and let someone else have the heartaches and the loss. Always remember: Construction is a business in which you are supposed to make a fair, reasonable profit.

REVIEW QUESTIONS

1. Where should office overhead include?

2. How should project risk affect the profit on a job?

3. What is the problem with "needing" or "wanting" a job?

CHAPTER TWENTY-ONE

OTHER ESTIMATING METHODS

21–1 OVERVIEW

For the most of this book, we have focused on the detailed estimate, identifying and pricing each of the components needed to build a project. Other estimating methods were introduced in Chapter 1. In this chapter, we will look at project comparison, square foot, and assembly estimating methods in more depth. Care must be taken when using these methods because they often produce less accurate results than a detailed estimate, but results that are accurate enough for the purpose of some estimates. The estimator should make sure that the estimating method used is appropriate for the purpose of the estimate. When ordering materials, there is no substitute for a good detailed estimate.

21–2 PROJECT COMPARISON METHOD

A project comparison estimate is prepared by comparing the proposed project to one or more similar projects that the company has built. The estimator starts with the costs for a similar project and then adjusts these costs for any differences between the comparison project and the proposed project. When comparing the project to multiple projects, the cost of the proposed project should be calculated for each comparison project, and then the estimator uses his best judgment as to which of the costs is most accurate.

When preparing a project comparison estimate, the estimator must make sure that the projects are very similar in both design and use. It is obvious that one cannot use the costs from a warehouse to determine the costs of a school. They are simply too different. But what about using the costs for an apartment complex to estimate the costs for a condominium project? Even if the design is the same, the use and clientele are different. It costs more to deal with many owners (in the case of the condominiums) rather than one owner (in the case of the apartments). If this comparison is made, the estimator needs to take into account the difference in the use and ownership. When comparing two projects, the estimator needs to consider the following:

- Size. The size (usually square footage) should be within 10 percent for the projects being compared. As the size of the building increases, the cost per square foot decreases. Conversely, as the size decreases, the cost per square foot increases. This is due to economies of scale.
- Height between floors. Increasing the height between floors increases the envelope cost without increasing the square footage of the building.
- Length of perimeter. Two buildings, a long skinny one and a square one, will have very different perimeters and envelope costs for the same square footage.
- Project location. Labor availability, materials availability, and government regulations vary by location.
- When the project was built. Inflation in labor, materials, and fuel costs; weather (summer versus winter).
- Type of structure: Concrete, steel, wood framed, block.
- Level of finishes. Quality of materials and workmanship.
- Utilization of the space. Some spaces cost more than others do. Bathrooms and kitchens cost more per square foot than bedrooms.
- Union versus nonunion labor.
- Soil conditions.

EXAMPLE 21-1 PROJECT COMPARISON

Last year your construction company built a 70,000-sf warehouse for $6,203,595. The owner wants to build another warehouse of similar size; this time they want to add 1,000 sf of office space in one corner of the building. It is estimated that the offices space will cost $100,000 and that costs have risen 3 percent during the last year. Using this information, prepare a preliminary estimate for the new warehouse.

$$\text{Cost (\$)} = \$6,203,595 \times 1.03 + \$100,000$$
$$= \$6,489,703 \text{ — Use } \$6,490,000 \quad \blacksquare$$

When collecting data for use in project comparison, it is important that any unusual conditions that would skew the costs higher or lower be documented. It is simply not enough to record just the cost. The estimator must know whether the cost is typical for the type of project and understand why the cost might have been higher or lower than the average.

21–3 SQUARE-FOOT ESTIMATING

Square-foot estimates are prepared by multiplying the square footage of a building by a cost per square foot and then by adjusting the price to compensate for differences in the design. Units other than square footage may be used; for example, a parking garage may be measured by the number of parking stalls. The cost per square foot or other unit may be determined by dividing the cost by the size of the building.

EXAMPLE 21-2   SQUARE-FOOT ESTIMATE

Last year your construction company built a 100,000-sf, 5-story parking garage for $5,395,621. The parking garage included one

elevator, which cost $126,345, and its costs are included in the $5,395,621. Using this information, prepare a preliminary cost estimate for a 95,000-sf, 5-story parking garage with two elevators. It is estimated that costs have risen 3 percent during the last year.

Determine the cost per sf for the parking garage, excluding the elevator.

$$\text{Cost per sf (\$/sf)} = (\$5,395,621 - \$126,345)/100,000 \text{ sf}$$
$$= \$52.69 \text{ per sf}$$

$$\text{Cost without elevators (\$)} = \$52.69 \text{ per sf} \times 95,000 \text{ sf}$$
$$= \$5,005,550$$

$$\text{Add for elevators (\$)} = \$126,345 \times 2 = \$252,690$$

Add costs and adjust for inflation.

$$\text{Total cost (\$)} = (\$5,005,550 + \$252,690) \times 1.03$$
$$\text{Total cost (\$)} = \$5,258,240 \times 1.03$$
$$= \$5,415,987 \text{ — Use } \$5,420,000 \quad \blacksquare$$

The square-foot method may be used for the entire project of just specific trades within the project. For example, we may prepare a square-foot estimate for the electrical to check our electrical pricing. In addition, the square-foot method may be applied to each cost code in a bid. This requires the estimator to think through each cost code and determine the best unit of measure to use each cost code. Some items may be best based on the square footage (for example, floor slabs and roofing) or the length of the perimeter of the building (for example, exterior brick), and others may be nearly fixed for a building unless a large change in size occurs (for example, elevators).

EXAMPLE 21-3   LINE-BY-LINE SQUARE-FOOT ESTIMATE

Last year your construction company built a 3-story, 26,000-sf apartment building for $1,915,071. The costs by cost code are shown in the second column of Figure 21.1. The perimeter of the building was 240′. Using this information, prepare an estimate for a 3-story, 24,000-sf apartment building with a perimeter of 200′. The apartments have the same room counts, kitchen sizes, and bathroom sizes.

Use the square-foot method   to adjust each line of the cost codes. The cost codes will be adjusted based upon the ratios of the areas, the ratio of the perimeters, or will be treated as a fixed cost. The ratio of the areas is determined by dividing the new area by the old area. Multiplying the total cost of the old project by the area ratio is equivalent to dividing the total cost of the old project by the old area to determine a cost per square foot for the old project and then multiplying the cost per square foot by the new area. This results in an estimated cost for the new building. The ratios for this building are as follows:

$$\text{Area Ratio} = 24,000 \text{ sf}/26,000 \text{ sf} = 0.923$$

$$\text{Perimeter Ratio} = 200′/240′ = 0.833$$

Items that change proportionally with the area of the building will be adjusted using the area ratio. The following will be adjusted using the area ratio: under-slab gravel, slab floor, light-weight

concrete, rough carpentry, lumber, trusses, finish carpentry, insulation, roofing, drywall, carpet and vinyl, paint, and wire shelving.

Items that deal with the exterior wall of the building will be adjusted using the perimeter ratio. The following will be adjusted using the perimeter ratio: footing and foundation, rebar, masonry, waterproofing, stucco, siding, and rain gutters.

Some items are a function of the number of rooms and the size of the kitchens and bathrooms. For example, a small increase in the square footage, without changing the number of rooms, will have little effect on the heating system other than adding or subtracting a few feet of duct; the door count would remain the same. Because the number of rooms and the size of the kitchen and bathrooms are the same, these items are treated as fixed costs and will not be adjusted. These items include miscellaneous steel, firesafing, metal doors and frames, wood doors, hardware, windows, window sills, appliances, cabinetry and counter tops, window treatments, plumbing, HVAC, electrical, and clean-up.

The adjustment factor is shown in the third column of Figure 21.1, and the adjusted cost estimate for the new apartment building is shown in the fourth column. From Figure 21.1, we see that the cost per square foot increased by $1.54 (2.1%) due to changes in the size of the building and length of the perimeter.

Div.	Item	Old Project	Adjustment	New Project
3	Under-Slab Gravel	8,989	0.923	8,297
3	Footing and Foundation	63,067	0.833	52,535
3	Slab Floor	32,364	0.923	29,872
3	Light-Weight Concrete	15,685	0.923	14,477
3	Rebar	8,528	0.833	7,104
4	Masonry	48,764	0.833	40,620
5	Misc. Steel	37,524	1.000	37,524
6	Rough Carpentry	213,596	0.923	197,149
6	Lumber	241,461	0.923	222,869
6	Trusses	37,011	0.923	34,161
6	Finish Carpentry	27,269	0.923	25,169
7	Waterproofing	4,707	0.833	3,921
7	Insulation	18,767	0.923	17,322
7	Firesafing	2,946	1.000	2,946
7	Stucco	45,173	0.833	37,629
7	Siding	66,127	0.833	55,084
7	Rain Gutters	5,138	0.833	4,280
7	Roofing	20,367	0.923	18,799
8	Metal Doors and Frames	18,532	1.000	18,532
8	Wood Doors	63,912	1.000	63,912
8	Hardware	20,846	1.000	20,846
8	Windows	25,493	1.000	25,493
9	Drywall	178,426	0.923	164,687
9	Carpet & Vinyl	70,195	0.923	64,790
9	Paint	36,085	0.923	33,306
9	Window Sills	2,024	1.000	2,024
10	Wire Shelving	8,299	0.923	7,660
11	Appliances	66,906	1.000	66,906
12	Cabinetry & Counter Tops	106,736	1.000	106,736
12	Window Treatments	5,632	1.000	5,632
15	Plumbing	135,093	1.000	135,093
15	HVAC	137,278	1.000	137,278
16	Electrical	135,677	1.000	135,677
17	Clean-Up	6,454	1.000	6,454
	Subtotal	1,915,071		1,804,784
	Square Footage	26,000		24,000
	Cost per Square Foot	73.66		75.20

FIGURE 21.1. Apartment Costs.

In *Square Foot Costs*, RS Means publishes square footage cost data for different types of construction projects. Figures 21.2 and 21.3 show the RS Means square foot data for a warehouse. Using this book, the estimated costs for a construction project may be determined by the following steps:

1. Find the correct page for the type of construction to be estimated. Figures 21.2 and 21.3 are for a warehouse.

2. Determine the base cost per square foot by finding the type of exterior wall and frame system in the left two columns and the area in the top row of Figure 21.2. For example, the base cost is $75.10 per sf for a 30,000-sf warehouse with a concrete block exterior that uses the exterior walls as bearing walls. When the size falls between two sizes on the page, the estimator should interpolate between the two sizes. For example, the base cost per square foot for a 33,000-sf warehouse with a

concrete block exterior that uses the exterior walls as bearing walls would be determined as follows:

$$\text{Cost (\$/sf)} = \frac{33{,}000 - 30{,}000}{35{,}000 - 30{,}000} (\$73.65 - \$75.10) + \$75.10$$

$$\text{Cost (\$/sf)} = \$74.23$$

3. Next, the base cost per square foot needs to be adjusted for differences in the perimeters. This is done by determining the difference between the proposed building's perimeter and the perimeter specified in the second row of the table. The differences are measured in 100' or a fraction thereof. For example, for a 30,000-sf warehouse, the perimeter is 700'; and if the proposed building's perimeter is 680', the difference in the perimeters would be 0.2 hundred feet, which would be a deduction because we are reducing the perimeter. This is multiplied

COMMERCIAL/INDUSTRIAL/ INSTITUTIONAL	**M.690**	**Warehouse**

Costs per square foot of floor area

Exterior Wall	S.F. Area	10000	15000	20000	25000	30000	35000	40000	50000	60000
	L.F. Perimeter	410	500	600	640	700	766	833	966	1000
Tiltup Concrete Panels	Steel Frame	98.05	89.30	85.25	81.30	79.15	77.60	76.55	75.00	73.00
Brick with Block Back-up	Bearing Walls	114.60	102.05	96.35	90.25	86.95	84.75	83.10	80.85	77.55
Concrete Block	Steel Frame	96.50	88.00	84.05	80.25	78.15	76.70	75.65	74.20	72.20
	Bearing Walls	93.75	85.10	81.10	77.25	**75.10**	73.65	72.55	71.05	69.10
Galvanized Steel Siding	Steel Frame	102.10	93.25	89.10	85.05	82.85	81.35	80.20	78.70	76.65
Metal Sandwich Panels	Steel Frame	104.30	94.35	89.70	85.10	82.55	80.80	79.55	77.85	75.40
Perimeter Adj., Add or Deduct	Per 100 L.F.	6.35	4.25	3.25	2.60	2.10	1.90	1.55	1.30	1.00
Story Hgt. Adj., Add or Deduct	Per 1 Ft.	.75	.60	.55	.45	.40	.40	.35	.40	.25

For Basement, add $23.55 per square foot of basement area

The above costs were calculated using the basic specifications shown on the facing page. These costs should be adjusted where necessary for design alternatives and owner's requirements. Reported completed project costs, for this type of structure, range from $30.13 to $120.30 per S.F.

Common additives

Description	Unit	$ Cost
Dock Leveler, 10 ton cap.		
6' x 8'	Each	6225
7' x 8'	Each	6225
Emergency Lighting, 25 watt, battery operated		
Lead battery	Each	265
Nickel cadmium	Each	770
Fence, Chain link, 6' high		
9 ga. wire	L.F.	20
6 ga. wire	L.F.	29
Gate	Each	293
Flagpoles, Complete		
Aluminum, 20' high	Each	1375
40' high	Each	3125
70' high	Each	9725
Fiberglass, 23' high	Each	1675
39'-5" high	Each	3225
59' high	Each	8025
Paving, Bituminous		
Wearing course plus base course	S.Y.	7.85
Sidewalks, Concrete 4" thick	S.F.	3.96

Description	Unit	$ Cost
Sound System		
Amplifier, 250 watts	Each	2125
Speaker, ceiling or wall	Each	174
Trumpet	Each	335
Yard Lighting, 20' aluminum pole	Each	2600
with 400 watt		
high pressure sodium		
fixture.		

Important: See the Reference Section for Location Factors

FIGURE 21.2. Warehouse Square Foot Costs.

by the perimeter adjustment, and the resultant is added or deducted from the base cost per square foot.

4. Next, the base cost per square foot needs to be adjusted for differences in the heights of the stories. This is done by determining the difference between the proposed building's story height and the story height used to develop the square foot costs. The story height is 24' for the warehouse and is found in the top left-hand corner of Figure 21.3. This difference is then multiplied by the story height adjustment, and the resultant is added or deducted from the base cost per square foot.

5. The square footage of the building is multiplied by the cost per square foot, including adjustments for the perimeter and story height.

6. Other costs are then added or subtracted from this price to account for differences in the design.

EXAMPLE 21-4   SQUARE-FOOT ESTIMATE USING RS MEANS

Using Figures 21.2 and 21.3, determine the cost for a 250' by 120', tilt-up concrete warehouse. The warehouse is 30' high and has a steel frame. Include four 7' × 8' dock levelers in the costs.

$$\text{Area} = 250' \times 120' = 30,000 \text{ sf}$$

$$\text{Perimeter} = 250' + 120' + 250' + 120' = 740'$$

From Figure 21.2, the cost per sf for a tilt-up concrete warehouse with an area of 30,000 sf is $79.15 per sf. From Figure 21.2, the base perimeter is 700 lf and an add of $2.10 per sf per 100 lf of perimeter is required.

Add for Perimeter ($/sf) = $(740' - 700')/100'$
$$\times \$2.10 \text{ per sf per 100 lf} = \$0.84 \text{ per sf}$$

From Figure 21.3, the base story height is 24'; and from Figure 21.2, the adjustment for story height is $0.40 per sf per foot of height.

Add for story height ($/sf) = $(30' - 24') \times \$0.40$ per sf per ft
$$= \$2.40 \text{ per sf}$$

Base cost ($/sf) = \$79.15 per sf + \$0.84 per sf + \$2.40 per sf

Base cost ($/sf) = \$82.39 per sf

Base cost ($) = 30,000 sf $\times$ \$82.39 per sf = \$2,471,700

From Figure 21.2, the added cost for a 7' $\times$ 8' dock leveler is $6,225 per leveler.

Dock levelers = 4 ea $\times$ \$6,225 per ea = \$24,900

Total cost = \$2,471,700 + \$24,900
$$= \$2,496,600 — \text{Use } \$2,500,000 \blacksquare$$

21–4 ASSEMBLY ESTIMATING

In assembly estimating, rather than bidding each component separately, the components are grouped into assemblies, consisting of items that would be installed together (but not necessarily by the same trade), and then the assembly is bid as a single component. For example, the cost per linear foot for an 8-foot-high interior wall (consisting of track, metal studs, insulation, drywall, and paint) may be determined. Then all 8-foot-high interior walls may be bid by the linear foot rather than by bidding the track, metal studs, drywall, insulation, and paint separately.

To create an assembly, the estimator determines the quantity of materials needed for one unit of the assembly, linear foot of wall in the case of the interior wall. From these quantities, the cost for materials, equipment, and labor to construct one unit of the assembly is then determined. This

Model costs calculated for a 1 story building with 24' story height and 30,000 square feet of floor area			Unit	Unit Cost	Cost Per S.F.	% Of Sub-Total
A. SUBSTRUCTURE						
1010	Standard Foundations	Poured concrete; strip and spread footings	S.F. Ground	1.22	1.22	
1030	Slab on Grade	5" reinforced concrete with vapor barrier and granular base	S.F. Slab	10.73	10.73	25.3%
2010	Basement Excavation	Site preparation for slab and trench for foundation wall and footing	S.F. Ground	.14	.14	
2020	Basement Walls	4' foundation wall	L.F. Wall	69	2.10	
B. SHELL						
B10 Superstructure						
1010	Floor Construction	Mezzanine: open web steel joists, slab form, concrete beams, columns 10% of area	S.F. Floor	18.00	1.80	13.8%
1020	Roof Construction	Metal deck, open web steel joists, beams, columns	S.F. Roof	5.95	5.95	
B20 Exterior Enclosure						
2010	Exterior Walls	Concrete block 95% of wall	S.F. Wall	10.00	5.32	
2020	Exterior Windows	N/A	—	—	—	11.1%
2030	Exterior Doors	Steel overhead, hollow metal 5% of wall	Each	2436	.89	
B30 Roofing						
3010	Roof Coverings	Built-up tar and gravel with flashing; perlite/EPS composite insulation	S.F. Roof	4.55	4.55	8.7%
3020	Roof Openings	Roof hatches and skylight	S.F. Roof	.35	.35	
C. INTERIORS						
1010	Partitions	Concrete block (office and washrooms) 100 S.F. Floor/L.F. Partition	S.F. Partition	7.63	.61	
1020	Interior Doors	Single leaf hollow metal 5000 S.F. Floor/Door	Each	815	.17	
1030	Fittings	N/A	—	—	—	
2010	Stair Construction	Steel gate with rails	Flight	10,950	.73	10.0%
3010	Wall Finishes	Paint	S.F. Surface	11.63	1.86	
3020	Floor Finishes	90% hardener, 10% vinyl composition tile	S.F. Floor	1.76	1.76	
3030	Ceiling Finishes	Suspended mineral tile on zee channels in office area 10% of area	S.F. Ceiling	4.71	.47	
D. SERVICES						
D10 Conveying						
1010	Elevators & Lifts	N/A	—	—	—	0.0%
1020	Escalators & Moving Walks	N/A	—	—	—	
D20 Plumbing						
2010	Plumbing Fixtures	Toilet and service fixtures, supply and drainage 1 Fixture/2500 S.F. Floor	Each	2300	.92	
2020	Domestic Water Distribution	Gas fired water heater	S.F. Floor	.21	.21	4.3%
2040	Rain Water Drainage	Roof drains	S.F. Roof	1.28	1.28	
D30 HVAC						
3010	Energy Supply	Oil fired hot water, unit heaters 90% of area	S.F. Floor	4.07	4.07	
3020	Heat Generating Systems	N/A	—	—	—	
3030	Cooling Generating Systems	N/A	—	—	—	8.7%
3050	Terminal & Package Units	Single zone unit gas, heating, electric cooling 10% of area	S.F. Floor	.81	.81	
3090	Other HVAC Sys. & Equipment	N/A	—	—	—	
D40 Fire Protection						
4010	Sprinklers	Sprinklers, ordinary hazard	S.F. Floor	2.73	2.73	4.9%
4020	Standpipes	N/A	—	—	—	
D50 Electrical						
5010	Electrical Service/Distribution	200 ampere service, panel board and feeders	S.F. Floor	.43	.43	
5020	Lighting & Branch Wiring	Fluorescent fixtures, receptacles, switches, A.C. and misc. power	S.F. Floor	4.30	4.30	9.1%
5030	Communications & Security	Alarm systems	S.F. Floor	.38	.38	
5090	Other Electrical Systems	N/A	—	—	—	
E. EQUIPMENT & FURNISHINGS						
1010	Commercial Equipment	N/A	—	—	—	
1020	Institutional Equipment	N/A	—	—	—	4.2%
1030	Vehicular Equipment	Dock boards, dock levelers	S.F. Floor	2.37	2.37	
1090	Other Equipment	N/A	—	—	—	
F. SPECIAL CONSTRUCTION						
1020	Integrated Construction	N/A	—	—	—	0.0%
1040	Special Facilities	N/A	—	—	—	
G. BUILDING SITEWORK	**N/A**					
			Sub-Total		56.15	100%
	CONTRACTOR FEES (General Requirements: 10%, Overhead: 5%, Profit: 10%)			25%	14.04	
	ARCHITECT FEES			7%	4.91	
			Total Building Cost		75.10	

Warehouse

FIGURE 21.3. Warehouse Square Foot Costs.

cost is then used to bid the assembly by multiplying the quantity of the assembly by the unit cost of the assembly.

EXAMPLE 21-5   WALL ASSEMBLY UNIT COST

Determine the quantities of materials needed and the costs for an 8-foot-high interior partition wall assembly. The wall consists of track and metal studs, insulation, 1/2″ drywall on both sides, and paint. The studs are 16″ on center. Openings in the partition will be handled as a separate assembly.

$$\text{Length of track} = 2 \times 1' = 2'$$

$$\text{Number of studs} = 1'/16'' = 12''/16'' = 0.75 \text{ each}$$

$$\text{Area of insulation} = 8' \times 1' \times 15''/16'' = 7.5 \text{ sf}$$

$$\text{Area of drywall} = 2 \text{ sides} \times 8' \times 1' = 16 \text{ sf}$$

$$\text{Area of paint} = 2 \text{ sides} \times 8' \times 1' = 16 \text{ sf}$$

The costs are as follows:

$$\text{Track (\$)} = 2' \times \$1.80 \text{ per lf} = \$3.60$$

$$\text{Studs (\$)} = 0.75 \text{ each} \times \$9.52 \text{ per each} = \$7.14$$

$$\text{Insulation (\$)} = 7.5 \text{ sf} \times \$0.81 \text{ per sf} = \$6.08$$

$$\text{Drywall (\$)} = 16 \text{ sf} \times \$1.62 \text{ per sf} = \$25.92$$

$$\text{Paint (\$)} = 16 \text{ sf} \times \$0.61 \text{ per sf} = \$9.76$$

$$\text{Total cost (\$/ft)} = \$52.50 \qquad \blacksquare$$

EXAMPLE 21-6   ASSEMBLY ESTIMATING USING INTERIOR WALL ASSEMBLY

Determine the cost of 251′ of an 8-foot-high interior wall using the unit cost from Example 21-5.

$$\text{Cost (\$)} = 251' \times \$52.50 = \$13,177 \qquad \blacksquare$$

In *Square Foot Costs,* RS Means publishes cost data for a number of assemblies that can be used to prepare an assembly estimate for an entire building. The costs for a steel joist and deck roof that bears on the exterior walls and steel columns are shown in Figure 21.4. The cost estimate is prepared by determining the costs for each assembly and then by adding these costs together.

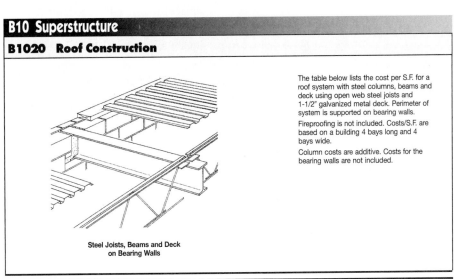

B10 Superstructure

B1020   Roof Construction

The table below lists the cost per S.F. for a roof system with steel columns, beams and deck using open web steel joists and 1-1/2″ galvanized metal deck. Perimeter of system is supported on bearing walls.

Fireproofing is not included. Costs/S.F. are based on a building 4 bays long and 4 bays wide.

Column costs are additive. Costs for the bearing walls are not included.

Steel Joists, Beams and Deck on Bearing Walls

B1020 108		Steel Joists, Beams & Deck on Columns & Walls						
	BAY SIZE (FT.)	SUPERIMPOSED LOAD (P.S.F.)	DEPTH (IN.)	TOTAL LOAD (P.S.F.)	COLUMN ADD	COST PER S.F.		
						MAT.	INST.	TOTAL
3000	25x25	20	18	40		3.37	1.21	4.58
3100					columns	.30	.08	.38
3200		30	22	50		3.84	1.43	5.27
3300					columns	.40	.10	.50
3400		40	20	60		4.03	1.42	5.45
3500					columns	.40	.10	.50
3600	25x30	20	22	40		3.60	1.17	4.77
3700					columns	.34	.09	.43
3800		30	20	50		3.92	1.51	5.43
3900					columns	.34	.09	.43
4000		40	25	60		4.25	1.35	5.60
4100					columns	.40	.10	.50
4200	30x30	20	25	42		3.95	1.27	5.22
4300					columns	.28	.07	.35
4400		30	22	52		4.33	1.37	5.70
4500					columns	.34	.09	.43
4600		40	28	62		4.50	1.41	5.91
4700					columns	.34	.09	.43
4800	30x35	20	22	42		4.12	1.32	5.44
4900					columns	.29	.08	.37
5000		30	28	52		4.35	1.38	5.73
5100					columns	.29	.08	.37
5200		40	25	62		4.72	1.48	6.20
5300					columns	.34	.09	.43
5400	35x35	20	28	42		4.14	1.32	5.46
5500					columns	.25	.07	.32

FIGURE 21.4. Steel Joists, Beams and Deck on Columns and Walls.

EXAMPLE 21-7   ASSEMBLY ESTIMATING USING RS MEANS

Determine the costs for 50′ × 100′ warehouse excluding site work. The costs for each assembly are as follows:

Footings (24″ × 12″) = 300′ × $30.95 per foot = $9,285

Spread footings (4′6″ × 4′6″) = 3 ea × $273 per each = $819

Floor slab (6″ reinforced) = 5,000 sf × $6.14 per sf = $30,700

Block walls (8″ × 8″ × 16″) = 25′ × 300′ × $8.14 per sf = $61,050

Steel columns (wide flanged) = 75′ × $25.45 per ft = $1,909

Joist, deck, and beams = 5,000 sf × $5.27 per sf = $26,350

Exterior doors (3′ × 6′8″) = 4 ea × $1,425 per door = $5,700

Overhead doors (12′ × 12′) = 8 ea × $3,575 per door = $28,600

Roof insulation (3″) = 5,000 sf × $1.37 per sf = $6,850

Roofing = 5,000 sf × $2.40 per sf = $12,000

Skylights (4′ × 4′) = 32 ea × 16 sf × $24 per sf = $12,288

Roof hatch (2′6″ × 3′) = 1 ea × $777 per hatch = $777

Bathroom (toilet and sink) = 1 ea × $2,400 per bathroom = $2,400

Fire sprinklers (ordinary hazard) = 5,000 sf × $3.18 per sf = $15,900

Electrical service (200 A) = 1 ea × $2,725 per ea = $2,725

Lighting (HID) = 5,000 sf × $4.46 per sf = $22,300

Total = $239,653 — Use $240,000   ∎

WEB RESOURCES

www.rsmeans.com

REVIEW QUESTIONS

1. How would you determine the costs for a project using the project comparison method?

2. How would you determine the costs for a project using the square-foot method and in-house cost data?

3. What are some of the ratios that you would use to prepare a line-by-line square-foot estimate?

4. Determine the following using Figure 21.2:

   a. Base cost per square foot for a 40,000-sf, tilt-up concrete warehouse.

   b. Base cost per square foot for a 57,000-sf, tilt-up concrete warehouse.

   c. Base perimeter for a 50,000-sf warehouse.

   d. Perimeter adjustment for a 25,000-sf warehouse.

   e. Story height adjustment for a 15,000-sf warehouse.

   f. Added cost for a 70-foot-high aluminum flagpole.

5. What components would you include in an exterior wall assembly for a residence?

DRAWINGS AND OUTLINE SPECIFICATIONS FOR SMALL COMMERCIAL BUILDING

The drawings that accompany Appendix A are included in the drawing packet that accompanies this textbook. These drawings are used in the example problems throughout the textbook and selected review questions.

OUTLINE SPECIFICATIONS

Section 1—Excavation

Excavation shall be carried 2 feet beyond all walls.

Excavations shall be carried to depths shown on the drawings.

All excess earth not needed for fill shall be removed from the premises.

Section 2—Concrete

Concrete shall develop a compressive strength of 2,500 pounds per square inch (psi) in 28 days.

Reinforcing steel shall be A-15 Intermediate Grade.

Reinforcing mesh shall be 6 × 6 10/10 welded wire mesh.

All concrete shall receive a machine-troweled finish.

Vapor barrier shall be 6-mil polyethylene.

Expansion joint filler shall be 1/2-inch × 4-inch asphalt-impregnated fiberboard.

Prestressed hollow core.

Section 3—Masonry

Concrete masonry units shall be 16 inches in length and 8 inches wide.

Masonry wall reinforcement shall be truss design and installed every third course.

Brick shall be standard size, red.

All mortar joints are 3/8 inch thick.

Section 4—Door Frames, Doors, Windows

Door frames shall be 16-gauge steel.

Doors shall be flush, solid core, seven-ply construction with a rotary birch veneer.

Prefinish interior doors with plastic laminate lacquer and exterior doors with polyurethane varnish.

Hardware for doors (except overhead)—allow $250 for the purchase of the hardware.

Hardware to be installed under this contract.

Section 5—Roofing

Roofing shall be three-ply, built-up roof, 20-year bond.

Cant strips shall be asphalt-impregnated fiberboard.

Insulation shall be fiberglass.

Gravel stop shall be 0.032-inch aluminum.

Section 6—Finishes

All exposed concrete shall receive one coat of concrete paint.

All exposed surfaces of gypsum board and plaster wall shall receive two coats of latex paint.

Floors—1/8 inch thick, 12 × 12 inches, vinyl, marbleized design, light colors.

Base—1/8 inch thick, 4 inches high, vinyl.

Addendum #1

1. Detail 16 on sheet S8.1 is a cross section of Detail 1 on sheet S8.1.

2. Detail 7 on sheet S8.1 is a cross section of Detail 16 on sheet S8.1.

3. The vertical rebar in Detail 13 on sheet S8.1 should be #5 in lieu of the #4 specified.

4. In Detail 13 on sheet S8.1, provide #5, L-shaped dowels at 11 o.c. to connect foundation to the footing.

5. Figure A.1 is to be used to construct the grade beam along the south wall of the building.

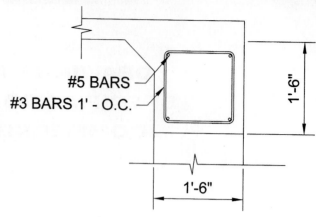

FIGURE A.1. Grade Beam Detail

DRAWINGS AND OUTLINE SPECIFICATIONS FOR RESIDENTIAL BUILDING PROJECT

The drawings that accompany Appendix B are included in the drawing packet that accompanies this textbook. These drawings are used for selected review questions.

OUTLINE SPECIFICATIONS

Section 1—Excavation

Topsoil excavation shall be carried 4 feet beyond all walls.

Excavation shall be carried to the depth shown on the drawings.

All excess earth, not needed for fill, shall be removed from the premises.

Existing grade is 106′6″.

Section 2—Concrete

Concrete for footings shall develop a compressive strength of 3,000 psi in 28 days.

Reinforcing bars shall be A-15 Intermediate Grade.

Section 3—Masonry

Concrete masonry shall be 16 inches in length.

Mortar joints shall be 3/8 inch thick, gray.

Brick shall be standard size, red.

Section 4—Framing

All wood framing shall be sized in accordance with the drawings and building code requirements.

All wood framing shall be kiln dried.

Metal bridging is required in the floor joists.

Roof ridge shall be 10×80.

Collar ties are required every third rafter.

Section 5—Door Frames, Doors, Windows

Door frames shall be ponderosa pine.

Interior doors shall be 1⅜ inches thick, birch, hollow core.

Exterior door—allow $200 for exterior doors.

Doors shall be of the sizes indicated on the drawings.

Windows shall be wood, of the size and shape shown on the drawings.

Hardware—allow $250.

Sliding glass doors shall have insulating glass, anodized aluminum frames, and a screen.

Section 6—Roofing

Roofing felt shall be 15 pounds.

Roofing shingles shall be 300 pounds per square, asphalt, Johns-Manville or equal. Roof slope shall be 4 in 12.

Section 7—Finishes

All exposed exterior wood finishes shall receive one coat of approved stain.

All interior drywall finishes shall receive two coats of approved alkyd paint.

Resilient tile shall be 3/32 inch, vinyl, 12×12 inches from Designer Essentials collection by Mannington.

Base shall be 1/8-inch vinyl, 4 inches high, wood base required in carpeted areas.

Carpeting shall be selected from the Premier Collection by Mohawk.

Insulate garage the same as the exterior walls.

Section 8—Drywall

Furnish and install 1/2-inch gypsum board in accordance with the drawings.

DRAWINGS AND OUTLINE SPECIFICATIONS FOR COMMERCIAL BUILDING PROJECT

The drawings that accompany Appendix C are included in the drawing packet that accompanies this textbook. These drawings are used for selected review questions.

OUTLINE SPECIFICATIONS

Section 1—Excavation

Topsoil excavation shall be carried 5 feet beyond all walls.

Excavations shall be carried to the depths shown on the drawings.

All excess earth, not needed for fill, shall be removed from the premises.

Existing grade is 104.2′.

Section 2—Concrete

Concrete for footings shall develop a compressive strength of 3,000 psi in 28 days.

All other concrete shall develop a compressive strength of 2,500 psi in 28 days unless otherwise specified.

Reinforcing bars shall be A-15 Intermediate Grade.

Reinforcing mesh shall be 4 × 4⅜ welded wire mesh.

Vapor barrier shall be 6-mil polyethylene.

Expansion joint filler shall be 1/2-inch × 4-inch asphalt-impregnated fiberboard.

Concrete lintels shall be precast, 5,000 psi in 28 days.

Section 3—Masonry

Concrete masonry units shall be 16 inches in length.

Masonry wall reinforcement shall be ladder design and installed every third course.

Stone shall be bluestone with stone thickness 3/4 to 2 inches. All mortar joints are 3/8 inches thick.

Exterior window sills shall be bluestone, 1¼ inches thick, slip-in type.

Section 4—Structural

Complete structural system as indicated on the drawings.

All required holes in the roof deck to be made by roof deck installer.

Furnish all required accessories as indicated on the drawings.

Furnish and install in compliance with all manufacturers' specifications.

Section 5—Door Frames, Doors, and Windows

Door frames 16-gauge steel.

Doors shall be flush, solid core, five-ply construction with a rotary birch veneer.

Prefinish interior doors with semigloss lacquer and exterior doors with polyurethane varnish.

Allow $500 for the purchase of hardware.

Install hardware in accordance with the manufacturer's specifications.

Windows shall be custom made, 1/8-inch-thick aluminum, 2-inch × 4-inch frames, glazing materials as indicated on the drawings.

Section 6—Roofing

Roofing shall be five-ply, 20-year bond.

Cant strips shall be asphalt-impregnated fiberboard.

Insulation shall be rigid, perlite.

Slag surface (400 lbs. per square).

Section 7—Finishes

All exposed exterior masonry finishes shall receive two coats of silicone spray.

All exposed interior finishes of drywall shall receive two coats of an approved oil base paint.

All vinyl tile shall be 1/8 inch thick, sheet vinyl.

Bases shall be 1/8 inch thick, 2½ inches high, color to be selected by the architect.

All concrete slabs shall be machine trowel finished.

Section 8—Drywall and Wetwall

Furnish and install complete systems with any required accessories as indicated on the drawings.

Addendum #1

1. Elevation of bottom of footing shall be 102′4″.
2. Interior footings shall be the same size and same reinforcing as the exterior footings.
3. Interior masonry shall extend up to elevation 11′4″.
4. The length of the parking lot is 100 feet.
5. All interior masonry walls are exposed masonry unless otherwise noted.
6. Precast concrete lintels at W3 and W4 shall be 12 × 12 inches.
7. Acoustical tile shall be 2 feet × 4 feet.
8. Steel joists shall be 14J7, 9.4 pounds per foot; bridging shall be one No. 4 bar in the middle of the span.
9. Apply two coats of masonry paint to all exterior exposed masonry (block) walls.
10. Apply two coats of semi-gloss paint to all door frames.
11. Lintel L1 shall be 10 inches × 12 inches reinforced concrete.

COMMON TERMS USED IN THE BUILDING INDUSTRY

Addenda   Statements or drawings that modify the basic contract documents after the latter have been issued to the bidders, but prior to the taking of bids.

Alternates   Proposals required of bidders reflecting amounts to be added to or subtracted from the basic proposal in the event that specific changes in the work are ordered.

Anchor Bolts   Bolts used to anchor structural members to concrete or the foundation.

Approved Equal   The term used to indicate that material or product finally supplied or installed must be equal to that specified and as approved by the architect (or engineer).

As-Built Drawings   Drawings made during the progress of construction, or subsequent thereto, illustrating how various elements of the project were actually installed.

Astragal   A closure between the two leafs of a double-swing or double-slide door to close the joint. This can also be a piece of molding.

Axial   Anything situated around, in the direction of, or along an axis.

Baseplate   A plate attached to the base of a column which rests on a concrete or masonry footing.

Bay   The space between column centerlines or primary supporting members, lengthwise in a building. Usually the crosswise dimension is considered the span or width module, and the lengthwise dimension is considered the bay spacing.

Beam   A structural member that is normally subjected to bending loads and is usually a horizontal member carrying vertical loads. (An exception to this is a purlin.) There are three types of beams:

a. Continuous Beam: A beam that has more than two points of support.

b. Cantilevered Beam: A beam that is supported at only one end and is restrained against excessive rotation.

c. Simple Beam: A beam that is freely supported at both ends, theoretically with no restraint.

Beam and Column   A primary structural system consisting of a series of beams and columns; usually arranged as a continuous beam supported on several columns with or without continuity that is subjected to both bending and axial forces.

Beam-Bearing Plate   Steel plate with attached anchors that is set on top of a masonry wall so that a purlin or a beam can rest on it.

Bearing   The condition that exists when one member or component transmits load or stress to another by direct contact in compression.

Benchmark   A fixed point used for construction purposes as a reference point in determining the various elevations of floor, grade, etc.

Bid   Proposal prepared by prospective contractor specifying the charges to be made for doing the work in accordance with the contract documents.

Bid Bond   A surety bond guaranteeing that a bidder will sign a contract, if offered, in accordance with their proposal.

Bid Security   A bid bond, certified check, or other forfeitable security guaranteeing that a bidder will sign a contract, if offered, in accordance with the proposal.

Bill of Materials   A list of items or components used for fabrication, shipping, receiving, and accounting purposes.

Bird Screen   Wire mesh used to prevent birds from entering the building through ventilators or louvers.

Bond   Masonry units interlocked in the face of a wall by overlapping the units in such a manner as to break the continuity of vertical joints.

Bonded Roof   A roof that carries a printed or written warranty, usually with respect to weather tightness, including repair and/or replacement on a prorated cost basis for a stipulated number of years.

Bonus and Penalty Clause   A provision in the proposal form for payment of a bonus for each day the project is completed prior to the time stated, and for a charge against the contractor for each day the project remains uncompleted after the time stipulated.

Brace Rods Rods used in roofs and walls to transfer wind loads and/or seismic forces to the foundation (often used to plumb building but not designed to replace erection cables when required).

Bridging The structural member used to give lateral support to the weak plane of a truss, joist, or purlin; provides sufficient stability to support the design loads, sag channels, or sag rods.

Built-Up Roofing Roofing consisting of layers of rag felt or jute saturated with coal tar pitch, with each layer set in a mopping of hot tar or asphalt; ply designation as to the number of layers.

Camber A permanent curvature designed into a structural member in a direction opposite to the deflection anticipated when loads are applied.

Canopy Any overhanging or projecting structure with the extreme end unsupported. It may also be supported at the outer end.

Cantilever A projecting beam supported and restrained only at one end.

Cap Plate A horizontal plate located at the top of a column.

Cash Allowances Sums that the contractor is required to include in the bid and contract amount for specific purposes.

Caulk To seal and make weathertight the joints, seams, or voids by filling with a waterproofing compound or material.

Certificate of Occupancy Statement issued by the governing authority granting permission to occupy a project for a specific use.

Certificate of Payment Statement by an architect informing the owner of the amount due a contractor on account of work accomplished and/or materials suitably stored.

Change Order A work order, usually prepared by the architect and signed by the owner or by the owner's agent, authorizing a change in the scope of the work and a change in the cost of the project.

Channel A steel member whose formation is similar to that of a C-section without return lips; may be used singularly or back to back.

Clip A plate or angle used to fasten two or more members together.

Clip Angle An angle used for fastening various members together.

Collateral Loads A load, in addition to normal live, wind, or dead loads, intended to cover loads that are either unknown or uncertain (sprinklers, lighting, etc.).

Column A main structural member used in a vertical position on a building to transfer loads from main roof beams, trusses, or rafters to the foundation.

Contract Documents Working drawings, specifications, general conditions, supplementary general conditions, the owner-contractor agreement, and all addenda (if issued).

Curb A raised edge on a concrete floor slab.

Curtain Wall Perimeter walls that carry only their own weight and wind load.

Datum Any level surface to which elevations are referred (see Benchmark).

Dead Load The weight of the structure itself, such as floor, roof, framing, and covering members, plus any permanent loads.

Deflection The displacement of a loaded structural member or system in any direction, measured from its no-load position, after loads have been applied.

Design Loads Those loads specified by building codes, state or city agencies, or owner's or architect's specifications to be used in the design of the structural frame of a building. They are suited to local conditions and building use.

Door Guide An angle or channel guide used to stabilize and keep plumb a sliding or rolling door during its operation.

Downspout A hollow section such as a pipe used to carry water from the roof or gutter of a building to the ground or sewer connection.

Drain Any pipe, channel, or trench for which waste water or other liquids are carried off, i.e., to a sewer pipe.

Eave The line along the side wall formed by the intersection of the inside faces of the roof and wall panels; the projecting lower edges of a roof, overhanging the walls of a building.

Equal (see Approved Equal).

Erection The assembly of components to form the completed portion of a job.

Expansion Joint A connection used to allow for temperature-induced expansion and contraction of material.

Fabrication The manufacturing process performed in the plant to convert raw material into finished metal building components. The main operations are cold forming, cutting, punching, welding, cleaning, and painting.

Fascia A flat, broad trim projecting from the face of a wall, which may be part of the rake or the eave of the building.

Field The job site or building site.

Field Fabrication Fabrication performed by the erection crew or others in the field.

Field Welding Welding performed at the job site, usually with gasoline/diesel-powered machines.

Filler Strip Preformed neoprene material, resilient rubber, or plastic used to close the ribs or corrugations of a panel.

Final Acceptance The owner's acceptance of a completed project from a contractor.

Fixed Joint A connection between two members in such a manner as to cause them to act as a single continuous member; provides for transmission of forces from one member to the other without any movement in the connection itself.

Flange That portion of a structural member normally projecting from the edges of the web of a member.

Flashing A sheet-metal closure that functions primarily to provide weather tightness in a structure and secondarily to enhance appearance; the metalwork that prevents leakage over windows, doors, around chimneys, and at other roof details.

Footing   That bottom portion at the base of a wall or column used to distribute the load into the supporting soil.

Foundation   The substructure that supports a building or other structure.

Framing   The structural steel members (columns, rafters, girts, purlins, brace rods, etc.) that go together to comprise the skeleton of a structure ready for covering to be applied.

Furring   Leveling up or building out of a part of wall or ceiling by wood, metal, or strips.

Glaze (Glazing)   The process of installing glass in window and door frames.

Grade   The term used when referring to the ground elevation around a building or other structure.

Grout   A mixture of cement, sand, and water used to solidly fill cracks and cavities; generally used under setting places to obtain a solid, uniform, full bearing surface.

Gutter   A channel member installed at the eave of the roof for the purpose of carrying water from the roof to the drains or downspouts.

Head   The top of a door, window, or frame.

Impact Load   The assumed load resulting from the motion of machinery, elevators, cranes, vehicles, and other similar moving equipment.

Instructions to Bidders   A document stating the procedures to be followed by bidders.

Insulation   Any material used in building construction for the protection from heat or cold.

Invitation to Bid   An invitation to a selected list of contractors furnishing information on the submission of bids on a subject.

Jamb   The side of a door, window, or frame.

Joist   Closely spaced beams supporting a floor or ceiling. They may be wood, steel, or concrete.

Kip   A unit of weight, force, or load that is equal to 1,000 pounds.

Lavatory   A bathroom sink.

Liens   Legal claims against an owner for amounts due those engaged in or supplying materials for the construction of the building.

Lintel   The horizontal member placed over an opening to support the loads (weight) above it.

Liquidated Damages   An agreed-to sum chargeable against the contractor as reimbursement for damages suffered by the owner because of contractor's failure to fulfill contractual obligations.

Live Load   The load exerted on a member or structure due to all imposed loads except dead, wind, and seismic loads. Examples include snow, people, movable equipment, etc. This type of load is movable and does not necessarily exist on a given member of structure.

Loads   Anything that causes an external force to be exerted on a structural member. Examples of different types are as follows:

a. Dead Load: in a building, the weight of all permanent constructions, such as floor, roof, framing, and covering members.

b. Impact Load: the assumed load resulting from the motion of machinery, elevators, craneways, vehicles, and other similar kinetic forces.

c. Roof Live Load: all loads exerted on a roof (except dead, wind, and lateral loads) and applied to the horizontal projection of the building.

d. Floor Live Loads: all loads exerted on a floor (except dead, wind, and lateral loads), such as people and furnishings.

e. Seismic Load: the assumed lateral load due to the action of earthquakes and acting in any horizontal direction on the structural frame.

f. Wind Load: the load caused by wind blowing from any horizontal direction.

Louver   An opening provided with one or more slanted, fixed, or movable fins to allow flow of air, but to exclude rain and sun or to provide privacy.

Mullion   The large vertical piece between windows. (It holds the window in place along the edge with which it makes contact.)

Nonbearing Partition   A partition that supports no weight except its own.

OSB   Panels made of short strands of wood fibers oriented parallel the surface of the sheet.

Parapet   That portion of the vertical wall of a building that extends above the roof line at the intersection of the wall and roof.

Partition   A material or combination of materials used to divide a space into smaller spaces.

Performance Bond   A bond that guarantees to the owner, within specified limits, that the contractor will perform the work in accordance with the contract documents.

Pier   A structure of masonry (concrete) used to support the bases of columns and bents. It carries the vertical load to a footing at the desired load-bearing soil.

Pilaster   A flat, rectangular column attached to or built into a wall masonry or pier; structurally, a pier, but treated architecturally as a column with a capital, shaft, and base. It is used to provide strength for roof loads or support for the wall against lateral forces.

Precast Concrete   Concrete that is poured and cast in some position other than the one it will finally occupy; cast either on the job site and then put into place, or away from the site to be transported to the site and erected.

Prestressed Concrete   Concrete in which the reinforcing cables, wires, or rods are tensioned before there is load on the member.

Progress Payments Payments made during progress of the work, on account, for work completed and/or suitably stored.

Progress Schedule A diagram showing proposed and actual times of starting and completion of the various operations in the project.

Punch List A list prepared by the architect or engineer of the contractor's uncompleted work or work to be corrected.

Purlin Secondary horizontal structural members located on the roof extending between rafters, used as (light) beams for supporting the roof covering.

Rafter A primary roof support beam usually in an inclined position, running from the tops of the structural columns at the eave to the ridge or the highest portion of the roof. It is used to support the purlins.

Recess A notch or cutout, usually referring to the blockout formed at the outside edge of a foundation, providing support and serving as a closure at the bottom edge of wall panels.

Reinforcing Steel The steel placed in concrete to carry the tension, compression, and shear stresses.

Retainage A sum withheld from each payment to the contractor in accordance with the terms of the owner-contractor agreement.

Rolling Doors Doors that are supported on wheels that run on a track.

Roof Overhang A roof extension beyond the end or the side walls of a building.

Roof Pitch The angle or degree of slope of a roof from the eave to the ridge. The pitch can be found by dividing the height, or rise, by the span; for example, if the height is 8 feet and the span is 16 feet, the pitch is 8/16 or 1/2 and the angle of pitch is 45 degrees (see Roof Slope).

Roof Slope The angle that a roof surface makes with the horizontal, usually expressed as a certain rise in 12 inches of run.

Sandwich Panel An integrated structural covering and insulating component consisting of a core material with inner and outer metal or wood skins.

Schedule of Values A statement furnished to the architect by the contractor, reflecting the amounts to be allotted for the principal divisions of the work. It is to serve as a guide for reviewing the contractor's periodic application for payment.

Sealant Any material that is used to close cracks or joints.

Separate Contract A contract between the owner and a contractor other than the general contractor for the construction of a portion of a project.

Sheathing Rough boarding (usually plywood) on the outside of a wall or roof over which is placed brick, siding, shingles, stucco, or other finish material.

Shim A piece of steel used to level or square beams or column base plates. A piece of wood used to level or plumb doors.

Shipping List A list that enumerates by part, number, or description each piece of material to be shipped.

Shop Drawings Drawings that illustrate how specific portions of the work shall be fabricated and/or installed.

Sill The lowest member beneath an opening such as a window or door; also, the horizontal framing members at floor level, such as sill girts or sill angles; the member at the bottom of a door or window opening.

Sill Lug A sill that projects into the masonry at each end of the sill. It must be installed as the building is being erected.

Sill Slip A sill that is the same width as the opening—it will slip into place.

Skylight An opening in a roof or ceiling for admitting daylight; also, the reinforced plastic panel or window fitted into such an opening.

Snow Load In locations subject to snow loads, as indicated by the average snow depth in the reports of the US Weather Bureau, the design loads shall be modified accordingly.

Soffit The underside of any subordinate member of a building, such as the undersurface of a roof overhang or canopy.

Soil Borings A boring made on the site in the general location of the proposed building to determine soil type, depth of the various types of soils, and water table level.

Soil Pressure The allowable soil pressure is the load per unit area a structure can safely exert on the substructure (soil) without exceeding reasonable values of footing settlements.

Spall A chip or fragment of concrete that has chipped, weathered, or otherwise broken from the main mass of concrete.

Span The clear distance between supports of beams, girders, or trusses.

Spandrel Beam A beam from column to column carrying an exterior wall and/or the outermost edge of an upper floor.

Specifications A statement of particulars of a given job as to size of building, quality and performance of workers and materials to be used. A set of specifications generally indicates the design loads and design criteria.

Square One hundred square feet (100 sf).

Stock A unit that is standard to its manufacturer; it is not custom made.

Stool A shelf across the inside bottom of a window.

Stud A vertical wall member to which exterior or interior covering or collateral material may be attached. Load-bearing studs are those that carry a portion of the loads from the floor, roof, or ceiling as well as the collateral material on one or both sides. Non-load-bearing studs are used to support only the attached collateral materials and carry no load from the floor, roof, or ceiling.

Subcontractor A separate contractor for a portion of the work; hired by the general (prime) contractor.

Substantial Completion For a project or specified area of a project, the date when the construction is sufficiently completed in accordance with the contract documents, as modified by any change orders agreed to by the parties, so that the owner can occupy the project or specified area of the project for the use for which it was intended.

Supplementary General Conditions   One of the contract documents, prepared by the architect, that may modify provisions of the general conditions of the contract.

Temperature Reinforcing   Lightweight deformed steel rods or wire mesh placed in concrete to resist possible cracks from expansion or contraction due to temperature changes.

Time of Completion   The number of days (calendar or working) or the actual date by which completion of the work is required.

Truss   A structure made up of three or more members, with each member designed to carry basically a tension or a compression force. The entire structure in turn acts as a beam.

Veneer   A thin covering of valuable material over a less expensive body, e.g., brick on a wood frame building.

Wainscot   Protective or decorative covering applied or built into the lower portion of a wall.

Wall Bearing   In cases where a floor, roof, or ceiling rests on a wall, the wall is designed to carry the load exerted. These types of walls are also referred to as load-bearing walls.

Wall Covering   The exterior wall skin consisting of panels or sheets and including their attachment, trim, fascia, and weather sealants.

Wall Nonbearing   Wall not relied upon to support a structural system.

Water Closet   More commonly known as a toilet.

Working Drawing   The actual plans (drawings and illustrations) from which the building will be built. They show how the building is to be built and are included in the contract documents.

CONVERSIONS

USEFUL CONVERSION FACTORS

Cement

1 Sack = 1 cf = 94 pounds

1 barrel = 4 sacks = 376 pounds

Water

1 U.S. gallon = 231 ci = 0.1337 cf = 8.35 pounds

1 cf = 7.5 U.S. gallons = 62.4 pounds

Aggregate (Approximate)

1 ton = 2,000 pounds = 19 cf = 0.70 cy

1 cy = 27 cf = 2,800 pounds = 1.4 tons

Ready-Mix Concrete (Approximate)

1 cy = 27 cf = 2 tons = 4,000 pounds

1 ton = 2,000 pounds = 0.50 cy = 13.5 cf

Concrete: Weight (Approximate)

1 cf = 145 pounds (heavyweight)

1 cf = 115 pounds (lightweight)

Lengths

12 inches = 1 foot

3 feet = 1 yard

1 mile = 5,280 ft

Areas

144 si = 1 sf

9 sf = 1 sy

100 sf = 1 square

1 acre = 43,560 sf

Volume

1,728 ci = 1 cf

27 cf = 1 cy

1 cf = 7.4850 gallons

1 gallon = 231 ci

1 gallon = 8.33 pounds

1 cf = 62.3 pounds

Metric to US Conversions Factors

1 square cm	= 0.155 sf
1 square m	= 10.764 sf = 1.196 sy
1 cubic cm	= 0.06 ci
1 cubic dm	= 61.02 ci
1 cubic m	= 1.308 cy
1 millimeter (mm)	= 0.0394 inch
1 centimeter (cm)	= 0.3937 inch
1 decimeter (dm)	= 3.937 inches
1 meter (m)	= 39.37 inches
1 meter (m)	= 1.1 yards
1 kilometer (km)	= 0.621 mile

Inches Reduced to Decimals

Inches	Decimal
1/2	0.041
1	0.083
1½	0.125
2	0.167
2½	0.209
3	0.250
3½	0.292
4	0.333
4½	0.375
5	0.417
5½	0.458
6	0.500
6½	0.542
7	0.583
7½	0.625
8	0.667
8½	0.708
9	0.750
9½	0.792

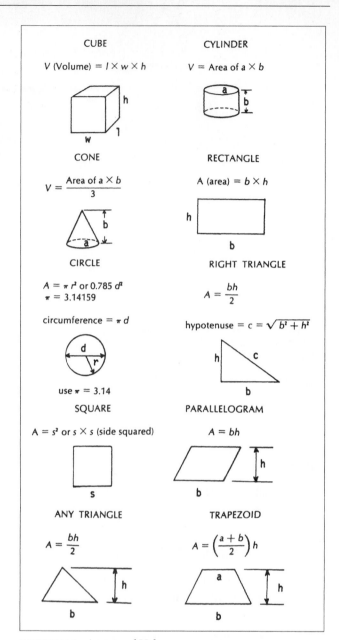

FIGURE E.1. Areas and Volumes.

BILLY'S C-STORE

T he drawings that accompany Appendix F are included in the drawing packet that accompanies this textbook. These drawings are used for selected review questions.

WinEst SOFTWARE AND SPREADSHEETS

The CD that is enclosed in this book contains an educational version of the WinEst software as well as several spreadsheets that were used in the previous chapters. The spreadsheets were generated using Microsoft Excel.

WinEst

The WinEst program provided with this book is one of many estimating software packages that are commercially available. There are many advantages to using such a system for estimating. First, they allow the estimate to be performed by assemblies (which facilitates project scheduling) and then priced out by unique item. The quantity by item is needed to request price quotes. Second, it simplifies the maintenance of pricing and productivity factors. This allows for changes in the productivity rates and prices to be automatically reflected throughout the estimate. There are many other advantages, such as error reductions and time savings.

When the WinEst CD is inserted in the CD drive, the autorun feature should automatically start the installation process, bringing up the screen shown in Figure G.1. Should the autorun feature fail to start, the user can start the installation by (1) double-clicking on the CD drive, (2) right-clicking on the CD drive and selecting Autoplay from the popup menu, or (3) opening the CD and double-clicking on

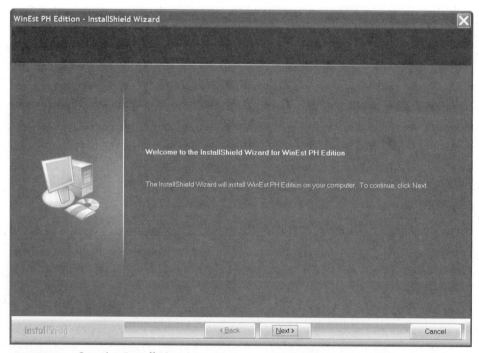

FIGURE G.1. Opening Installation Screen.

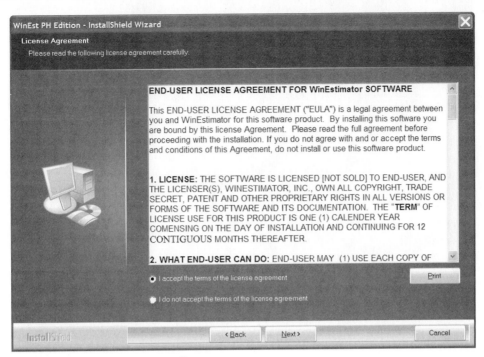

FIGURE G.2. License Agreement Screen.

setup.exe. To cancel the installation process, the user clicks on the Cancel button, clicks on the Yes button to confirm that she wants to exit, and then clicks on the Finish button to exit the installation. To continue the installation, the user clicks on the Next button to bring up the License Agreement screen shown in Figure G.2. If the user agrees to abide by the license agreement, he selects the I accept the terms of the license agreement radio button and clicks on the Next button to bring up the Choose Destination Location screen shown in Figure G.3. The user can change the location where the files

are located by clicking on the Change . . . button. When the file location is acceptable, the user clicks on the Next button to bring up the Ready to Install the Program screen shown in Figure G.4. The user clicks on the Install button to install WinEst. When the program has been installed, the InstallSheild Wizard Complete screen shown in Figure G.5 is displayed. The user clicks on the Finish button to close the installation software. The installation software will place the WinEst PH Edition icon, shown in Figure G.6, and the WinEst Calculator icon, shown in Figure G.7 on the desktop.

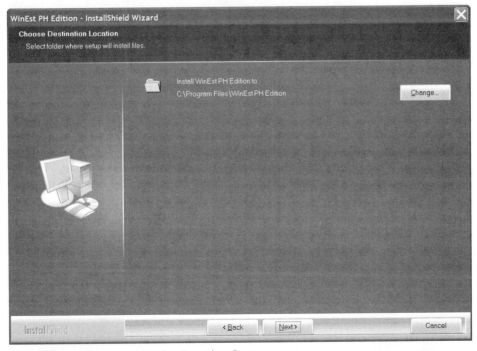

FIGURE G.3. Choose Destination Location Screen.

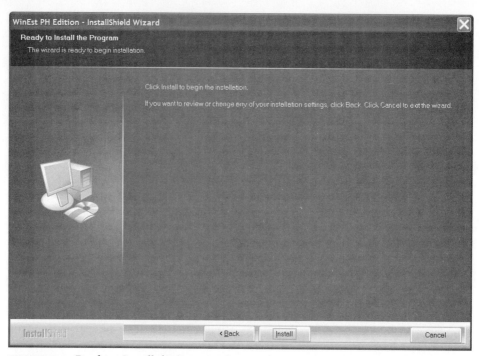

FIGURE G.4. Ready to Install the Program Screen.

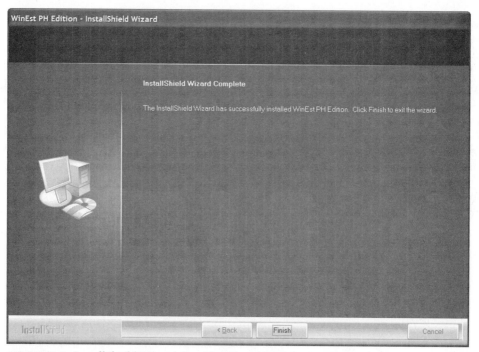

FIGURE G.5. InstallSheild Wizard Complete Screen.

The user clicks on the WinEst PH Edition icon, shown in Figure G.6, to open WinEst. WinEst will open to the Take-off screen shown in Figure G.8. The columns on the Takeoff screen have a striking resemblance to the ones that were used throughout this book.

Items are added to the estimate by right-clicking on the lower portion of the screen and by selecting Add Items . . . from the popup menu to bring up the Database Browser shown in Figure G.9. Items are selected in the top half of the Database Browser and appear in the Takeoff Items box after

FIGURE G.6. WinEst PH Edition Icon.

FIGURE G.7. WinEst Calculator Icon.

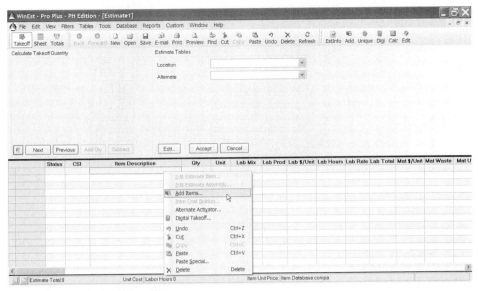

FIGURE G.8. WinEst Takeoff Screen.

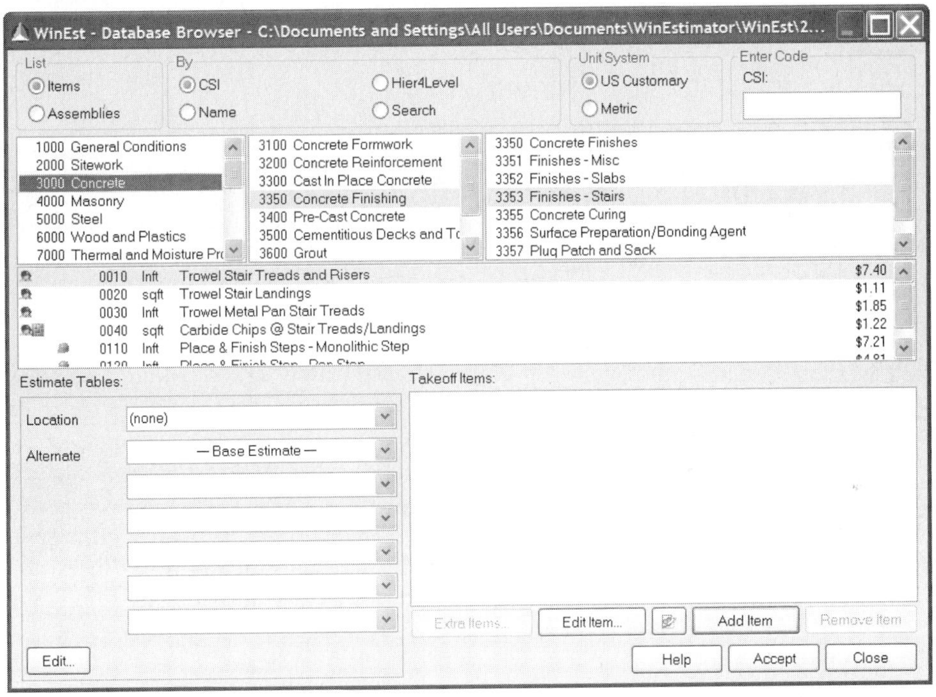

FIGURE G.9. WinEst Database Browser.

they are selected. The user clicks on the Accept button to send the selected items to the Takeoff screen.

Another feature of the WinEst software is the selection of assemblies from the Database Browser. The user switches to assemblies by selecting the Assemblies radio button at the top left corner of the Database Browser. This feature allows the estimator to select component parts of the project rather than specific items. For example, slab on grade would be an assembly. With that assembly, the estimator would provide dimensions and specifications. The software would then generate all of the items within that assembly and their associated quantities based on the provided dimensions. The assemblies that are provided with the student edition are

sparse; however, a good estimator will always be working on developing and enhancing assemblies, as these items speed up the estimating process and can ultimately be passed to a scheduling software package.

For more information on how to use WinEst, see the user's manual that is installed with the software. The user's manual is in PDF format and requires Adobe Acrobat Reader to open. The user manual is opened by selecting WinEst User's Manual from the WinEst 2006 PH Edition folder under the Start menu.

WinEst includes a calculator that can be accessed by clicking on the WinEst Calculator icon shown in Figure G.7. The WinEst calculator is shown in Figure G.10.

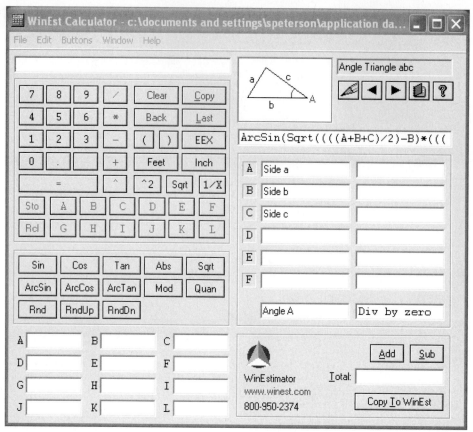

FIGURE G.10. WinEst Calculator.

LOADING THE SPREADSHEETS

Before modifying the spreadsheets, they need to be loaded on a local computer's hard drive. This is done by inserting the disk into the CD drive. If the Autorun starts the installation process, the user must cancel the installation process by clicking on the Cancel button, by clicking on the Yes button to confirm that he wants to exit, and by clicking on the Finish button to exit the installation. The user then opens My Computer, right-clicks on the CD drive, and selects Explore from the popup menu to bring up the window shown in Figure G.11. The user then copies the Excel Files folder to a local computer. The user may now access these files from the local computer.

The table below contains the file name, description, and figure reference for the spreadsheets from the CD. These spreadsheets are provided as guides and may need to be revised to meet specific needs.

File Name	Description	Figure
Bid-Tab.xls	Subcontractor bid tabulation	Figure G.12
Cut-Fill.xls	Cut and fill worksheet	Figure G.13
Est-Summary.xls	Project estimate summary sheet	Figure G.14
Rebar.xls	Worksheet to calculate reinforcing bars	Figure G.15
Recap.xls	Estimate summary sheet	Figure G.16
Slab-Rebar.xls	Worksheet to calculate reinforcing bars in slabs	Figure G.17
Steel.xls	Worksheet for quantifying structural steel	Figure G.18
Warehouse.xls	Warehouse Model	Figure G.19
Work-Sheet.xls	Generic estimate work up sheet	Figure G.20

Note: The highlighted cells contain formulas and do not require any input.

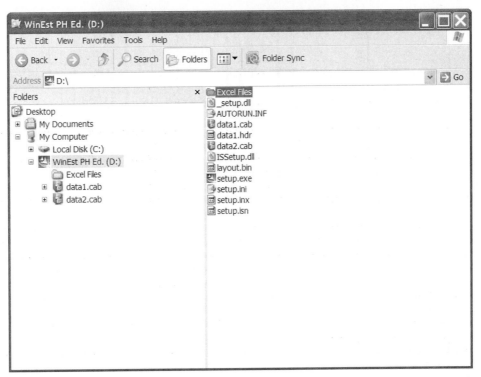

FIGURE G.11. Explore CD Window.

BID TABULATION

Project: _____
Location: _____
Architect: _____
Subcontract Package: _____

Estimate No. _____
Sheet No. _____
Date: _____
By: _____   Checked: _____

Scope of Work	Subcontractor 1	Subcontractor 2	Subcontractor 3	Subcontractor 4	Subcontractor 5
Base Bid					
Adjustments					
1					
2					
3					
4					
5					
6					
7					
8					
9					
10					
11					
Adjusted Bids					

Comments

FIGURE G.12. Subcontractor Bid Tabulation.

ESTIMATE WORK SHEET

Project: _____
Location _____
Architect _____
Items

CUT & FILL

Estimate No. _____
Sheet No. _____
Date _____
By _____ Checked _____

Grid	Fill									Cut								
	Fill At Intersections					Points	Average	Area	Total	Cut At Intersections					Points	Average	Area	Total
	1	2	3	4	5					1	2	3	4	5				
1						0	0		0						0	0		0
2						0	0		0						0	0		0
3						0	0		0						0	0		0
4						0	0		0						0	0		0
5						0	0		0						0	0		0
6						0	0		0						0	0		0
7						0	0		0						0	0		0
8						0	0		0						0	0		0
9						0	0		0						0	0		0
10						0	0		0						0	0		0
11						0	0		0						0	0		0
12						0	0		0						0	0		0
13						0	0		0						0	0		0
14						0	0		0						0	0		0
15						0	0		0						0	0		0
16						0	0		0						0	0		0
17						0	0		0						0	0		0
18						0	0		0						0	0		0
19						0	0		0						0	0		0
20						0	0		0						0	0		0
21						0	0		0						0	0		0
22						0	0		0						0	0		0
23						0	0		0						0	0		0
24						0	0		0						0	0		0
25						0	0		0						0	0		0
26						0	0		0						0	0		0
27						0	0		0						0	0		0
28						0	0		0						0	0		0
29						0	0		0						0	0		0
30						0	0		0						0	0		0
31						0	0		0						0	0		0
32						0	0		0						0	0		0
33						0	0		0						0	0		0
34						0	0		0						0	0		0
35						0	0		0						0	0		0
36						0	0		0						0	0		0
37						0	0		0						0	0		0
38						0	0		0						0	0		0
39						0	0		0						0	0		0
40						0	0		0						0	0		0
41						0	0		0						0	0		0
42						0	0		0						0	0		0

TOTAL FILL - Compacted Cubic Feet	0	TOTAL CUT - Bank Cubic Feet	0
Compacted Cubic Yards	0	Bank Cubic Yards	0
Shrinkage Factor		Swell Factor	
Required Bank Cubic Yards of Fill	0	Loose Cubic Yards of Cut to Haul	0
Net Bank Cubic Yards to Purchase	0		

FIGURE G.13. Cut and Fill Worksheet.

PROJECT ESTIMATE SUMMARY

PROJECT: _____
LOCATION: _____
ARCHITECT: _____

Estimate Number _____
Date _____
By _____
Checked _____

DIV. DESCRIPTION	WORK HOURS	LABOR $	MATERIAL $	EQUIPMENT $	SUBCONTRACT $	TOTAL $
DIRECT FIELD COSTS						
2 SITEWORK						0
3 CONCRETE						0
4 MASONRY						0
5 METALS						0
6 WOODS & PLASTICS						0
7 MOISTURE - THERMAL CONTROL						0
8 DOORS, WINDOWS & GLASS						0
9 FINISHES						0
10 SPECIALTIES						0
11 EQUIPMENT						0
12 FURNISHINGS						0
13 SPECIAL CONSTRUCTION						0
14 CONVEYING SYSTEMS						0
15 MECHANICAL						0
16 ELECTRICAL						0
						0
						0
						0
						0
TOTAL DIRECT FIELD COSTS	0	0	0	0	0	0
INDIRECT FIELD COSTS						
FIELD STAFF						0
TEMPORARY OFFICES						0
TEMPORARY FACILITIES						0
TEMPORARY UTILITIES						0
REPAIRS & PROTECTION						0
CLEANING						0
PERMITS						0
PROFESSIONAL SERVICES						0
BONDS						0
INSURANCE						0
MISC. EQUIPMENT						0
						0
						0
						0
LABOR BURDENS (STAFF)						0
LABOR BURDENS (CRAFT)						0
SALES TAX						0
TOTAL INDIRECT FIELD COSTS	0	0	0	0	0	0
HOME OFFICE COSTS						0
LAST MINUTE CHANGES						0
						0
						0
						0
						0
						0
						0
TOTAL LAST MINUTE CHANGES	0	0	0	0	0	0
PROFIT						0
TOTAL PROJECT COST	0	0	0	0	0	0

COMMENTS

_____
_____
_____
_____
_____
_____

FIGURE G.14. Project Estimate Summary Sheet.

ESTIMATE SUMMARY SHEET
REINFORCING STEEL

Project										Estimate No.
Location										Sheet No.
Architect										Date
Items										By
										Checked

Cost Code	Description	Dimensions			Count	Bar Size	Linear Feet	Pounds/ Foot	Quantity	Unit
		L ft	W ft	Space /ft						

FIGURE G.15. Worksheet to Calculate Reinforcing Bars.

ESTIMATE SUMMARY SHEET

Project _____
Location _____
Architect _____
Items _____

Estimate No. _____
Sheet No. _____
Date _____
By _____   Checked _____

Cost Code	Description	Q.T.O.	Waste Factor	Purch. Quan.	Unit	Crew	Prod. Rate	Wage Rate	Labor Hours	Unit Cost Labor	Unit Cost Material	Unit Cost Equipment	Labor	Material	Equipment	Total
									0.00				$0	$0	$0	$0

FIGURE G.16. Estimate Summary Sheet.

ESTIMATE WORK SHEET

Project: _____
Location _____
Architect _____
Items _____

Estimate No. _____
Sheet No. _____
Date _____
By _____   Checked _____

Reinforcing Steel

Cost Code	Description	Slab Width		Bar Spacing	Pcs	Slab Length		Cover age	Bar Length		Place an X in the coorresponding bar size						Bar Weight						Quantity	Unit
		Ft	In	In.–O.C.		Ft.	In.	In.	Ea	Total	3	4	5	6	7	8	3	4	5	6	7	8		
																	0.376	0.668	1.043	1.502	2.044	2.67		
																							0	Pound
																							0	Pound
																							0	Pound
																							0	Pound
																							0	Pound
																							0	Pound
																							0	Pound
																							0	Pound
																							0	Pound
																							0	Pound
																							0	Pound
																							0	Pound
																							0	Pound
																							0	Pound
																							0	Pound
																							0	Pound
																							0	Pound
																							0	Pound
																							0	Pound
																							0	Pound
																							0	Pound
																							0	Pound
																							0	Pound
																							0	Pound
																							0	Pound
																							0	Pound
																							0	Pound
																							0	Pound
																							0	Pound
																							0	Pound
																							0	Pound
																							0	Pound
																							0	Pound
																							0	Pound
																							0	Pound
																							0	Pound
																							0	Pound

FIGURE G.17. Worksheet to Calculate Reinforcing Bars in Slabs.

ESTIMATE WORK SHEET
STRUCTURAL STEEL

Project: _____

Location _____

Architect _____

Items _____

Estimate No. _____

Sheet No. _____

Date _____

By _____   Checked _____

Cost Code	Description	Designation	Pounds / Foot	Length Ft.	Length In.	Length Ft.	Count				Quantity	Unit
						0					0	Pounds
						0					0	Pounds
						0					0	Pounds
						0					0	Pounds
						0					0	Pounds
						0					0	Pounds
						0					0	Pounds
						0					0	Pounds
						0					0	Pounds
						0					0	Pounds
						0					0	Pounds
						0					0	Pounds
						0					0	Pounds
						0					0	Pounds
						0					0	Pounds
						0					0	Pounds
						0					0	Pounds
						0					0	Pounds
						0					0	Pounds
						0					0	Pounds
						0					0	Pounds
						0					0	Pounds
						0					0	Pounds
						0					0	Pounds
						0					0	Pounds
						0					0	Pounds
						0					0	Pounds
						0					0	Pounds
						0					0	Pounds
						0					0	Pounds
						0					0	Pounds
						0					0	Pounds
						0					0	Pounds
						0					0	Pounds
						0					0	Pounds

FIGURE G.18. Worksheet for Quantifying Structural Steel.

Building Parameters

Building Length	100	ft
Number of Bays Long	5	ea
Bay Length	20	ft
Building Width	60	ft
Number of Bays Wide	3	ea
Bay Width	20	ft
Wall Height (above grade)	25	ft
Depth to Top of Footing	24	in
Floor Thickness	5	in
Wire Mesh in Slab?	Yes	
Roof Hatch	1	ea
Personnel Doors	2	ea
10' x 10' OH door	-	ea
12' x 12' OH door	8	ea
14' x 14' OH door	-	ea
12' x 16' OH door	-	ea
2' x 4' Skylight	-	ea
4' x 4' Skylight	15	ea
4' x 6' Skylight	-	ea
Fire sprinklers?	Yes	
Bathrooms?	Unisex	

	Quantity	Unit Price	Total Cost	Notes
Concrete			27,596	
Masonry			81,704	
Steel, Joist, & Deck			53,980	
Roofing			23,450	
Doors & Skylights			29,381	
Finishes			9,200	
Accessories			250	
Fire Sprinklers			15,480	
Plumbing			11,000	
HVAC			18,000	
Electrical			22,750	
Excavation			**13,836**	
Subtotal			306,627	
Building Permit			2,153	
Overhead			50,000	
Profit			24,530	
Total			383,310	

FIGURE G.19. Warehouse Model.

ESTIMATE WORK SHEET

Project: _____
Location _____
Architect _____
Items _____

Estimate No. _____
Sheet No. _____
Date _____
By _____ Checked _____

Cost Code	Description	Comments / Calculations	Sub Totals	Quantity	Unit

FIGURE G.20. Generic Estimate Work Up Sheet.

INDEX